Stability and Control of Nonlinear Time-varying Systems

Shuli Guo · Lina Han

Stability and Control of Nonlinear Time-varying Systems

Springer

Shuli Guo
Key Laboratory of Complex System
  Intelligent Control and Decision,
  School of Automation
Beijing Institute of Technology
Beijing
China

Lina Han
Department of Cardiovascular Internal
  Medicine of Nanlou Branch,
  National Clinical Research Center
  for Geriatric Diseases
Chinese PLA General Hospital
Beijing
China

ISBN 978-981-10-8907-7   ISBN 978-981-10-8908-4 (eBook)
https://doi.org/10.1007/978-981-10-8908-4

Library of Congress Control Number: 2018935849

© Springer Nature Singapore Pte Ltd. 2018
This work is subject to copyright. All rights are reserved by the Publisher, whether the whole or part of the material is concerned, specifically the rights of translation, reprinting, reuse of illustrations, recitation, broadcasting, reproduction on microfilms or in any other physical way, and transmission or information storage and retrieval, electronic adaptation, computer software, or by similar or dissimilar methodology now known or hereafter developed.
The use of general descriptive names, registered names, trademarks, service marks, etc. in this publication does not imply, even in the absence of a specific statement, that such names are exempt from the relevant protective laws and regulations and therefore free for general use.
The publisher, the authors and the editors are safe to assume that the advice and information in this book are believed to be true and accurate at the date of publication. Neither the publisher nor the authors or the editors give a warranty, express or implied, with respect to the material contained herein or for any errors or omissions that may have been made. The publisher remains neutral with regard to jurisdictional claims in published maps and institutional affiliations.

Printed on acid-free paper

This Springer imprint is published by the registered company Springer Nature Singapore Pte Ltd. part of Springer Nature
The registered company address is: 152 Beach Road, #21-01/04 Gateway East, Singapore 189721, Singapore

We sincerely thank the Ministry of Science and Technology of P.R. China, and the Administrative Center for China's Agenda 21 for their supports through National Key Research and Development Program of China (2017YFF0207400).

We sincerely thank Prof. Irene Moroz.

We love our daughter, a little girl, Guo Fusu.

Preface

We ever have a dream about writing our work on "Stability and Control of Nonlinear Time-varying Systems". Since March 2012, our sincere friend, Prof. Irene Moroz from Oxford University has ever given us so many helps and advices about our work. Professor Han and I ever discuss every detail and try our best to finish them and keep our dream come true as quickly as possible.

Here are some of our motives to summary our work. For examples, we organize all material in the most-minded, straightforward manner, and we edit our research work into different parts or chapters based on different nonlinear fields or similar fields, and we try our best to help all readers into these deep fields for open research.

This book presents special systems that come from industrial models, including the complex saturation nonlinear systems, fuzzy uncertain differential systems, Lurie nonlinear systems with time-varying delay feedback, neutral delay differential equations, differential inclusions with nonlinear integral delays, and more industrial 2K-H planetary gear transmission systems and three-stage chaotic communication systems. And simultaneously, it also presents several typical methods, such as the classical Lyapunov function methods, integral inequalities methods, 0–1 algebra-geometry structure equations on saturation nonlinear function, solvable matrix inequalities, fuzzy observer, fuzzy controller design, etc. This book is structurally divided into four parts, such as stability and control of linear time-varying systems subject to actuator saturations, stability analysis for several types of nonlinear time-varying systems, integral inequalities and their applications in time-varying nonlinear systems, and applications in nonlinear mechanical and electrical systems.

In Part 1, we outline four chapters as follows.

Chapter 1 is dedicated to presenting a new method of saturation nonlinear fields, which can be used to tackle problems of analysis and synthesis for linear systems subject to actuator saturations. Noting that new systematic techniques which can be formally presented, its objective is to show not only the recent methods but also their practical applications. Its focus is the so-called 0–1 algebra-geometry structure equations. The some constructive qualitative and quantitative methods for linear systems with saturated inputs in both global and local contexts are attempted to provided. In Chap. 2, commutative matrices of multiple-input multiple-output (MIMO) linear systems are considered. The existence of the feedback matrices of a commutative state matrix set in the MIMO closed loops is reduced to the existence of an invariant subspace of a matrix A. The existence of feedback matrices in systems in open loop is equivalent to the existence of the solution of matrix equations denoted by Kronecker products. Based on 0–1 algebra-geometry structure equations, the relationship between equilibrium points is discussed for a linear system with a single saturated input. In Chap. 3, a new method for control system with saturation inputs is presented. By defining new equilibrium points, the relationship criterion among equilibrium points is discussed for linear system with saturated inputs. The asymptotical stability of the origin of the linear system in the presence of a single saturation input is analyzed, and the existence equations of closed trajectory are also considered for the same control systems. Finally, characteristics of commutative matrices of MIMO linear systems are considered. And at the same time, we present the simple criterion equations of asymptotic stability of second-order control systems with single saturated input when $ABK = BKA$. Chapter 4 is dedicated to presenting equilibrium points for second-order differential systems with single saturated input, and at the same time, the relationship between equilibrium points is discussed for second-order linear systems with a single saturated input. Moreover, many interesting examples, including their corresponding stability and possible limited cycles, are effectively shown to illustrate the above results.

In Part 2, we outline three chapters as follows.

In Chap. 5, fuzzy state observer and fuzzy controller are developed for a type of uncertain nonlinear system, which is represented by more general fuzzy modeling. Many interesting results are obtained as follows: First, by constructing Lyapunov function approaches, the adaptive observer laws including Riccati equations, two differentiators, and many solvability conditions about the above Riccati equations are presented. Second, based on Lyapunov function approaches, the proposed controllers are designed to guarantee the stability of the overall closed-loop system, and many solvability conditions on the proposed controllers are analyzed, too. More importantly, we give the structure of the common stable matrices for this kind of control problem, including their disturbance structures and Lie-algebraic conditions. In Chap. 6, the stability problems of Lurie time-varying nonlinear system with time-varying delay feedback are presented. First, absolute stability criteria, which we define as solvable matrix inequalities, are outlined by constructing a Lyapunov–Razumikhin functional. Second, we analyze these solvable matrix inequalities and give their sufficient and necessary solvable conditions that are easily computed in practical systems. Third, based on our norm definition, solvable conditions and time-delay bound estimations can be obtained from two cases of solvable norm inequalities. Moreover, by analyzing all obtained procedures, we present the optimal combination method for a pair of (τ, Q). In Chap. 7, it is very difficult in directly analyzing the robust stability of uncertain differential inclusions. If all points of the convex polyhedron are found, we can easily analyze the stability of convex points to obtain the stability of the uncertain differential inclusions. The new method of constructing polyhedron Lyapunov functional for given differential inclusions with nonlinear integral delays is presented. Algebraic criteria of asymptotical stable of the zero solution of a class of differential inclusion with nonlinear integral delays are outlined too. Moreover, the above conclusions are similarly generalized to show by the Riccati matrix inequalities.

In Part 3, we outline three chapters as follows.

In Chap. 8, several new integral inequalities are presented, which are effective in dealing with the integrodifferential inequalities whose variable exponents are greater than one. Compared with existed integral inequalities, those proposed here can be applied to more complicated differential equations. In Chap. 9, the notions of uniform Lipschitz stability are generalized, and the relations between these notions are analyzed. Several sufficient conditions about uniform Lipschitz asymptotic stability of nonlinear systems are established by the proposed integral inequalities. These sufficient conditions can be similarly generalized to linearly perturbed differential systems that appear in the existed literatures. In Chap. 10, the notions of (ω, Ω) stable, (ω, Ω) asymptotically stable, and especially (c_1, c_1) stable, (c_1, c_1) asymptotically stable, are presented. Several sufficient conditions on (c_1, c_1) stable, (c_1, c_1) asymptotically stable of time-varying delay neutral differential equations are established. The above sufficient conditions may similarly be generalized to delay neutral differential linear time-varying equations.

In Part 4, we outline two chapters as follows.

In order to improve the transmission and safety performance of the single-stage chaotic system, such as chaotic masking, chaotic shift keying, and chaotic modulation, we present a new three-stage chaotic communication system in Chap. 10. The unified chaotic systems are consist of the transmitter and the receiver end, which has only one parameter to set. By adjusting the parameters in the linear part and the interference system, we can obtain tens types of different transmission signals, which make it more difficult for the illegal receiver to decode the encrypted signal. The nonlinear dynamical modeling of 2K-H planetary gear transmission, which takes into consideration of clearance, is presented in this Chap. 11. Various chaotic characteristics of the dynamical system, such as the largest Lyapunov exponent, Poincaré section, FFT spectrum, and the phase locus are considered. Because of the existence of clearance, the Jacobi matrix of the above system does not exist. A new discrete method to calculate the largest Lyapunov exponent of the system is deduced based on the definition of classical Lyapunov exponent. Numerical simulation of the periodic movement, quasi-periodic movement, and chaotic movement is carried out under different parameters.

It is well known that nonlinear sciences are difficult and heavy subjects. Our styles are to make this book a litter lighter and easily read as poems and songs. And we wish all readers to enjoy it. Every effort has been extended to make every subject move rapidly and to make the sealing-in from one topic to the next as seamless as possible.

This book will help practitioners to have more concise model systems for modern nonlinear techniques, which have the potential functions for future applications.

This book will be a valuable guide for researcher and graduate students in the fields of mathematics, control, and engineering, too.

Beijing, China Shuli Guo
December 2017 Lina Han

Acknowledgements

We wish to thank the many professors who over the years have given us comments concerning these topics on this book and those encouraged us to carry it out. It is difficult to do anything in life without the friends' help, and many of my friends have contributed much to this book.

Our sincere gratitude goes especially to Academician of the Chinese Academy of Sciences, Prof. Huang Lin from Peking University, Prof. Irene Moroz from Oxford University, President and Prof. Fan Li from PLA General Hospital, Academician of the Chinese Academy of Engineering, Prof. Chen Jie from Beijing Institute of Technology, President and Prof. Fu Mengyin from Nanjing University of Science & Technology, Prof. Si Ligeng from Inner Mongolia Normal University, President and Prof. Wang Chenghong from Chinese Association of Automation, Prof. Xu Li from Akita Prefectural University, Prof. Ji Linhong from Tsinghua University, Prof. Yan Shaoze from Tsinghua University, Prof. Ma Wanbiao from University of Science and Technology, Beijing, Director and Prof. Hu Fujun from Chinese Hua Ling Press, Prof. Cui Weiqun from National Institute of Metrology, China, and Director and Prof. Lin Jie from Beijing Institute of Technology Press.

We wish to thank our colleagues and friends Prof. Wang Junzheng, Prof. Xia Yuanqing, Prof. Liu Xiangdong, Prof. Wu Qinghe, Prof. Zhang Baihai, Prof. Wang Zhaohua, Prof. Qu Dacheng, Prof. Dai Yaping, Prof. Yao Xiaolan, Prof. Xu Xiangyang, and Prof. Song Chunlei.

We wish to thank our graduate students Mr. Feng Xianjia, Mrs. Xin Wenfang, Mr. Liu Jinlei, Miss Lu Beibei, Miss Guo Yang, Miss Hao Xiaoting, Mr. Yuan Zhenbin, Mr. Gui Xinzhe, Mr. Chen Qimin, Mr. Zhang Yitong, and Mr. Zhang He.

We wish to thank the Ministry of Science and Technology of China and The Administrative Center for China's Agenda 21 for their supports through National Key Research and Development Program of China (2017YFF0207400).

We also wish to take this opportunity to thank Prof. Wang Long from Peking University and Prof. Xia Yuanqing for critically checking the entire manuscript and offering detail improvements on our manuscript.

We are truly indebted to Mr. Hu Yin for working with me for 3 months and for taking care of the typing and preparation of this book's manuscript.

Lastly, this book will be dedicated to Vinoth. S and his colleagues for their active efforts.

Beijing, China
December 2017

Shuli Guo
Lina Han

Contents

Part I Stability and Control of Linear Time-Varying Systems Subject to Actuator Saturations

1 Mathematical Modeling and Stability of Linear Uncertain Systems with Actuator Saturations 3
   1.1   Introduction 3
   1.2   Problem Statements, Mathematical Modeling, and Equilibrium Points 4
   1.3   Stability and Boundedness of Linear System with Saturation Inputs 19
   1.4   Stability of Linear System with Lipschitz Nonlinearity 21
   1.5   Robust Stability and Linear Matrix Inequalities 23
       1.5.1   Polytopic Uncertainty 23
       1.5.2   Norm Boundary Uncertainty 23
   1.6   Example Analysis and Simulink 24
   1.7   Conclusion 29
   References 29

2 Equilibrium Points of Linear Systems with Single Saturated Input Under Commuting Conditions 31
   2.1   Introduction 31
   2.2   Properties of Commuting Matrices 35
   2.3   The Existence of Feedback Matrices in MIMO/SISO Control Systems 37
       2.3.1   Closed-Loop Control 38
       2.3.2   Open-Loop Control 39
   2.4   Equilibrium Points of SISO Control Systems with Single Input 41

	2.5	Example Analysis and Simulink	44
	2.6	Conclusion	52
	References		53

3 Stability and Closed Trajectory for Second-Order Control Systems with Single-Saturated Input 55
- 3.1 Introduction 55
- 3.2 Criteria for Stability Analysis 56
- 3.3 Criteria to the Closed Trajectory 60
- 3.4 The Commutative Case 64
- 3.5 Example and Simulation 68
- 3.6 Conclusion 71
- References 72

4 Equilibrium Points of Second-Order Linear Systems with Single Saturated Input 73
- 4.1 Introduction 73
- 4.2 Problem Statements, Equilibrium Points 74
- 4.3 Some Discussions and Numerical Simulation 78
  - 4.3.1 $x_{eq,1,0}$ Being Stable Focus 79
  - 4.3.2 $x_{eq,1,0}$ Being Unstable Focus (or Center) 83
  - 4.3.3 $x_{eq,1,0}$ Being Genuine Stable Node 87
  - 4.3.4 $x_{eq,1,0}$ Being Unstable Node 91
  - 4.3.5 $x_{eq,1,0}$ Being Saddle 96
- 4.4 Conclusion 100
- References 101

Part II Stability Analysis for Three Types of Nonlinear Time-Varying Systems

5 Fuzzy Observer, Fuzzy Controller Design, and Common Hurwitz Matrices for a Class of Uncertain Nonlinear Systems 105
- 5.1 Introduction 105
- 5.2 Problem Statement 107
- 5.3 Design the Fuzzy Observer 109
- 5.4 Design the Fuzzy Controller 114
- 5.5 Structures of the Common Hurwitz Matrices 117
- 5.6 Numerical Simulation 121
- 5.7 Conclusion 129
- References 129

6 Stability of Lurie Time-Varying Systems with Time-Varying Delay Feedbacks 131
- 6.1 Introduction 131
- 6.2 Problem Formulation and Preliminaries 132

6.3	Some Results on Absolute Stability	135
6.4	Some Discussions on Solvable Conditions	140
6.5	Numerical Examples	146
6.6	Conclusion	148
References		149

7 Stability Criteria on a Type of Differential Inclusions with Nonlinear Integral Delays 151

7.1	Introduction	151
7.2	Algebraic Criteria of Asymptotic Stability	154
7.3	Numerical Example Analysis	158
7.4	Conclusion	165
References		165

Part III Integral Inequalities and Their Applications in Time-Varying Nonlinear Systems

8 Several Integral Inequalities 169

8.1	Introduction	169
8.2	Two Integral Inequalities	169
8.3	Generalization of Two Integral Inequalities	173
8.4	Conclusion	177
References		177

9 Lipschitz Stability Analysis on a Type of Nonlinear Perturbed System 179

9.1	Introduction	179
9.2	Basic Notions and General Results	179
9.3	Lipschitz Stability on Nonlinear Perturbed Systems	184
9.4	Example Analysis	190
9.5	Conclusion	190
References		190

10 (c_1, c_1) Stability of a Class of Neutral Differential Equations with Time-Varying Delay 193

10.1	Introduction	193
10.2	Stability Criteria	194
10.3	Numerical Examples	204
10.4	Conclusion	208
References		209

Part IV Applications in Nonlinear Mechanical and Electrical Systems

11 The Chaos Synchronization, Encryption for a Type of Three-Stage Communication System 213
    11.1 Introduction 213
    11.2 Problem Statement 214
    11.3 Proof of the Feasibility of the Recovery Signal 218
    11.4 Numerical Simulation 219
    11.5 Conclusion 222
    References 223

12 The Numerical Solutions and Their Applications in 2K-H Planetary Gear Transmission Systems 227
    12.1 Introduction 227
    12.2 The Numerical Integration Algorithms 228
    12.3 Nonlinear Dynamics of 2K-H Planetary Gear Transmission Systems 231
    12.4 Numerical Simulation 244
    12.5 Conclusion 250
    References 251

Appendix A: Some Theory about Lyapunov Stability 253

Glossary 259

About the Authors

Shuli Guo was born in 1973 in Inner Mongolia, P.R. China. He received his Bachelor Degree (1995) and Master Degree (1998) from Inner Mongolia Normal University, and his Ph.D. Degree from Peking University (2001). Dr. Guo worked as a postdoctoral researcher from Jan. 2002 to Apr. 2004 at Tsinghua University, as a researcher fellow from January 2008 to September 2008 at Akita Prefectural University, and as a postdoctoral researcher from September 2008 to September 2009 at Oxford University. He has been worked at Automatic School, Beijing Institute Technology since 2004. His research interests are in the area of control theory and stability of nonlinear systems, wireless sensor networks, and cardiovascular mathematics.

Lina Han was born in 1973 in Jilin Province, P.R. China. She received her Med. Bachelor Degree (1995), Med. Master Degree (2000), and Ph.D. Degree (2003) from Jilin University. Dr. Han worked as a postdoctoral researcher from January 2003 to April 2005 at Chinese PLA General Hospital, and as a research fellow from August 2008 to September 2009 at Kyoto University. And now, she works as a research fellow in Department of Cardiovascular Internal Medicine, National Clinical Research Center for Geriatric Diseases, Nanlou Branch of Chinese PLA General Hospital, and PLA School of Medicine. Her research interests focus on many 3D modeling problems on cardiovascular systems and their medical solutions.

Acronyms

(Λ, Λ^*)	0–1 algebra-geometry structure
MIMO	Multiple-input multiple-output
SISO	Single-input single-output
SIMO	Single-input multiple-output
$\otimes$	Kronecker products
LMIs	Linear matrix inequalities
GAS	Global asymptotic stability
US	Unstable
S	Stable
Ge	Genuine
Sp	Spurious
MagLev	Magnetic levitation
LMI	Linear matrix inequality
FFT	Fast Fourier transformation
y^p	The superscript "p" denotes a predicted value
y^c	The superscript "c" denotes a correct value
ODE	Ordinary differential equation
ϕ	The empty set

Part I
Stability and Control of Linear Time-Varying Systems Subject to Actuator Saturations

Chapter 1
Mathematical Modeling and Stability of Linear Uncertain Systems with Actuator Saturations

1.1 Introduction

Input saturation is encountered in all control system implementations because of the physical limits of actual actuators. The analysis and synthesis of controllers for dynamic systems subject to actuator saturation have been attracting increasingly more attention. Common examples of such limits are the deflection limits in aircraft actuators, the voltage limits in electrical actuators, the limits on flow volume or rate in hydraulic actuators, and so on. Although such limits obviously restrict the performance achievable in the systems of which they are part, if these limits are not treated carefully and if the relevant controllers do not account for these limits appropriately, peculiar and pernicious behavior may be observed. In particular, actuator saturation or rate limits have been implicated in various aircraft crashes and so on.

Global asymptotic stability (GAS) of the origin of linear systems with saturated inputs has been extensively studied by many researchers [1–6, 10–14]. More recently, efforts have been made to extend these results to nonlinear systems and delay systems [1, 6]. Lee and Hedrick [6] put forward some control problems about the GAS of the origin when an existing linear control law, which was designed to stabilize a linear plant, saturation has to occur because the required inputs exceed the physical limits of the actuator. Studying this aspect of the saturation problem is of concern because it is very common way in which control saturation occurs in practice. In practice, control bounds are usually ignored in the initial design. Available methods, such as linear quadratic or input–output linearization, are used to obtain control laws. For example, if the resulting control signal happens to exceed the limit, the control is then allowed to saturate.

Lee and Hedrick [6] used the hypercube in R^n to obtain GAS of the origin. The same method is used in references [1, 11], and some results about GAS of the origin are obtained. It is fact that the control is allowed to saturate when the resulting control signal exceeds the limit. We find that there exist nonzero equilibrium points when saturation happens. It is necessary to learn how the other equilibrium points except the origin function in analyzing the control systems with saturated inputs.

In this chapter, a new mathematical modeling method about linear systems with actuator saturation is presented, which are sufficient and necessary to describe linear systems with actuator saturation compared with the appeared results. Moreover, a method about equilibrium points is defined, which is very different from the classical equilibrium methods. The main contents of this chapter are organized as follows: First, a new mathematical model about linear systems with actuator saturation is presented, and the corresponding phase space is defined, too. The spatial structure and the position relations among the intervals in phase space are analyzed too. Based on the above discussions, new equilibrium points are defined, and the criterion of equilibrium points is presented too. Second, some sufficient conditions for determining asymptotical stable and boundedness of an equilibrium point under control saturation are presented. In addition, we try to extend the stability results to Lipschitz nonlinear systems, polytopic uncertainty systems, and norm boundary uncertainty. Many Riccati inequations or linear matrix inequalities (LMIs) as sufficient conditions are proposed when taking into account these uncertainties or nonlinearities. Finally, we present an example to illustrate the obtained results.

1.2 Problem Statements, Mathematical Modeling, and Equilibrium Points

The continuous-time single-input single-output (SISO) plant with a state feedback of the following form in [1, 6, 11]:

$$\begin{cases} \dot{x}(t) = Ax(t) + Bu_s(t), \\ y(t) = C^T x(t), \\ u_s = sat(-Kx(t)), \end{cases} \tag{1.1}$$

is considered, where $x \in \mathbb{R}^n$ is the state vector, $y \in \mathbb{R}$ is the output vector, and $u_s \in \mathbb{R}$ is the control input. A, B, C, and K are constant matrices of appropriate dimensions, and the system is minimal.

The saturated input u_s can be expressed as

$$sat(-Kx(t)) = \begin{cases} u_{\lim} & if \ -Kx \geq u_{\lim}, \\ -Kx(t) & if \ |-Kx| < u_{\lim}, \\ -u_{\lim} & if \ -Kx \leq -u_{\lim}, \end{cases} \tag{1.2}$$

where $u_{\lim}$ represents the saturation limit of the feedback system and $|u_s| \leq u_{\lim}$.

On the common method, the saturated input u_s in [1, 6, 11] is converted into the following form:

$$u_s = sat(-Kx) = -\mu(x)Kx, \tag{1.3}$$

1.2 Problem Statements, Mathematical Modeling, and Equilibrium Points

where

$$\mu(x) = \begin{cases} 1, & if \quad |-Kx| < u_{\lim}, \\ \frac{u_{\lim}}{|-Kx|}, & otherwise \ |-Kx| \geq u_{\lim}. \end{cases} \quad (1.4)$$

Thus, $\mu(x) \in [0, 1]$. Clearly, $\mu(x) = 1$ if saturation does not occur and $\mu(x) \in (0, 1)$ if saturation occurs.

Lee and Hedrick [6] proved the following lemma.

Lemma 1.1 *Given an interval* $\mathbb{E} = [a, b] \subset \mathbb{R}^1$, $M(\mu) = M_0 + \mu M_1$, *where* M_0, $M_1 \in \mathbb{R}^{n \times n}$, $M(\mu) = M^T(\mu)$, $\mu \in \mathbb{E}$, *then*

$$\max_{\mu \in E} \lambda_{\max}(M(\mu)) = \max(\lambda_{\max}(M(a)), \lambda_{\max}(M(b))),$$

where $\lambda_{\max}(M(\mu))$ *denotes the maximum eigenvalue of* $M(\mu)$.

So system (1.1) and (1.2) can be written as [1–3, 6]

$$\dot{x}(t) = Ax(t) - \mu BKx(t) = (A - \mu BK)x(t),$$

where $\mu = 0$ and $\mu = 1$ represent two vertexes of the interval matrices $A - \mu BK$. At the same option, we omit these discussions or results using Lemma 1.1.

Similarly, MIMO systems with many saturating inputs are represented by

$$\begin{cases} \dot{x}(t) = Ax(t) + Bu_s(t), \\ y = C^T x(t), \\ u_s(s) = \begin{pmatrix} u_{s,1} & u_{s,2} & \ldots & u_{s,m}, \end{pmatrix} \\ u_{s,i} = sat(-K_i x), \ , i = 1, 2, \ldots, n, \\ K^T = \begin{pmatrix} K_1 & K_2 & \ldots & K_m, \end{pmatrix}^T \end{cases} \quad (1.5)$$

where $x \in \mathbb{R}^n$ is the state variable; $y \in \mathbb{R}^m$ is the output variable; $u_{s,1} \in \mathbb{R}^1$ is the control input; A, B, C, and K are constant matrices of appropriate dimensions; and the system is minimal.

The saturated input can be expressed as

$$u_{s,i} = sat(-K_i x) = \begin{cases} u_{\lim,i}, & if \ -K_i x \geq u_{\lim,i}, \\ -K_i x, & if \ |-K_i x| < u_{\lim,i}, \\ -u_{\lim,i}, & if \ -K_i x \leq -u_{\lim,i}, \end{cases}$$

where $u_{\lim,i}$ ($i = 1, 2, \ldots, m$) represents the saturation limit of the feedback system. The systems in (1.5) can be written into the following form ([1, 6, 11]):

$$\begin{cases} \dot{x}(t) = (A - BG_{\mu_0}K)x(t), \\ y(t) = Cx(t), \end{cases} \quad (1.6)$$

where for any $x \in \mathbb{E}$, elements of vector $\mu(x)$ admit a lower bound

$$\mu_{0,i} = \min\{\mu_i(x) : \forall x \in \mathbb{E}\},$$

where G_{μ_0} is a diagonal matrix of positive scalars $\mu_i(x)$ for $i = 1, \ldots, m$ which arbitrarily take the value 1 or $\mu_{0,i}$. But we can obtain some special points in the following examples.

Example 1.1 We demonstrate some special points in the following saturating systems:
$$\dot{x}(t) = Ax(t) + Bu_s(t),$$

where $A = \begin{pmatrix} -1 & 1 & 0 \\ 0 & -2 & 1 \\ 0 & 0 & -4 \end{pmatrix}$, $B = \begin{pmatrix} 1 \\ 0 \\ 0 \end{pmatrix}$, and u_s denotes a saturated state feedback

$$u_s = \begin{cases} 4, & -[3\ 3\ 1]x \geq 4, \\ -[3\ 3\ 1]x, & |-[3\ 3\ 1]x| < 4, \\ -4, & -[3\ 3\ 1]x \leq -4, \end{cases}$$

where $K = [3\ 3\ 1]$.

It is clear that $ABK = BKA$, $-1 < 0$, $-1 + (-3) < 0$, $\begin{pmatrix} -2 & 1 \\ 0 & -4 \end{pmatrix} \in Hurwitz$, then
$$\begin{cases} A \in Hurwitz, \\ A - BK \in Hurwitz. \end{cases}$$

While there exist the following special equilibrium points:
$$x_{eq,1} = 0,\ x_{eq,2} = [4\ 0\ 0]^T,\ x_{eq,3} = [-4\ 0\ 0]^T,$$

which correspondingly satisfy
$$\begin{aligned} \dot{x}(t) &= (A - BK)x(t), \\ \dot{x}(t) &= Ax(t) + 4B, \\ \dot{x}(t) &= Ax(t) - 4B. \end{aligned}$$

Let
$$\begin{aligned} \mathbb{D}_1 &= \{x\,|-4 < -[3\ 3\ 1]x < 4\}, \\ \mathbb{D}_2 &= \{x\,|-[3\ 3\ 1]x \geq 4\}, \\ \mathbb{D}_3 &= \{x\,|-[3\ 3\ 1]x \leq -4,\,\}. \end{aligned}$$

Clearly,
$$\begin{aligned} x_{eq,1} &= 0 \in \mathbb{D}_1, \\ x_{eq,2} &= [4\ 0\ 0]^T \notin \mathbb{D}_2, \\ x_{eq,3} &= -[4\ 0\ 0]^T \notin \mathbb{D}_3. \end{aligned}$$

1.2 Problem Statements, Mathematical Modeling, and Equilibrium Points

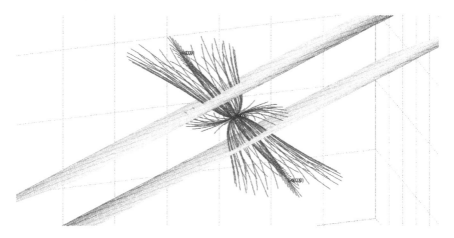

Fig. 1.1 Phase trajectories

Figure 1.1 shows that the solution of Example 1.1 which start from regions $\mathbb{D}_1$, $\mathbb{D}_2$ and $\mathbb{D}_3$ (including $x_{eq,2}, x_{eq,3}$) converges to $x_{eq,1}$.

Example 1.2 We demonstrate some special points in the following saturating systems:

$$\dot{x}(t) = Ax(t) + Bu_s(t),$$

where $A = \begin{pmatrix} 1 & 1 & 0 \\ 0 & -2 & 1 \\ 0 & 0 & -4 \end{pmatrix}$, $B = \begin{pmatrix} 1 \\ 0 \\ 0 \end{pmatrix}$, and where u_s denotes a saturated state feedback

$$u_s = \begin{cases} 4, & -[15\ 5\ 1]x \geq 4, \\ -[15\ 5\ 1]x, & \left| -[15\ 5\ 1]x \right| < 4, \\ -4, & -[15\ 5\ 1]x \leq -4, \end{cases}$$

where $K = [15\ 5\ 1]$.

It is clear that $ABK = BKA$, $1 > 0$, $1 + (-15) < 0$, $\begin{pmatrix} -2 & 1 \\ 0 & -4 \end{pmatrix} \in Hurwitz$, then

$$\begin{cases} A \notin Hurwitz, \\ A - BK \in Hurwitz, \end{cases}$$

While there exist the following special equilibrium points:

$$x_{eq,1} = 0, x_{eq,2} = [-4\ 0\ 0]^T, x_{eq,3} = [4\ 0\ 0]^T,$$

which correspondingly satisfy

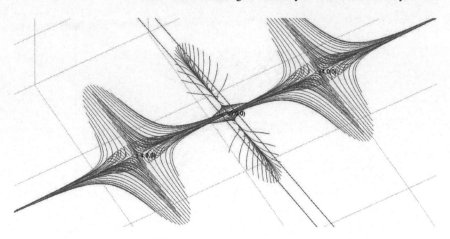

Fig. 1.2 Phase trajectories

$$\dot{x}(t) = (A - BK)x(t),$$
$$\dot{x}(t) = Ax(t) + 4B,$$
$$\dot{x}(t) = Ax(t) - 4B.$$

Let

$$\mathbb{D}_1 = \{x \mid -4 < -\begin{bmatrix} 15 & 5 & 1 \end{bmatrix} x < 4\},$$
$$\mathbb{D}_2 = \{x \mid -\begin{bmatrix} 15 & 5 & 1 \end{bmatrix} x \geq 4\},$$
$$\mathbb{D}_3 = \{x \mid -\begin{bmatrix} 15 & 5 & 1 \end{bmatrix} x \leq -4\}.$$

Clearly,

$$x_{eq,1} = 0 \in \mathbb{D}_1,$$
$$x_{eq,2} = [-4\ 0\ 0]^T \in \mathbb{D}_2,$$
$$x_{eq,3} = [4\ 0\ 0]^T \in \mathbb{D}_3.$$

Figure 1.2 shows that the solutions of Example 1.2 which start near $x_{eq,1}$ will converge to $x_{eq,1}$ and some solutions of Example 1.2 which start near $x_{eq,2}$ or $x_{eq,3}$ will converge to $x_{eq,1}$, but others will diffuse to be infinite.

Example 1.3 We demonstrate some special points in the following saturating systems:

$$\dot{x}(t) = Ax(t) + Bu_s(t),$$

where $A = \begin{pmatrix} -1 & 1 & 0 \\ 0 & -2 & 1 \\ 0 & 0 & -4 \end{pmatrix}$, $B = \begin{pmatrix} 1 \\ 0 \\ 0 \end{pmatrix}$, and u_s denotes a saturated state feedback

$$u_s = \begin{cases} 4, & -[-3\ -3\ -1]x \geq 4, \\ -[-3\ -3\ -1]x, & \left|-[-3\ -3\ -1]x\right| < 4, \\ -4, & -[-3\ -3\ -1]x \leq -4, \end{cases}$$

1.2 Problem Statements, Mathematical Modeling, and Equilibrium Points

where $K = [-3\ -3\ -1]$.

It is clear that $ABK = BKA$, $-1 < 0$, $-1 - (-3) > 0$, $\begin{pmatrix} -2 & 1 \\ 0 & -4 \end{pmatrix} \in Hurwitz$, then

$$\begin{cases} A \in Hurwitz, \\ A - BK \notin Hurwitz. \end{cases}$$

While there exist the following special equilibrium points:

$$x_{eq,1} = 0,\ x_{eq,2} = [4\ 0\ 0]^T,\ x_{eq,3} = [-4\ 0\ 0]^T,$$

which correspondingly satisfy

$$\dot{x}(t) = (A - BK)x(t),$$
$$\dot{x}(t) = Ax(t) + 4B,$$
$$\dot{x}(t) = Ax(t) - 4B.$$

Let

$$\mathbb{D}_1 = \{x \mid -4 < -[-3\ -3\ -1]x < 4\},$$
$$\mathbb{D}_2 = \{x \mid -[-3\ -3\ -1]x \geq 4\},$$
$$\mathbb{D}_3 = \{x \mid -[-3\ -3\ -1]x \leq -4\}.$$

Clearly,

$$x_{eq,1} = 0 \in \mathbb{D}_1,$$
$$x_{eq,2} = [4\ 0\ 0]^T \in \mathbb{D}_2,$$
$$x_{eq,3} = [-4\ 0\ 0]^T \in \mathbb{D}_3.$$

Figure 1.3 shows that the solutions of Example 3, which start in $\mathbb{D}_1$, $\mathbb{D}_2$ and $\mathbb{D}_3$ will converge to $x_{eq,2}$ or $x_{eq,3}$.

Example 1.4 We demonstrate some special points in the following saturating systems:

$$\dot{x}(t) = Ax(t) + Bu_s(t)$$

where

$$A = \begin{pmatrix} 1 & 1 & 0 \\ 0 & -2 & 1 \\ 0 & 0 & -4 \end{pmatrix},\ B = \begin{pmatrix} 1 \\ 0 \\ 0 \end{pmatrix}$$

and where u_s denotes a saturated state feedback

$$u_s = \begin{cases} 4, & -[-15\ -5\ -1]x \geq 4, \\ -[-15\ -5\ -1]x, & |-[-15\ -5\ -1]x| < 4, \\ -4, & -[-15\ -5\ -1]x \leq -4, \end{cases}$$

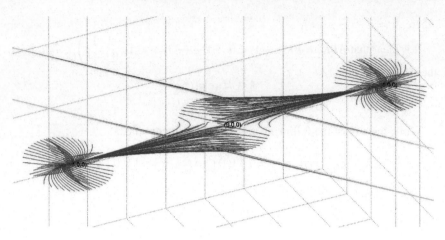

Fig. 1.3 Phase trajectories

and $K = [-15 \ -5 \ -1]$.

It is clear that $ABK = BKA$, $1 > 0$, $1 - (-15) > 0$, $\begin{pmatrix} -2 & 1 \\ 0 & -4 \end{pmatrix} \in Hurwitz$,

$$\begin{cases} A \notin Hurwitz, \\ A - BK \notin Hurwitz. \end{cases}$$

While there exist the following special equilibrium points:

$$x_{eq,1} = 0, x_{eq,2} = [-4 \ 0 \ 0]^T, x_{eq,3} = [4 \ 0 \ 0]^T$$

which correspondingly satisfy

$$\dot{x}(t) = (A - BK)x(t),$$
$$\dot{x}(t) = Ax(t) + 4B,$$
$$\dot{x}(t) = Ax(t) - 4B.$$

Let

$$\mathbb{D}_1 = \{x \mid -4 < -[-15 \ -5 \ -1]x < 4\},$$
$$\mathbb{D}_2 = \{x \mid -[-15 \ -5 \ -1]x \geq 4\},$$
$$\mathbb{D}_3 = \{x \mid -[-15 \ -5 \ -1]x \leq -4\}.$$

Clearly,

$$x_{eq,1} = 0 \in \mathbb{D}_1,$$
$$x_{eq,2} = [-4 \ 0 \ 0]^T \notin \mathbb{D}_2,$$
$$x_{eq,3} = [4 \ 0 \ 0]^T \notin \mathbb{D}_3.$$

1.2 Problem Statements, Mathematical Modeling, and Equilibrium Points

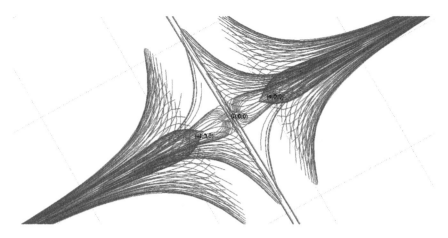

Fig. 1.4 Phase trajectories

Figure 1.4 shows that the solutions of Example 4 which start in $\mathbb{D}_1, \mathbb{D}_2, \mathbb{D}_3$ diffuse to be infinite.

It demonstrates that it is the fact that there exist nonzero equilibrium points in the classical sense (see Examples 1.2 and 1.3), but some points in the above systems (appearing Examples 1.1, 1.2, 1.3 and 1.4) are not equilibrium points which define in the classical sense. How these special points that appear Examples 1.1, 1.2, 1.3 and 1.4 affect the control system under nonlinear saturation is an unsolvable problem [1–15].

According to the saturating definition in (1.2), we give out the following $0-1$ algebra-geometry type structure equations that convert the SISO system (1.1) into the following control system:

$$\begin{cases} \dot{x}(t) = (A - \lambda BK)x(t) + \lambda^* Bu_{\lim}, \\ y(t) = Cx(t), \end{cases} \quad (1.7)$$

where λ correspondingly demonstrates 0 or 1, and λ^* is derived from the following restrictions:

$$\begin{cases} (\lambda, \lambda^*) = (0, 1), & \leftrightarrow \quad -Kx - u_{\lim} \geq 0, \\ (\lambda, \lambda^*) = (0, -1), & \leftrightarrow \quad -Kx + u_{\lim} \leq 0, \\ (\lambda, \lambda^*) = (1, 0), & \leftrightarrow \quad |-Kx| < u_{\lim}, \end{cases} \quad (1.8)$$

If $A - \lambda BK$ is nonsingular, then the equilibrium point under the restriction (λ, λ^*) is

$$x_{eq, \lambda, \lambda^*} = -\lambda^*(A - \lambda BK)^{-1} Bu_{\lim}. \quad (1.9)$$

Simultaneously, we define the following intervals in accord with (λ, λ^*) as

$$\mathbb{D}_{0,1} = \{x \mid -Kx \geq u_{\lim}\},$$
$$\mathbb{D}_{0,-1} = \{x \mid |-Kx \leq -u_{\lim}\},$$
$$\mathbb{D}_{1,0} = \{x \mid |-Kx| < u_{\lim}\}.$$

According to the above method, we convert MIMO system (1.5) into the following system:

$$\begin{cases} \dot{x}(t) = (A - B\Lambda K)x(t) + B\Lambda^* u_{\lim}, \\ y(t) = Cx(t), \end{cases} \quad (1.10)$$

where Λ demonstrates diagonal matrix whose elements can take 1 or 0, the diagonal matrix Λ^* is defined along Λ whose elements change according to (1.8). For example, when the elements of diagonal matrix Λ take 1, the corresponding elements of Λ^* take 0; when the elements of diagonal matrix Λ take 0, the corresponding ones of Λ^* take 1 or -1, where -1 denotes the interval $-K_i x \leq -u_{\lim,i}$ and 1 corresponds to the interval $-K_i x \geq u_{\lim,i}$.

Clearly, there are 2^m possibilities for Λ and 3^m possibilities for (Λ, Λ^*), that is to say

$$C_m^0 + C_m^1 2^1 + C_m^2 2^2 + \cdots + C_m^{m-1} 2^{m-1} + C_m^m 2^m = 3^m.$$

Thus, we can obtain 3^m linear differential equations. From the above method, we can see that Eq. (1.5) can be easily converted into Eq. (1.10), while Eq. (1.10) does not represent Eq. (1.5). But when we explain the corresponding elements of matrices (Λ, Λ^*) with (1.8), the solutions of system (1.10) have to be consistent with the ones of (1.5).

The following definitions will be instrumental to study the above special equilibrium points.

Definition 1.1 For MIMO system (1.10), if the elements 0 of diagonal matrix appear in $l_1 th, l_2 th, \ldots, l_p th$ elements, the other elements are 1, $p < m$, then we call matrix Λ with saturation order $l_1 th, l_2 th, \ldots, l_p th$.

Clearly, when $\Lambda = I$ holds, we can find that $\Lambda^* = 0$ holds, while when $\Lambda = 0$ holds, we can find that there exist 2^m possibilities for Λ^* such as

$$\Lambda^* = \begin{pmatrix} \pm 1 & 0 & \cdots & 0 & 0 \\ 0 & \pm 1 & \cdots & 0 & 0 \\ \vdots & \vdots & \ddots & 0 & 0 \\ 0 & 0 & \cdots & \pm 1 & 0 \\ 0 & 0 & \cdots & 0 & \pm 1 \end{pmatrix}.$$

Definition 1.2 For MIMO system (1.10), the whole space $\mathbb{R}^n$ is divided into 3^m intervals $\mathbb{D}_{\Lambda,\Lambda^*}$ according to the definitions of (Λ, Λ^*) as follows:

1.2 Problem Statements, Mathematical Modeling, and Equilibrium Points

$$\mathbb{D}_{\Lambda,\Lambda^*} = \left\{ x \left| \begin{array}{l} |-K_i x| < u_{\lim,i},\ i = l_1, l_2, \ldots, l_p, \\ -K_j x \geq u_{\lim,j},\ j = l_{p+1}, l_{p+2}, \ldots, l_{p+r}, \\ -K_z x \geq -u_{\lim,z},\ z = l_{p+r+1}, l_{p+r+2}, \ldots, l_{p+r+k}, \\ p+r+k = m. \end{array} \right. \right\}. \quad (1.11)$$

Definition 1.3 For MIMO system (1.10), if the intervals

$$\mathbb{D}_{\Lambda_1,\Lambda_1^*} \cap \mathbb{D}_{\Lambda_2,\Lambda_2^*} = \phi,\ \inf d(x_1, x_2) = 0,$$

where $x_1 \in D_{\Lambda_1,\Lambda_1^*}$, $x_2 \in D_{\Lambda_2,\Lambda_2^*}$, $d(.,.)$ denotes the distance between two points, then we call that they are neighboring; otherwise, they are apart.

It is clear in Example 1.1 that $\mathbb{D}_{I,0}, \mathbb{D}_{0,I}$ or $\mathbb{D}_{I,0}, \mathbb{D}_{0,-I}$ are neighboring and $\mathbb{D}_{0,I}, \mathbb{D}_{0,-I}$ are apart. Clearly, $\mathbb{D}_{I,0}$ is neighboring with every intervals such as $\mathbb{D}_{0,I}, \mathbb{D}_{0,-I}$.

Theorem 1.1 *(1) Given $\Lambda \neq I$, the intervals derived by some saturation order of Λ are apart;*
(2) For two different matrices Λ_1^*, Λ_2^*, if there is only one different element on the same position where the elements of Λ_1^*, Λ_2^* are ones such as $[0, 1]$, $[0, -1]$, $[1, 0]$, $[-1, 0]$, intervals $\mathbb{D}_{\Lambda_1,\Lambda_1^*}, \mathbb{D}_{\Lambda_2,\Lambda_2^*}$ are neighboring; otherwise, they are apart;
(3) The whole plane is divided into 3^m different intervals by all saturated planes;
(4) $\mathbb{D}_{I,0}$ is neighboring with every intervals such as $\mathbb{D}_{\Lambda_1,\Lambda_1^*}$, where $\Lambda_1 \neq I$.

Proof (1) Given any $\Lambda \neq I$ and suppose the saturation nonlinear happens on the i_{th} position of Λ, we may generally define Λ_i as

$$\Lambda_i = \begin{pmatrix} 1_1 & \cdots & 0 & 0 & \cdots & \cdots & 0 \\ \vdots & \ddots & \vdots & 0 & \ddots & \vdots & 0 \\ 0 & \cdots & 1_{i-1} & 0 & \cdots & \cdots & 0 \\ 0 & \cdots & 0 & 0_i & \cdots & \cdots & 0 \\ 0 & \cdots & 0 & 0 & 1_{i+1} & \cdots & 0 \\ \vdots & \vdots & \vdots & \vdots & \vdots & \ddots & \vdots \\ 0 & \cdots & 0 & 0 & 0 & \cdots & 1_m \end{pmatrix} = \begin{pmatrix} I_{i-1} & 0 & 0 \\ 0 & 0 & 0 \\ 0 & 0 & I_{m-i} \end{pmatrix}.$$

Its corresponding derived matrix is

$$\Lambda_i^* = \begin{pmatrix} 0_1 & \cdots & 0 & 0 & 0 & 0 & 0 \\ \vdots & \ddots & \vdots & 0 & 0 & 0 & 0 \\ 0 & \cdots & 0_{i-1} & 0 & 0 & 0 & 0 \\ 0 & \cdots & 0 & \pm 1_i & 0 & 0 & 0 \\ 0 & \cdots & 0 & 0 & 0_{i+1} & \cdots & 0 \\ \vdots & \vdots & \vdots & \vdots & \vdots & \ddots & \vdots \\ 0 & \cdots & 0 & 0 & 0 & \cdots & 0_m \end{pmatrix} = \begin{pmatrix} 0_{i-1} & 0 & 0 \\ 0 & \pm 1_i & 0 \\ 0 & 0 & 0_{m-i} \end{pmatrix}.$$

That is to say that $\mathbb{D}_{\Lambda_i,\Lambda_i^*}$ denotes the following two corresponding intervals:

$$\mathbb{D}_{\Lambda_i,\Lambda_{i+}^*} = \left\{ x \,\big|\, \big||-K_j x\big| < u_{\lim,j}, j = 1, \ldots, i-1, i+1, \ldots, m, -K_i x \geq u_{\lim,i} \right\},$$
$$\mathbb{D}_{\Lambda_i,\Lambda_{i-}^*} = \left\{ x \,\big|\, \big||-K_j x\big| < u_{\lim,j}, j = 1, \ldots, i-1, i+1, \ldots, m, -K_i x \leq -u_{\lim,i} \right\}.$$

Clearly, $\mathbb{D}_{\Lambda_i,\Lambda_{i+}^*}$, $\mathbb{D}_{\Lambda_i,\Lambda_{i-}^*}$ are apart. Using the same methods, we may obtain more general results when many saturations happen.

(2) From the above method, we see that for different Λ_1^*, Λ_2^*, if there is only one different element on the same position of Λ_1^*, Λ_2^*, whose combination pairs are such as $[0, 1]$, $[1, 0]$, $[0, -1]$, $[-1, 0]$, $[1, -1]$, $[-1, 1]$. Clearly, the combination pairs $[1, -1]$, $[-1, 1]$ mean the intervals $\mathbb{D}_{\Lambda_1,\Lambda_1^*}$, $\mathbb{D}_{\Lambda_2,\Lambda_2^*}$ apart. While the combination pairs $[0, 1]$,$[1, 0]$, $[0, -1]$, $[-1, 0]$ mean $\mathbb{D}_{\Lambda_1,\Lambda_1^*}$,$\mathbb{D}_{\Lambda_2,\Lambda_2^*}$ neighboring.

(3) The planes (lines) $-K_i x = u_{\lim,i}$, $-K_j x = u_{lim,j}$, $i \neq j, i, j = 1, 2, \ldots, m$ are intersecting, and $-K_i x = u_{\lim,i}$, $-K_i x = -u_{\lim,i}$, $i = 1, 2, \ldots, m$ are parallel. According to the assumptions that the whole system is minimal, we can obtain the fact that the rank of K means that these m vectors are independent. That is to say, these 2^m planes divide the whole space R^n into 3^m intervals.

The problem about the multiple equilibriums is considered below.

For system (1.10), if matrix $A - B\Lambda K$ is nonsingular, then the equilibrium point is under the restriction (Λ, Λ^*) by

$$x_{eq,\Lambda,\Lambda^*} = -(A - B\Lambda K)^{-1} B\Lambda^* u_{\lim}. \tag{1.12}$$

Definition 1.4 (1) For the given system (1.10), if $x_{eq,\Lambda,\Lambda^*} \in \mathbb{D}_{\Lambda,\Lambda^*}$, then we call that x_{eq,Λ,Λ^*} is a genuine equilibrium point; and if $A - B\Lambda K$ is stable (unstable), then x_{eq,Λ,Λ^*} is a genuine stable (unstable) equilibrium point;

(2) For system (1.10), if $x_{eq,\Lambda,\Lambda^*} \notin \mathbb{D}_{\Lambda,\Lambda^*}$, then we call that x_{eq,Λ,Λ^*} is a false equilibrium point, if $A - B\Lambda K$ is stable (unstable), then we call x_{eq,Λ,Λ^*} is a spurious stable (unstable) equilibrium point.

Note 1.1 In Example 1.1, from Definition 1.4, we know that $x_{eq,1}$ is a genuine equilibrium point, $x_{eq,2}$, $x_{eq,3}$ are spurious equilibrium points. Figure 1.1 shows that the solution of Example 1.1 which start from regions $\mathbb{D}_1$, $\mathbb{D}_2$, and $\mathbb{D}_3$ (including $x_{eq,2}$, $x_{eq,3}$) converges to $x_{eq,1}$. So, $x_{eq,1}$ is the genuine stable equilibrium point, and $x_{eq,2}$ and $x_{eq,3}$ are stable spurious equilibria.

Note 1.2 In Example 1.2, from Definition 1.4, we know that $x_{eq,1}$ is a genuine equilibrium point, $x_{eq,2}$, $x_{eq,3}$ are genuine equilibrium points. Figure 1.2 shows that the solutions of Example 1.2 which start near $x_{eq,1}$ will converge to $x_{eq,1}$ and some solutions of Example 1.2 which start near $x_{eq,2}$ or $x_{eq,3}$ will converge to $x_{eq,1}$, but others will diffuse to be infinite.

Note 1.3 In Example 1.3, from Definition 1.4, we know that $x_{eq,1}$ is a genuine unstable equilibrium point, and $x_{eq,2}, x_{eq,3}$ are stable genuine equilibrium points. Figure 1.3 shows that the solutions of Example 1.3, which start in $\mathbb{D}_1$, $\mathbb{D}_2$, and $\mathbb{D}_3$ will converge to $x_{eq,2}$ or $x_{eq,3}$. In conclusion, $x_{eq,1}$ is a genuine unstable equilibrium point, $x_{eq,2}$, $x_{eq,3}$ are stable genuine equilibrium points.

1.2 Problem Statements, Mathematical Modeling, and Equilibrium Points

Note 1.4 In Example 1.4, from Definition 1.4, we know that $x_{eq,1}$ is an unstable genuine equilibrium point and $x_{eq,2}$, $x_{eq,3}$ are unstable spurious equilibrium points. Figure 1.4 shows that the solutions of Example 1.4 which start in $\mathbb{D}_1$, $\mathbb{D}_2$, $\mathbb{D}_3$ diffuse to be infinite. Therefore, $x_{eq,1}$ is an unstable genuine equilibrium point, and $x_{eq,2}$, $x_{eq,3}$ are unstable spurious equilibrium points.

Theorem 1.2 *For system (1.10), if $x_{eq,\Lambda,\Lambda^*} = -(A - B\Lambda K)^{-1} B\Lambda^* u_{\lim}$ is a stable (unstable) genuine equilibrium point, then $x_{eq,\Lambda,-\Lambda^*} = (A - B\Lambda K)^{-1} B\Lambda^* u_{\lim}$ is a stable (unstable) genuine equilibrium point too, especially $x = 0$ is a genuine equilibrium point. Similarly, if $x_{eq,\Lambda,\Lambda^*} = -(A - B\Lambda K)^{-1} B\Lambda^* u_{\lim}$ is a stable (unstable) spurious equilibrium point, then $x_{eq,\Lambda,-\Lambda^*} = (A - B\Lambda K)^{-1} B\Lambda^* u_{\lim}$ is a stable (unstable) spurious equilibrium one too.*

The proof is obtained using Definition 1.2, and we omit these discussions.

The results below are obtained by Definition 1.4 and Lyapunov basic theorem (see Reference [9]).

Theorem 1.3 *For system (1.10), let $A - B\Lambda K$ be nonsingular, suppose the following are genuine:*

(1) If x_{eq,Λ,Λ^*} is a stable genuine equilibrium point, then there exist $P = P^T > 0$, ellipsoids $\mathbb{S}_{\Lambda,\Lambda^*} \in \mathbb{D}_{\Lambda,\Lambda^*}$ and for any $x_0 \in \mathbb{S}_{\Lambda,\Lambda^*}$, $t \geq t_0$ such that

$$(x(t,t_0,x_0) - x_{eq,\Lambda,\Lambda^*})^T ((A - B\Lambda K)^T P + P(A - B\Lambda K))(x(t,t_0,x_0) - x_{eq,\Lambda,\Lambda^*})$$

being negative definite quadratic, where $\mathbb{S}_{\Lambda,\Lambda^*}$ is the maximum invariant set in $\mathbb{D}_{\Lambda,\Lambda^*}$.

(2) If x_{eq,Λ,Λ^*} is a spurious stable equilibrium point, then for $x_0 \notin \mathbb{D}_{\Lambda,\Lambda^*}$, $x(t,t_0,x_0)$ reaches the boundary of $\mathbb{D}_{\Lambda,\Lambda^*}$ in finite time, that is to say, for any $x_0 \in \mathbb{S}_{\Lambda,\Lambda^*}$, $t \geq t_0$,

$$(x(t,t_0,x_0) - x_{eq,\Lambda,\Lambda^*})^T ((A - B\Lambda K)^T P + P(A - B\Lambda K))(x(t,t_0,x_0) - x_{eq,\Lambda,\Lambda^*}) < 0$$

does not always hold in $\mathbb{D}_{\Lambda,\Lambda^*}$.

Proof (1) According to Lyapunov basic theorem, if x_{eq,Λ,Λ^*} is a stable genuine equilibrium point, that is to say, $x_{eq,\Lambda,\Lambda^*} \in \mathbb{D}_{\Lambda,\Lambda^*}$ and there exists a positive definite symmetric Lyapunov matrix $P = P^T > 0$ and at least an ellipsoid $\mathbb{S}_{\Lambda,\Lambda^*} \in \mathbb{D}_{\Lambda,\Lambda^*}$ such that

$$\begin{cases} V = x(t,t_0,x_0)^T P x(t,t_0,x_0), \\ -\dot{V} = -x(t,t_0,x_0)^T ((A - B\Lambda K)^T P + P(A - B\Lambda K))x(t,t_0,x_0), \end{cases}$$

are both positive definite.

(2) Since x_{eq,Λ,Λ^*} is a stable spurious equilibrium point, that is to say, for any $x_0 \in \mathbb{D}_{\Lambda,\Lambda^*}$, $x_{eq,\Lambda,\Lambda^*} \notin \mathbb{D}_{\Lambda,\Lambda^*}$, t_i, t_0 and $P = P^T > 0$, then from

$$(x(t_1,t_0,x_0) - x_{eq,\Lambda,\Lambda^*})^T ((A - B\Lambda K)^T P + P(A - B\Lambda K))(x(t_1,t_0,x_0) - x_{eq,\Lambda,\Lambda^*}) < 0,$$

we obtain $t_i - t_0 < \infty$.

Otherwise, we get

$$\lim_{t_1 \to \infty} x(t_1, t_0, x_0)$$
$$= \lim_{t_1 \to \infty} e^{(A-BAK)(t_1-t_0)} x_0 + \lim_{t_1 \to \infty} [e^{(A-BAK)(t_1-t_0)} - I](A - BAK)^{-1} BA^* u_{\lim}$$
$$= -(A - BAK)^{-1} BA^* u_{\lim}$$
$$= x_{eq,\Lambda,\Lambda^*} \in \mathbb{S}_{\Lambda,\Lambda^*} \subset \mathbb{D}_{\Lambda,\Lambda^*}.$$

It is in contradiction with the fact $x_{eq,\Lambda,\Lambda^*} \notin \mathbb{D}_{\Lambda,\Lambda^*}$.

From the above discussions, we can easily get the results below.

Theorem 1.4 *If x_{eq,Λ,Λ^*} is a stable genuine equilibrium point, then it is attractive; and if x_{eq,Λ,Λ^*} is stable spurious one, then it does not be attractive.*

For MIMO system (1.10), the result below can be easily obtained.

Equivalence proposition *For any equilibrium points $x_{eq,\Lambda,\Lambda^*} = -(A - BAK)^{-1} BA^* u_{\lim}$ and the following inequality:*

$$\begin{cases} \left| -e_i K(A - BAK)^{-1} B(\Lambda_1 - \Lambda_2) u_{\lim} \right| < u_{\lim,i}, i = l_1, l_2, \ldots, l_p, \\ e_j K(A - BAK)^{-1} B(\Lambda_1 - \Lambda_2) u_{\lim} \geq u_{\lim,j}, j = l_{p+1}, l_{p+2}, \ldots, l_{p+r}, \\ e_z K(A - BAK)^{-1} B(\Lambda_1 - \Lambda_2) u_{\lim} \leq -u_{\lim,z}, z = l_{p+r+1}, l_{p+r+2}, \ldots, l_{p+r+k}, \\ p + r + k = m, \end{cases}$$
(1.13)

where $e_i, i = 1, 2, \ldots, m$ is the $i - th$ elemental vector, $\Lambda, \Lambda_1, \Lambda_2$ are diagonal matrices, Λ_1 denotes $m \times m$ order matrix whose $l_{p+1}th, l_{p+2}th, \ldots, l_{p+r}th$ elements are 1, others are 0, Λ_2 denotes $m \times m$ order matrix whose $l_{p+r+1}th, l_{p+r+2}th, \ldots, l_{p+r+k}th$ elements are 1, others are 0, and $\Lambda + \Lambda_1 + \Lambda_2 = I$.

If (1.13) holds, then the equilibrium point x_{eq,Λ,Λ^*} is genuine. If at least one inequality in (1.13) does not hold, then the equilibrium point x_{eq,Λ,Λ^*} is spurious. Inversely, if x_{eq,Λ,Λ^*} is genuine, it satisfies (1.13) and if x_{eq,Λ,Λ^*} is spurious, it does not satisfy (1.13).

Proof From system (1.5) or (1.10),

$$-K_i x_{eq,\Lambda,\Lambda^*}$$
$$= -e_i K[-(A - BAK)^{-1} BA^* u_{lim}] \qquad (1.14)$$
$$= e_i K(A - BAK)^{-1} BA^* u_{lim}.$$

Since Λ is a matrix with saturation order $l_1, l_2, \ldots, l_p$, its derived $m \times m$ matrix Λ^* is the matrix whose $l_{p+1}, l_{p+2}, \ldots, l_{p+r}$ elements are 1, $l_{p+r+1}, l_{p+r+2}, \ldots, l_{p+r+k}$ elements are -1, other elements are 0, so $\Lambda^* = \Lambda_1 - \Lambda_2$, Λ_1, Λ_2 are defined as above. Since Definition 1.2 on $\mathbb{D}_{\Lambda,\Lambda^*}$, if $x_{eq,\Lambda,\Lambda^*} \in \mathbb{D}_{\Lambda,\Lambda^*}$, then

1.2 Problem Statements, Mathematical Modeling, and Equilibrium Points

$$\begin{aligned}
|-K_i x_{eq,\Lambda,\Lambda^*}| &= |-e_i K[-(A - B\Lambda K)^{-1} B\Lambda^* u_{lim}]| \\
&= |e_i K (A - B\Lambda K)^{-1} B\Lambda^* u_{lim}| \\
&= |e_i K (A - B\Lambda K)^{-1} B(\Lambda_1 - \Lambda_2) u_{lim}| \\
&< u_{lim,i}.
\end{aligned} \quad (1.15)$$

Similarly, the other two inequalities are demonstrated. Inversely, if x_{eq,Λ,Λ^*} satisfies (1.13), it certainly satisfies Definition 1.2, hence $x_{eq,\Lambda,\Lambda^*} \in \mathbb{D}_{\Lambda,\Lambda^*}$. If there exists at least one inequality in (1.13) does not hold, we obtain x_{eq,Λ,Λ^*} is spurious.

We demonstrate how the above definitions and theorems can be used by analyzing the following example.

Example 1.5 Consider the following control systems:

$$\begin{cases} \dot{x} = \begin{pmatrix} -1 & 2 & 0 & 0 \\ 1 & -2 & 1 & 0 \\ 0 & 1 & -2 & 1 \\ 0 & 0 & 1 & -2 \end{pmatrix} x + \begin{pmatrix} 1.2 & 0 \\ 0.6 & 1.7 \\ 0.3 & 0.1 \\ 1.1 & 0 \end{pmatrix} u_s, \\ y = Cx, \end{cases}$$

the saturation state feedback $u = sat(-Kx)$ is

$$K = \begin{pmatrix} 9.55 & 12.5 & 7.675 & 3.75 \\ 8.35 & 11 & 6.625 & 3.1 \end{pmatrix} = \begin{pmatrix} K_1 \\ K_2 \end{pmatrix},$$

where

$$u_{s,1} = \begin{cases} 4, & -[9.55\ 12.5\ 7.675\ 3.75\,]x \geq 4, \\ -[9.55\ 12.5\ 7.675\ 3.75\,]x, & |-[9.55\ 12.5\ 7.675\ 3.75\,]x| < 4, \\ -4, & -[9.55\ 12.5\ 7.675\ 3.75\,]x \leq -4, \end{cases}$$

$$u_{s,2} = \begin{cases} 5, & -[8.35\ 11\ 6.625\ 3.1\,]x \geq 5, \\ -[8.35\ 11\ 6.625\ 3.1\,]x, & |-[8.35\ 11\ 6.625\ 3.1\,]x| < 5, \\ -5, & -[8.35\ 11\ 6.625\ 3.1\,]x \leq -5, \end{cases}$$

$$u_s = \begin{pmatrix} u_{s,1} \\ u_{s,2} \end{pmatrix} = \begin{pmatrix} sat(-K_1 x) \\ sat(-K_2 x) \end{pmatrix}, \quad u_{lim} = \begin{pmatrix} 4 \\ 5 \end{pmatrix},$$

and the following equilibrium points:

$$x_{e_1} = [0.8202\ 2.2014\ -0.6655\ -2.7266],$$
$$x_{e_2} = -[0.8202\ 2.2014\ -0.6655\ -2.7266],$$
$$x_{e_3} = [1.2902\ -1.7549\ 0.7837\ 2.5918],$$
$$x_{e_4} = -[1.2902\ -1.7549\ 0.7837\ 2.5918],$$
$$x_{e_5} = 0,$$
$$x_{e_6} = -[50.1\ -210\ -15.7\ 5.65],$$
$$x_{e_7} = [50.1\ -210\ -15.7\ 5.65],$$
$$x_{e_8} = [2.9\ -0.95\ 1.3\ 2.85],$$
$$x_{e_9} = -[2.9\ -0.95\ 1.3\ 2.85],$$

correspondingly satisfy equations as follows:

$$\dot{x}(t) = (A - B\begin{pmatrix}1 & 0 \\ 0 & 0\end{pmatrix}K)x(t) + B\begin{pmatrix}0 & 0 \\ 0 & 1\end{pmatrix}u_{\lim},$$

$$\dot{x}(t) = (A - B\begin{pmatrix}1 & 0 \\ 0 & 0\end{pmatrix}K)x(t) + B\begin{pmatrix}0 & 0 \\ 0 & -1\end{pmatrix}u_{\lim},$$

$$\dot{x}(t) = (A - B\begin{pmatrix}0 & 0 \\ 0 & 1\end{pmatrix}K)x(t) + B\begin{pmatrix}1 & 0 \\ 0 & 0\end{pmatrix}u_{\lim},$$

$$\dot{x}(t) = (A - B\begin{pmatrix}0 & 0 \\ 0 & 1\end{pmatrix}K)x(t) + B\begin{pmatrix}-1 & 0 \\ 0 & 0\end{pmatrix}u_{\lim},$$

$$\dot{x}(t) = (A - B\begin{pmatrix}1 & 0 \\ 0 & 1\end{pmatrix}K)x(t),$$

$$\dot{x}(t) = (A - B\begin{pmatrix}0 & 0 \\ 0 & 0\end{pmatrix}K)x(t) + B\begin{pmatrix}1 & 0 \\ 0 & 1\end{pmatrix}u_{\lim},$$

$$\dot{x}(t) = (A - B\begin{pmatrix}0 & 0 \\ 0 & 0\end{pmatrix}K)x(t) + B\begin{pmatrix}-1 & 0 \\ 0 & -1\end{pmatrix}u_{\lim},$$

$$\dot{x}(t) = (A - B\begin{pmatrix}0 & 0 \\ 0 & 0\end{pmatrix}K)x(t) + B\begin{pmatrix}1 & 0 \\ 0 & -1\end{pmatrix}u_{\lim},$$

$$\dot{x}(t) = (A - B\begin{pmatrix}0 & 0 \\ 0 & 0\end{pmatrix}K)x(t) + B\begin{pmatrix}-1 & 0 \\ 0 & 1\end{pmatrix}u_{\lim}.$$

Let

$$\mathbb{D}_1 = \{x\,||-K_1x| < 4, -K_2x \geq 5\},$$
$$\mathbb{D}_2 = \{x\,||-K_1x| < 4, -K_2x \leq -5\},$$
$$\mathbb{D}_3 = \{x\,|-K_1x \geq 4, |-K_2x| < 5\},$$
$$\mathbb{D}_4 = \{x\,|-K_1x \leq -4, |-K_2x| < 5\},$$
$$\mathbb{D}_5 = \{x\,||-K_1x| < 4, |-K_2x| < 5\},$$
$$\mathbb{D}_6 = \{x\,|-K_1x \geq 4, -K_2x \geq 5\},$$
$$\mathbb{D}_7 = \{x\,|-K_1x \leq -4, -K_2x \leq -5\},$$
$$\mathbb{D}_8 = \{x\,|-K_1x \geq 4, -K_2x \leq -5\},$$
$$\mathbb{D}_9 = \{x\,|-K_1x \leq -4, -K_2x \geq 5\}.$$

1.2 Problem Statements, Mathematical Modeling, and Equilibrium Points

It is clear that $x_{e_6} \in \mathbb{D}_6$, $x_{e_7} \in \mathbb{D}_7$, $0 \in \mathbb{D}_5$, and the other equilibrium points do not possess the above property.

Remark If the open-loop system is unstable, then there may exist genuine unstable nonzero equilibrium points, especially when $A - BK$ is stable, then the origin is only locally asymptotic stable. For example, x_{e_6}, x_{e_7} are genuine unstable equilibrium points and 0 is genuine stable.

1.3 Stability and Boundedness of Linear System with Saturation Inputs

The stability property of the MIMO control system with saturation inputs is considered below.

Theorem 1.5 *Given (1.5), if there exists a positive matrix $P = P^T > 0$ and the constant $\varepsilon > 0$ such that*

$$(A - B\Lambda K)^T P + P(A - B\Lambda K) + \varepsilon P B B^T + \frac{1}{\varepsilon} K^T (I - \Lambda) K < 0, \forall \Lambda \neq I,$$
$$(A - BK)^T P + P(A - BK) < 0,$$
(1.16)

then the following statements are genuine:
(1) $x_{eq,\Lambda,\Lambda^*} = -(A - B\Lambda K)^{-1} B \Lambda^* u_{\lim} \neq 0$ are stable spurious points;
(2) $x = 0$ is globally asymptotically stable.

Proof Given a Lyapunov function candidate $V(x) = x^T P x$, $x \in \mathbb{R}^n$, $\dot{V}$ becomes

$$\begin{aligned}
\left.\frac{dV}{dt}\right|_{(1.5)} &= \dot{x}^T P x + x^T P \dot{x} \\
&= [(A - B\Lambda K)x(t) + B\Lambda^* u_{\lim}]^T P x + x^T P[(A - B\Lambda K)x(t) + B\Lambda^* u_{\lim}] \\
&= x^T(t)[(A - B\Lambda K)^T P + P(A - B\Lambda K)]x(t) + [B\Lambda^* u_{\lim}]^T P x + x^T P B \Lambda^* u_{\lim}.
\end{aligned}$$
(1.17)

From the definition of $\mathbb{D}_{\Lambda,\Lambda^*}$, it is clear that

$$\mathbb{R}^n = \cup \mathbb{D}_{\Lambda,\Lambda^*}, \quad \mathbb{D}_{\Lambda_1,\Lambda_1^*} \cap \mathbb{D}_{\Lambda_2,\Lambda_2^*} = \phi, \ (\Lambda_1 \neq \Lambda_2).$$

From the inequalities

$$2\alpha\beta \leq \varepsilon\alpha^2 + \frac{1}{\varepsilon}\beta^2, \ \forall \varepsilon > 0.$$

(1.17) is converted into

$$\begin{aligned}
\left.\frac{dV}{dt}\right|_{(1.5)} &\leq x^T[(A - B\Lambda K)^T P + P(A - B\Lambda K) + \varepsilon P B B^T P]x + \frac{1}{\varepsilon} u^T_{\lim}(I - \Lambda^*\Lambda^*) u_{\lim} \\
&= x^T[(A - B\Lambda K)^T P + P(A - B\Lambda K) + \varepsilon P B B^T P]x + \frac{1}{\varepsilon} u^T_{\lim}(I - \Lambda) u_{\lim}.
\end{aligned}$$
(1.18)

By the assumptions of Theorem 1.5, it follows

$$x^T[(A - B\Lambda K)^T P + P(A - B\Lambda K) + \varepsilon PBB^T P + \frac{1}{\varepsilon} K^T(I - \Lambda)K]x < 0,$$

for $x \in \mathbb{D}_{\Lambda,\Lambda^*} \neq \mathbb{D}_{I,0}$.

Further, it follows from the definition of $x \in \mathbb{D}_{\Lambda,\Lambda^*} \neq \mathbb{D}_{I,0}$ that

$$\begin{aligned}
0 > &\max_{x \in \mathbb{D}_{\Lambda,\Lambda^*}} x^T[(A - B\Lambda K)^T P + P(A - B\Lambda K) + \varepsilon PBB^T P + \tfrac{1}{\varepsilon} K^T(I - \Lambda)K]x \\
\geq &\max_{x \in \mathbb{D}_{\Lambda,\Lambda^*}} x^T[(A - B\Lambda K)^T P + P(A - B\Lambda K) + \varepsilon PBB^T P]x + \max_{x \in \mathbb{D}_{\Lambda,\Lambda^*}} \tfrac{1}{\varepsilon} x^T K^T(I - \Lambda)Kx \\
\geq &\max_{x \in \mathbb{D}_{\Lambda,\Lambda^*}} x^T[(A - B\Lambda K)^T P + P(A - B\Lambda K) + \varepsilon PBB^T P]x + \max_{x \in \partial \mathbb{D}_{\Lambda,\Lambda^*}} \tfrac{1}{\varepsilon} x^T K^T(I - \Lambda)Kx \\
\geq &\max_{x \in \mathbb{D}_{\Lambda,\Lambda^*}} x^T[(A - B\Lambda K)^T P + P(A - B\Lambda K) + \varepsilon PBB^T P]x + \max_{x \in \partial \mathbb{D}_{\Lambda,\Lambda^*}} \tfrac{1}{\varepsilon} x^T K^T \Lambda^* \Lambda^* K x \\
\geq &\max_{x \in \mathbb{D}_{\Lambda,\Lambda^*}} x^T[(A - B\Lambda K)^T P + P(A - B\Lambda K) + \varepsilon PBB^T P]x + \max_{x \in \partial \mathbb{D}_{\Lambda,\Lambda^*}} \tfrac{1}{\varepsilon} (\Lambda^* u_{\lim})^T (\Lambda^* u_{\lim}) \\
= &\max_{x \in \mathbb{D}_{\Lambda,\Lambda^*}} x^T[(A - B\Lambda K)^T P + P(A - B\Lambda K) + \varepsilon PBB^T P]x + \tfrac{1}{\varepsilon} u^T_{\lim}(I - \Lambda) u_{\lim}.
\end{aligned}$$
(1.19)

Combining (1.18) with (1.19), for any $x \in \mathbb{D}_{\Lambda,\Lambda^*}$, it follows that

$$\frac{dV(x(t))}{dt}\bigg|_{(1.5)} < 0.$$

It implies that

$$x_{eq,\Lambda,\Lambda^*} = -(A - B\Lambda K)^{-1} B\Lambda^* u_{\lim}(\neq 0) \notin \mathbb{D}_{\Lambda,\Lambda^*},$$

and $A - B\Lambda K$ is stable, while for any $\forall x \in \mathbb{D}_{I,0}$, it follows that

$$\frac{dV}{dt}\bigg|_{(1.5)} = x^T(t)[(A - BK)^T P + P(A - BK)]x(t) < 0.$$

Hence, $x_{eq,I,0} = 0$ is the only genuine stable equilibrium point and the origin is globally asymptotically stable.

Theorem 1.6 *For the given system (1.5), if A is stable, then there exists a feedback $u = -Kx$ such that all solutions of saturation system (1.5) are bounded.*

Proof Since matrix A is stable, then there exists $P^T = P > 0$ such that

$$A^T P + PA < 0.$$

Assume the linear input $u = -k_0 B^T P$, where k_0 is any positive constants, it is clear that

$$A - B\Lambda K = A - k_0 B\Lambda B^T P.$$

So

$$(A - B\Lambda K)^T P + P(A - B\Lambda K) = A^T P + PA - 2k_0 PB\Lambda B^T P < 0.$$

1.3 Stability and Boundedness of Linear System with Saturation Inputs

Further, there exists a $T(\varepsilon) > 0$ such that

$$\left\| e^{(A-B\Lambda K)(t-t_0)} x_0 + [e^{(A-B\Lambda K)(t-t_0)} - I](A - B\Lambda K)^{-1} B\Lambda^* u_{\lim} \right\| - \left\| (A - B\Lambda K)^{-1} B\Lambda^* u_{\lim} \right\|$$
$$< \varepsilon, \ (t \geq t_0 + T(\varepsilon)).$$

By the assumption of Theorem 1.6 and for any $x_0 \in \mathbb{R}^n$, $\mathbb{R}^n = \cup \mathbb{D}_{\Lambda,\Lambda^*}$, $\cap \mathbb{D}_{\Lambda,\Lambda^*} = \phi$, it follows that

$$\begin{aligned}
&\|x(t, t_0, x_0)\| \\
&\leq \sup_{\forall x_0 \in \mathbb{R}^n, \forall t \in [0,\infty)} \left\{ \left\| e^{(A-B\Lambda K)(t-t_0)} x_0 + [e^{(A-B\Lambda K)(t-t_0)} - I](A - B\Lambda K)^{-1} B\Lambda^* u_{\lim} \right\| \right\} \\
&= \max_{x_0 \in \mathbb{R}^n, \forall \varepsilon(x_0)>0} \begin{cases} \sup_{\forall x_0 \in \mathbb{R}^n, \forall t \in [0,T(\varepsilon)]} \left\{ e^{(A-B\Lambda K)(t-t_0)} x_0 + [e^{(A-B\Lambda K)(t-t_0)} - I](A - B\Lambda K)^{-1} B\Lambda^* u_{\lim} \right\} \\ \max_{(\Lambda,\Lambda^*, \forall \varepsilon(x_0)>0)} \left\{ \left\| (A - B\Lambda K)^{-1} B\Lambda^* u_{\lim} \right\| + \varepsilon \right\} \end{cases} \\
&< \infty.
\end{aligned}$$

It implies that all solutions of (1.5) are bounded.

Theorem 1.7 *For the given system (1.5) with single saturation input, let A be stable, then all solutions of (1.5) are bounded.*

Because A is asymptotically stable, there exist $P = P^T > 0$ such that

$$A^T P + PA = -I.$$

Let V be the Lyapunov function candidate where

$$V = x^T P x,$$

then

$$\begin{aligned}
\left.\frac{dV}{dt}\right|_{(1.5)} &= -x^T x \pm 2x^T P B K u_{\lim} - \|PBKu_{\lim}\|^2 + \|PBKu_{\lim}\|^2 \\
&\leq -\|x \pm PBKu_{\lim}\|^2 + \|PBKu_{\lim}\|^2,
\end{aligned}$$

and when $\|x\|$ is large enough, the above inequality is negative.

1.4 Stability of Linear System with Lipschitz Nonlinearity

The stability results obtained from linear systems are extended to nonlinear systems. Consider the following nonlinear system:

$$\begin{cases} \dot{x} = Ax + Bu_s + f(t, x), \\ u(s) = \begin{pmatrix} u_{s,1} & u_{s,2} & \cdots & u_{s,m} \end{pmatrix}, \\ u_{s,i} = sat(-K_i x), \ i = 1, 2, \ldots, m, \end{cases} \quad (1.20)$$

where $x \in \mathbb{R}^n$ is the state vector, $y \in \mathbb{R}^1$ is the output vector, and $u_{s,i} \in \mathbb{R}^1$, ($i = 1, 2, \cdot, m$) is the control input. A, B, C, and K are constant matrices of appropriate dimensions, and the system is minimal.

It is known that $f(t, 0) \equiv 0$, and $f(t, x)$ is global Lipschitz in x with Lipschitz constants $k_i, i = 1, 2, \ldots, m$, i.e.,

$$|f_i(t, x) - f_j(t, y)| \le k_i |x - y|, \quad k_i > 0.$$

From the above discussions, we know that

$$\dot{x} = (A - B\Lambda K)x + B\Lambda^* u_{\lim} + f(t, x).$$

Let $V = x^T P x$, then

$$\begin{aligned}
\dot{V}|_{(1.20)} &= \dot{x}^T P x + x^T P \dot{x} \\
&= [Ax + Bu + f(t,x)]^T P x + x^T P[Ax + Bu + f(t,x)] \\
&= [(A - B\Lambda K)x + B\Lambda^* u_{\lim} + f(t,x)]^T P x + x^T P[(A - B\Lambda K)x \\
&\quad + B\Lambda^* u_{\lim} + f(t,x)] \\
&= x^T(t)[(A - B\Lambda K)^T P + P(A - B\Lambda K)]x(t) \\
&\quad + [B\Lambda^* u_{\lim}]^T P x + x^T P B\Lambda^* u_{\lim} + f^T(t,x) P x + x^T P f(t,x).
\end{aligned} \quad (1.21)$$

Let $x \in \mathbb{D}_{I,0}$, then (1.21) transforms to

$$\dot{V}|_{(1.20)} \le x^T[(A - BK)^T P + P(A - BK) + \varepsilon_1 \text{diag}\left(k_1^2 \ k_2^2 \ \ldots \ k_n^2\right) + \frac{1}{\varepsilon_1} P^2]x < 0.$$

Let $x \in \mathbb{D}_{\Lambda, \Lambda^*} \ne \mathbb{D}_{I,0}$, then (1.21) further transforms to

$$\begin{aligned}
\dot{V}|_{(1.20)} &\le x^T[(A - B\Lambda K)^T P + P(A - B\Lambda K) + \varepsilon_1 \text{diag}(k_1^2 \ k_2^2 \ \ldots \ k_n^2) \\
&\quad \frac{1}{\varepsilon_1} P^2]x + \varepsilon_2 x^T B^T B x + \frac{1}{\varepsilon_2} x^T P^2 x \\
&= x^T \left[(A - B\Lambda K)^T P + P(A - B\Lambda K) + \varepsilon_1 \text{diag}(k_1^2 \ k_2^2 \ \ldots \ k_n^2) \right. \\
&\quad \left. + \varepsilon_2 B^T B + \left(\frac{1}{\varepsilon_1} + \frac{1}{\varepsilon_2}\right) P^2 \right] x \\
&< 0,
\end{aligned} \quad (1.22)$$

The above results are made in the theorem below.

Theorem 1.8 *For the given system (1.20), if there exists a matrix $P^T = P > 0$ and any constants $\varepsilon_1, \varepsilon_2$ such that*

$$(A - BK)^T P + P(A - BK) + \varepsilon_1 \text{diag}\left(k_1^2 \ k_2^2 \ \ldots \ k_n^2\right) + \frac{1}{\varepsilon_1} P^2 < 0,$$
$$(A - B\Lambda K)^T P + P(A - B\Lambda K) + \varepsilon_1 \text{diag}(k_1^2 \ k_2^2 \ \ldots \ k_n^2) + \varepsilon_2 B^T B + (\frac{1}{\varepsilon_1} + \frac{1}{\varepsilon_2}) P^2 < 0,$$

then $x = 0$ of (1.20) is globally asymptotically stable.

1.5 Robust Stability and Linear Matrix Inequalities

We consider more generally the continuous-time system

$$\begin{cases} \dot{x} = A(t)x(t) + B(t)u(t), \\ y = C^T x(t), \\ u(s) = \begin{pmatrix} u_{s,1} & u_{s,2} & \cdots & u_{s,m} \end{pmatrix}, \\ u_{s,i} = sat(-K_i x), i = 1, 2, \ldots, m, \end{cases} \quad (1.23)$$

where $x \in \mathbb{R}^n$ is the state vector, $y \in \mathbb{R}^1$ is the output vector, $u_{s,i} \in \mathbb{R}^1 (i = 1, 2, \ldots, m)$ is the control input. A, B, C, and K are constant matrices of appropriate dimensions, and the system is minimal.

In order to represent uncertainties that may affect system (1.23), we restrict time-varying systems matrices to

$$A(t) \in \mathbb{A}, \quad B(t) \in \mathbb{B},$$

where $\mathbb{A}$, $\mathbb{B}$ are compact sets.

In the sequel, we will consider two different types of uncertainty, namely, polytopic uncertainty and norm boundary uncertainty.

1.5.1 Polytopic Uncertainty

That is to say that the above system matrices belong to matrices polytopics

$$\mathbb{A} = convex\ hull\{A_1\ A_2\ \ldots\ A_{n_A}\},$$
$$\mathbb{B} = convex\ hull\{B_1\ B_2\ \ldots\ B_{n_B}\}.$$

Moreover, we assume that a robust stabilizing state feedback control matrix K is built for (1.23) by some design method, polytopic uncertainty conditions as follows:

$$\begin{aligned} & (A_i - B_i \Lambda K)^T P + P(A_i - B_i \Lambda K) + \varepsilon P B_i B_i^T P + \frac{1}{\varepsilon} K^T (I - \Lambda) K < 0, \ \Lambda \neq I, \\ & (A_i - B_i K)^T P + P(A_i - B_i K) < 0, \end{aligned} \quad (1.24)$$

hold, then $x = 0$ of (1.23) is globally asymptotically stable.

1.5.2 Norm Boundary Uncertainty

We assume

$$A(t) = \{A_0 + DF(t)E_1, \|F(t)\|_2 \leq 1\},$$
$$B(t) = \{B_0 + DF(t)E_2, \|F(t)\|_2 \leq 1\},$$

for the given nominal matrices A_0, B_0 and constant matrices D, E_1, E_2.

So, the system (1.23) is transformed into

$$\dot{x}(t) = A_{\Lambda,\Lambda^*} x(t) + B_{\Lambda,\Lambda^*} u_{\lim}, \tag{1.25}$$

where

$$A_{\Lambda,\Lambda^*} = convex\ hull\{A(t) + B(t)\Lambda K\},$$
$$B_{\Lambda,\Lambda^*} = convex\ hull\{B(t)\Lambda^*\},$$

Λ, Λ^* are defined as the above, and then we obtain the following norm boundary uncertainty conditions:

$$\begin{bmatrix} (A_0 + B_0 K)^T P + P(A_0 + B_0 K) & PD & E_1^T + K^T E_2^T \\ D^T P & -I & 0 \\ E_1 + E_2 K & 0 & -I \end{bmatrix} < 0,$$

$$\begin{bmatrix} (A_0 + B_0 \Lambda K)^T P + P(A_0 + B_0 \Lambda K) + & PD & E_1^T + K^T E_2^T \\ \varepsilon(I - A)[B_0 B_0^T + E_2 E_2^T] + \frac{1}{\varepsilon}[P^2 + D^T D] & & \\ D^T P & -I & 0 \\ E_1 + E_2 K & 0 & -I \end{bmatrix} < 0, \tag{1.26}$$

and $x = 0$ of (1.23) is globally asymptotically stable.

1.6 Example Analysis and Simulink

We give one example to demonstrate the results of Theorem 1.5, which is one important result in the chapter.

Example 1.6 We consider the following second-order control system with single saturated input:

$$\begin{cases} \dot{y}_1(t) = -4y_1(t) + 2y_2(t) + u_s, \\ \dot{y}_2(t) = -3y_2(t), \\ u_s = sat(-0.5y_1(t) - y_2(t)), \end{cases} \tag{1.27}$$

saturated input u_s is defined as

$$sat(-0.5y_1(t) - y_2(t)) = \begin{cases} 1, & if\ -0.5y_1(t) - y_2(t) \geq 1, \\ -0.5y_1(t) - y_2(t), & if\ |-0.5y_1(t) - y_2(t)| < 1, \\ -1, & if\ -0.5y_1(t) - y_2(t) \leq -1, \end{cases}$$

1.6 Example Analysis and Simulink

and
$$A = \begin{bmatrix} -4 & 2 \\ 0 & -3 \end{bmatrix}, B = \begin{bmatrix} 1 \\ 0 \end{bmatrix}, K = \begin{bmatrix} -0.5 \\ -1 \end{bmatrix}^T,$$

choosing $P = I$, we have

$$(A - BK)^T P + P(A - BK)$$

$$= \left[\begin{bmatrix} -4 & 2 \\ 0 & -3 \end{bmatrix} - \begin{bmatrix} 0.5 & 1 \\ 0 & 0 \end{bmatrix} \right] + \left[\begin{bmatrix} -4 & 2 \\ 0 & -3 \end{bmatrix} - \begin{bmatrix} 0.5 & 1 \\ 0 & 0 \end{bmatrix} \right]^T$$

$$= \begin{bmatrix} -9 & 1 \\ 1 & -3 \end{bmatrix} < 0,$$

$$A^T P + PA + PBB^T P + K^T K$$

$$= \begin{bmatrix} -4 & 2 \\ 0 & -3 \end{bmatrix} + \begin{bmatrix} -4 & 2 \\ 0 & -3 \end{bmatrix}^T + \begin{bmatrix} 1 & 0 \\ 0 & 0 \end{bmatrix} + \begin{bmatrix} \frac{1}{4} & \frac{1}{2} \\ \frac{1}{2} & 1 \end{bmatrix}$$

$$= \begin{bmatrix} -\frac{27}{4} & \frac{5}{2} \\ \frac{5}{2} & -5 \end{bmatrix} < 0, \varepsilon = 1,$$

$$A^T P + PA + 2PBB^T P + \tfrac{1}{2} K^T K$$

$$= \begin{bmatrix} -4 & 2 \\ 0 & -3 \end{bmatrix} + \begin{bmatrix} -4 & 2 \\ 0 & -3 \end{bmatrix}^T + 2\begin{bmatrix} 1 & 0 \\ 0 & 0 \end{bmatrix} + \tfrac{1}{2}\begin{bmatrix} \frac{1}{4} & \frac{1}{2} \\ \frac{1}{2} & 1 \end{bmatrix}$$

$$= \begin{bmatrix} -\frac{47}{8} & \frac{9}{4} \\ \frac{9}{4} & -\frac{11}{2} \end{bmatrix} < 0, \varepsilon = 2,$$

$$A^T P + PA + \tfrac{7}{2} PBB^T P + \tfrac{2}{7} K^T K$$

$$= \begin{bmatrix} -4 & 2 \\ 0 & -3 \end{bmatrix} + \begin{bmatrix} -4 & 2 \\ 0 & -3 \end{bmatrix}^T + \tfrac{7}{2}\begin{bmatrix} 1 & 0 \\ 0 & 0 \end{bmatrix} + \tfrac{2}{7}\begin{bmatrix} \frac{1}{4} & \frac{1}{2} \\ \frac{1}{2} & 1 \end{bmatrix}$$

$$= \begin{bmatrix} -\frac{31}{7} & \frac{15}{7} \\ \frac{15}{7} & -\frac{40}{7} \end{bmatrix} < 0, \varepsilon = 3.5.$$

It follows that the origin of (1.27) is globally asymptotical stability.

Its phase and movement trajectories of (1.27) are denoted passing many initial values (the red dots) by MATLAB as follows:

(i) Phase portrait $y_2(y_1)$ and trajectories $y_1(t)$, $y_2(t)$ passing the initial points, which satisfy $-0.5y_1 - y_2 \leq -1$ and $(y_1 - 2)^2 + (y_2 - 2)^2 = 1$ may see Figs. 1.5, 1.6 and 1.7.

(ii) Phase portrait $y_2(y_1)$ and trajectories $y_1(t)$, $y_2(t)$ passing initial the points, which satisfy $|-0.5y_1 - y_2| < 1$ and $y_1^2 + y_2^2 = 0.64$ may see Figs. 1.8, 1.9 and 1.10.

Fig. 1.5 Phase trajectories starting on the circle of $(y_1 - 2)^2 + (y_2 - 2)^2 = 1$

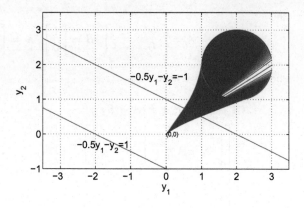

Fig. 1.6 Trajectories of $y_1(t)$

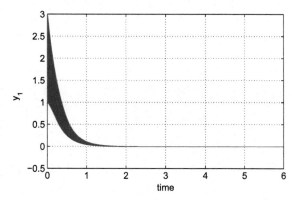

Fig. 1.7 Trajectories of $y_2(t)$

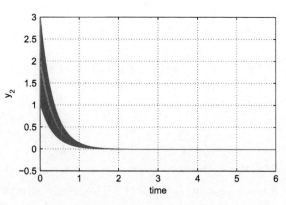

(iii) Phase portrait $y_2(y_1)$ and trajectories $y_1(t)$, $y_2(t)$ passing initial the points, which satisfy $-0.5y_1 - y_2 \geq 1$ and $(y_1 + 2)^2 + (y_2 + 2)^2 = 1$ may see Figs. 1.11, 1.12 and 1.13.

1.6 Example Analysis and Simulink

Fig. 1.8 Phase trajectories starting on the circle of $y_1^2 + y_2^2 = 0.64$

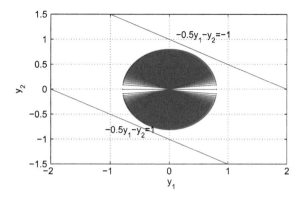

Fig. 1.9 Trajectories of $y_1(t)$

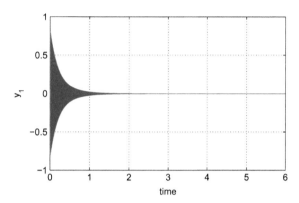

Fig. 1.10 Trajectories of $y_2(t)$

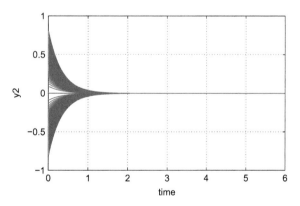

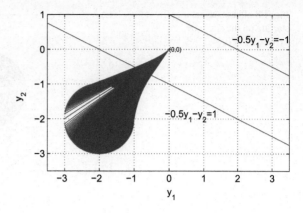

Fig. 1.11 Phase trajectories starting on the circle of $(y_1 + 2)^2 + (y_2 + 2)^2 = 1$

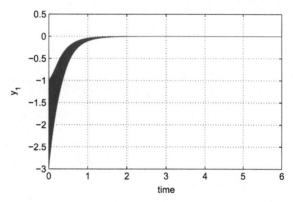

Fig. 1.12 Trajectories of $y_1(t)$

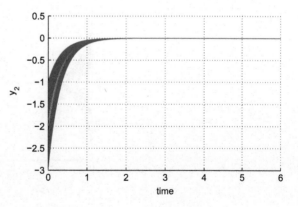

Fig. 1.13 Trajectories of $y_2(t)$

1.7 Conclusion

This chapter is dedicated to presenting a new method of saturation nonlinear fields which can be used to tackle problems of analysis and synthesis for linear systems subject to actuator saturations. Noting that new systematic techniques which can be formally presented in a chapter, our main objective is to show not only the recent methods but also their practical applications. The focus of this chapter is on the so-called $0 - 1$ algebra-geometry type structure equations. The chapter attempt to provide constructive qualitative analysis methods and stability methods for linear systems with saturated inputs in both global and local contexts. Our hope is also that this chapter will enable practitioners to have more concise model systems to modern saturation nonlinear techniques and that this will encourage future applications. Such novel mathematical model presents different technical and mathematical points of view to those traditionally presented in the saturation nonlinear literatures but is sure to keep the field alive with new and adventurous ideas.

Acknowledgements This work is supported by National Key Research and Development Program of China (2017YFF0207400).

References

1. Dugard L, Verrist EI, editors. Stability and control of time-delay systems. Berlin: Springer; 1998. p. 303–31.
2. Henrion D, Tarbouriech S. LMI relaxations for robust stability of linear systems with saturating controls. Automatica. 1999;35:1599–604.
3. Horisberger HP, Belanger PR. Regulator for linear, time invariant plants with uncertain parameters. IEEE Trans Autom Control. 1976;42:705–8.
4. Huang L. Stability theory. Beijing: Peking University Press; 1992. p. 235–83.
5. Kapoor N, Daoutidis P. An observer-based anti-windup scheme for nonlinear systems with input constraints. Int J Control. 1999;72(1):18–29.
6. Lee AW, Hedrick JK. Some new results on closed-loop stability in the presence of control saturation. Int J Control. 1995;62(3):619–51.
7. Liberzon D, Hespanta JP, Morse AS. Stability of switched systems:a Lie-algebraic condition. Syst Control Lett. 1999;37:117–22. North-Holland
8. Mareada KS, Balakrishan J. A common Lyapunov function for stable LTI systems with commuting A-martices. IEEE Trans Autom Control. 1994;39(12):2469–71.
9. Mancilla-Aguilar JL, Garcia RA. A converse Lyapunov theorem for nonlinear switched systems. Syst Control Lett. 2000;41(1):69–71.
10. Molchanov AP, Pyatnitskiy YS. Criteria of asymptotic stability of differential and difference inclusions encountered in control theory. Systems and Control Letters, North-Holland. 1989;13:59–64. North-Holland
11. Polanski K. On absolute stability analysis by polyhedric Lypunov functions. Automatica. 2000;36:573–8.
12. Scibile L, Kouvaritakis BIF. Stability region for a class of open-loop unstable linear systems:theory and application. Automatica. 2000;36(1):37–44.
13. Sontag ED, et al., Sussmann HJ. Nonlinear output feedback design for linear systems with saturated control. Proc 29th Conf Decis Control. 1990;45(5):719–21.

14. Tatsushi, Yasuyuki F. Two conditions concerning common quadratic lyapunov functions for linear systems. IEEE Trans Autom Control. 1997;42(5):750–61.
15. Tarbouriech S, Gomes da JM. Synthesis of controllers for continuous-times delay systems with saturated control via LMIs. IEEE Trans Autom Control. 2000;45(1):105–11.

Chapter 2
Equilibrium Points of Linear Systems with Single Saturated Input Under Commuting Conditions

2.1 Introduction

Global asymptotic stability (GAS) of the origin of linear systems with saturated inputs has been extensively studied by many researchers [1–16]. More recently, efforts have been made to extend these results to nonlinear and delay systems [1, 6, 16]. Lee and Hedrick [6] proposed conditions for the GAS of the origin under a linear control law, which was designed to stabilize a linear plant. Saturation occurs because the required inputs exceed the physical limits of the actuator. Studying this aspect of the saturation problem is of concern because it is a very common way in which control saturation occurs in practice, where control bounds are usually ignored in the initial design. Available methods, such as linear, quadratic, or input–output linearization, are used to obtain control laws. For example, if the resulting control signal happens to exceed the limit, the control is then allowed to saturate.

Lee and Hedrick [6] used the hypercube in $\mathbb{R}^n$ to obtain GAS of the origin. The same method was used in [1–3, 5–16], where some results about GAS of the origin were obtained. Henrion and Tarbouriech [2] pointed out that the problem of finding the set of initial conditions that steer the dynamics to the origin is a very challenging and a largely unsolved problem in general. When the control is allowed to saturate, the resulting control signal exceeds the limit and new, possibly spurious nonzero equilibrium points appear.

It is necessary to learn how such equilibrium points, apart from the origin, function in the analysis of control systems with saturated inputs.

The main contribution of this chapter is to present a new method for analysing nonlinear systems by investigating these new equilibrium points. If these equilibria also appear in the unsaturated system, we term them "genuine"; if they do not, we term them "spurious".

The outline of this chapter is as follows. First, the new equilibrium points are defined. Second, the structure of the commutative matrices is considered. The existence of feedback matrices of the commutative state matrices, set in the MIMO closed-loop systems, is considered. The existence of feedback matrices in the systems in open loop is equivalent to the existence of the solution to a matrix equation denoted by a Kronecker product. Third, from definitions of the equilibrium points for the control systems with saturated inputs, different kinds of equilibrium points are analyzed when the state matrices commute. By defining new equilibrium points, the relationship between equilibrium points is discussed for linear systems with a single saturated input. Four criteria for equilibrium points are outlined for such linear systems. Finally, four interesting examples, including their corresponding simulation plots, are shown to illustrate the above results. We find that all equilibrium points, except for the origin, are spurious but stable when analyzing the GAS of the origin of the control systems with saturated inputs. When the state matrix is unstable, it is possible that one or several pairs of nonzero stable genuine equilibrium points exist under certain feedback laws.

The continuous-time system of [1, 6, 13] is considered:

$$\begin{aligned} \dot{x}(t) &= Ax(t) + Bu_s(t), \\ y(t) &= C^T x(t), \\ u_s(t) &= sat(-Kx), \end{aligned} \quad (2.1)$$

where $x(t) \in \mathbb{R}^n$ is the state vector, $y(t) \in \mathbb{R}^1$ is the output vector, $u_s(t) \in \mathbb{R}^1$ is the control input, A, B, C, K are matrices of suitable dimensions, and the system is minimal. We define the saturated input $sat(-Kx)$ as

$$sat(-Kx) = \begin{cases} u_{\lim}, & if - Kx \geq u_{\lim}, \\ -Kx, & if |-Kx| < u_{\lim}, \\ -u_{\lim}, & if - Kx \leq -u_{\lim}, \end{cases} \quad (2.2)$$

where $u_{\lim}$ is the saturation limit of the feedback system and $\|u_s\| \leq u_{\lim}$.

From (2.1), (7.6), it is possible to rewrite the single-input single-output (SISO) system in the following form:

$$\begin{cases} \dot{x}(t) = (A - \lambda BK)x(t) + \lambda^* Bu_{\lim}, \\ y = C^T x(t), \end{cases} \quad (2.3)$$

where $\lambda = -1, 0, 1$, λ^* is obtained from λ as follows:

$$\begin{cases} (\lambda, \lambda^*) = (0, 1) \iff -Kx - u_{\lim} \geq 0, \\ (\lambda, \lambda^*) = (1, 0) \iff |-Kx| < u_{\lim}, \\ (\lambda, \lambda^*) = (0, -1) \iff -Kx + u_{\lim} \leq 0. \end{cases} \quad (2.4)$$

If $A - \lambda BK$ is nonsingular, then (2.3) gives the equilibrium state

$$x_{eq,\lambda,\lambda^*} = -\lambda^*(A - \lambda BK)^{-1} Bu_{\lim}.$$

If we introduce

$$\mathbb{D}_{0,1} = \{x \,|\, -Kx \geq u_{\lim}\},$$

2.1 Introduction

$$\mathbb{D}_{0,-1} = \{x \mid -Kx \leq -u_{\lim}\},$$

$$\mathbb{D}_{1,0} = \{x \mid |-Kx| < u_{\lim}\},$$

then we have the following definitions.

Definition 2.1 (I) For system (2.1), some pair (λ, λ^*) and its corresponding x_{eq,λ,λ^*} satisfying (2.3), if $x_{eq,\lambda,\lambda^*} \in \mathbb{D}_{\lambda,\lambda^*}$, we call x_{eq,λ,λ^*} a genuine equilibrium point. For such an equilibrium, if $(A - \lambda BK)$ is a stable (unstable) matrix, then x_{eq,λ,λ^*} is a genuine stable (unstable) equilibrium point.

(II) For system (2.1), some pair (λ, λ^*) and its corresponding x_{eq,λ,λ^*} satisfying (2.3), if $x_{eq,\lambda,\lambda^*} \notin \mathbb{D}_{\lambda,\lambda^*}$, then we call x_{eq,λ,λ^*} a spurious equilibrium point. For such spurious equilibria, if $(A - \lambda BK)$ is stable (unstable), then we say that x_{eq,λ,λ^*} is a spurious stable (unstable) equilibrium point.

Theorem 2.1 *If x_{eq,λ,λ^*} is a genuine stable (unstable) equilibrium point for system (2.1), then so is $x_{eq,\lambda,-\lambda^*}$. Similarly, if x_{eq,λ,λ^*} is a spurious stable (unstable) equilibrium point, then so is $x_{eq,\lambda,-\lambda^*}$.*

The proof, which is readily obtained via Definition 2.1 and noting that $x_{eq,\lambda,-\lambda^*} = -x_{eq,\lambda,\lambda^*}$, is omitted here.

For unsaturated system $\dot{x} = Ax$, it is clear that $x = 0$ is always a equilibrium point. And for system (2.1), when $(\lambda, \lambda^*) = (1, 0)$ happens, it means there is no saturation and $\dot{x} = (A - BK)x$. It is clearly $x_{eq} = 0$ and $|-Kx_{eq}| = |-K * 0| = 0 < u_{lim}$. So the origin $x = 0$ is always an equilibrium point for both the unsaturated and the saturated systems.

From the above discussion, we easily obtain the following facts.

Note 2.1 We know that, for system (2.1),

(i) if there exists genuine nonzero equilibrium point, the stability of any equilibrium point is local stable;

(ii) when we consider GAS of the origin, it is necessary that $A - \lambda BK$ are Hurwitz for all λ and nonzero equilibrium points are spurious.

Note 2.2 For system

$$\dot{x} = \begin{bmatrix} 5 & -1 \\ 6 & -2 \end{bmatrix} x + \begin{bmatrix} 4 \\ 2 \end{bmatrix} u, \ K = \begin{bmatrix} 3 & 2 \end{bmatrix}$$

$$u = sat(-Kx) = \begin{cases} 1, & if -Kx \geq 1, \\ -Kx, & if |-Kx| < 1, \\ -1, & if -Kx \leq -1, \end{cases}$$

we can get that origin is genuine stable equilibrium point and other equilibrium points $x_{eq,2} = (-1.5, -3.5), x_{eq,3} = (1.5, 3.5)$ are genuine unstable equilibrium points. It shows that the connection between $x_{eq,2}$ and $x_{eq,3}$ does not coincide with the attraction region of the origin from Fig. 2.1 (the green region denotes the attraction region of the origin, the blue dot denotes $x_{eq,2}$, and the yellow dot denotes $x_{eq,3}$).

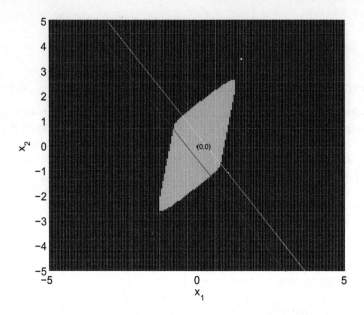

Fig. 2.1 Attraction region of the origin

While the [17] showed a special case.

Note 2.3 When $n = 2$ in (2.1), if k_1, k_2 are the eigenvalues for matrix $(A - \lambda BK)$, and an equilibrium point $x_{eq,\lambda,\lambda^*} \in \mathbb{D}_{\lambda,\lambda^*}$, we outline the following classifications.

(1) if k_1 and k_2 are negative real numbers, we call the equilibrium point as the genuine stable node.

(2) if k_1 and k_2 are positive real numbers, we call the equilibrium point as the genuine unstable node.

(3) if $k_1 = \bar{k}_2$, $\text{Re}(k_1) < 0$ and $\text{Re}(k_2) < 0$, we call the equilibrium point as the genuine stable spiral point.

(4) if $k_1 = \bar{k}_2$, $\text{Re}(k_1) > 0$ and $\text{Re}(k_2) > 0$, we call the equilibrium point as the genuine unstable spiral point.

(5) if k_1 and k_2 are real numbers and $k_1 \cdot k_2 < 0$, we call the equilibrium point as the genuine saddle point.

Similarly for spurious equilibrium points.

Note 2.4 According to our methods, if x_{eq} is a genuine equilibrium point, it is a equilibrium point in the traditional sense. That is to say that it too is classified into genuine stable equilibrium points (such as a genuine stable node, a genuine stable spiral point, etc.) and genuine unstable equilibrium points (such as a genuine unstable node, a genuine unstable spiral point, a genuine saddle point, etc.), which are the same definition as traditional equilibrium points methods. Similarly for the spurious stable (unstable) equilibrium points.

From the above notes, we can easily obtain the following results.

2.1 Introduction

Theorem 2.2 *In system (2.1), for some pair (λ, λ^*), and the corresponding x_{eq,λ,λ^*} satisfying (2.3),*
(1) If $\forall x_{eq,\lambda,\lambda^*}$ is a genuine stable equilibrium point, then

$$\lim_{t \to \infty} \| x(t, t_0, x_{eq,\lambda,\lambda^*}) - x_{eq,\lambda,\lambda^*} \| = 0;$$

(2) Similarly, if $\forall x_{eq,\lambda,\lambda^*}$ is a spurious stable equilibrium point, then

$$\lim_{t \to \infty} \| x(t, t_0, x_{eq,\lambda,\lambda^*}) - x_{eq,\lambda,\lambda^*} \| \neq 0.$$

The proof, which is readily obtained via Definition 2.1 and Notes 2.3 and 2.4, is omitted here.

It is interesting for us to discuss these genuine (or spurious) equilibrium points. Here, we will discuss the cases when $ABK = BKA$ holds for system (2.1). So, we first consider the properties of commuting matrices.

2.2 Properties of Commuting Matrices

The following theorems give the properties of commuting matrices.

Theorem 2.3 *Let $A, B \in \mathbb{R}^{n \times n}$, with $rank(A) = n$. Then $AB = BA$, where $rank(B) = m < r$, is equivalent to saying that there exist nonsingular matrices Q_1, R_1 such that*

$$\begin{aligned} Q_1^{-1} A Q_1 &= \begin{pmatrix} L_{11} & L_{12} \\ 0 & L_{22} \end{pmatrix}, \\ R_1 A R_1^{-1} &= \begin{pmatrix} L_{11} & 0 \\ D_{21} & D_{22} \end{pmatrix}, \end{aligned} \quad (2.5)$$

where $L_{11} \in \mathbb{R}^{m \times m}$; $L_{12}, L_{21}, D_{21} \in \mathbb{R}^{m \times (n-m)}$; $D_{22}, L_{22} \in \mathbb{R}^{(n-m) \times (n-m)}$; and B is defined as

$$B = Q_1 \begin{pmatrix} I_m & 0 \\ 0 & 0 \end{pmatrix} R_1,$$

where I_m is the identity matrix of rank m. Thus

$$A = Q_1 \begin{pmatrix} L_{11} & L_{12} \\ 0 & L_{22} \end{pmatrix} Q_1^{-1}, \quad B = Q_1 \begin{pmatrix} C_{11} & C_{12} \\ 0 & 0 \end{pmatrix} Q_1^{-1}, \quad (2.6)$$

where $Q_1 = R_1^{-1} C$, for the matrix $C = \begin{pmatrix} C_{11} & C_{12} \\ C_{21} & C_{22} \end{pmatrix}$.

The above results are obtained using the nonsingular decomposition theorem, whose proof we omit here.

We also have the following theorem.

Theorem 2.4 Let $A, B \in \mathbb{R}^{n \times n}$, where $rank(A) = n$. If $AB = BA$, where $rank(B) = m < r$, then, from (2.6) in Theorem 2.3,

(I)
$$A - B = Q_1 \begin{pmatrix} L_{11} - C_{11} & L_{12} - C_{12} \\ 0 & L_{22} \end{pmatrix} Q_1^{-1}; \qquad (2.7)$$

(II) Using (2.7), $AB = BA$ is equivalent to

$$C_{11}L_{11} = L_{11}C_{11}, \quad C_{11}L_{12} + C_{12}L_{22} = L_{11}C_{12},$$

(III) Using (2.5), we obtain

$$\begin{cases} C_{11}L_{11} = L_{11}C_{11}, \\ C_{21}L_{11} = D_{21}C_{11} + D_{22}C_{21}, \end{cases}$$

where $L_{11}, C_{11} \in \mathbb{R}^{m \times m}$; $L_{12}, D_{21}, C_{12} \in \mathbb{R}^{m \times (n-m)}$; $C_{22}, L_{22}, D_{22} \in \mathbb{R}^{(n-m) \times (n-m)}$.

In particular, when $m = 1$ and $rank(B) = 1$ in Theorem 2.2, we obtain the following result.

Theorem 2.5 $AB = BA$, when the $rank(A) = n$ is true if and only if there exist nonsingular matrices Q_1, C such that

$$A = Q_1 \begin{pmatrix} l_{11} & L_{12} \\ 0 & L_{22} \end{pmatrix} Q_1^{-1}, \quad B = Q_1 \begin{pmatrix} c_{11} & C_{12} \\ 0 & 0 \end{pmatrix} Q_1^{-1},$$

where l_{11}, c_{11} are constants, and $L_{12}, C_{12} \in \mathbb{R}^{1 \times (n-1)}$.

From Theorems 2.4 and 2.5, we obtain the following result.

Theorem 2.6 The existence of B in the commuting matrix equation $AB = BA$ is equivalent to the existence of the solution X_1, X_2 to the matrix equation

$$L_{11}X_1 - X_1 L_{11} = 0, \quad X_1 L_{12} = L_{11}X_2 - X_2 L_{22}, \qquad (2.8)$$

where $(A, B) \in \left\{ \left(Q_1 \begin{pmatrix} L_{11} & L_{12} \\ 0 & L_{22} \end{pmatrix} Q_1^{-1}, Q_1 \begin{pmatrix} X_1 & X_2 \\ 0 & 0 \end{pmatrix} Q_1^{-1} \right) \right\}$, and Q_1 is any nonsingular matrix. This follows directly from (2.6) and (2.8) if we write $C_{11} = X_1$ and $C_{21} = X_2$.

Definition 2.2 For $A = (a_{ij})^{n \times n}$, we define the row space $rs(A)$ and the column space $cs(A)$ as

$$rs(A) = [a_{11} \, a_{12} \, \ldots \, a_{1n} \, a_{21} \, a_{22} \, \ldots \, a_{2n} \, \ldots \, a_{m1} \, a_{m2} \, \ldots \, a_{mn}],$$

and

$$cs(A) = [a_{11} \, a_{21} \, \ldots \, a_{m1} \, a_{12} \, a_{22} \, \ldots \, a_{m2} \, \ldots \, a_{1n} \, a_{2n} \, \ldots \, a_{mn}].$$

2.2 Properties of Commuting Matrices

From the definition of Kronecker product and the theory of linear algebraic equations, we have the following results for each matrix equation in (2.8).

Theorem 2.7 *For $L_{11}X_1 - X_1L_{11} = 0$, there exist r linearly independent solutions which are all row vectors of X_1.*
Since the eigenvalues of $I_r \otimes L_{11} - L_{11}^T \otimes I_r$ are always zero, there are r linearly independent vectors for the Kronecker product equation

$$(I_r \otimes L_{11} - L_{11}^T \otimes I_r)cs(X_1) = 0,$$

where I_r is the identity matrix with rank r.

Theorem 2.8 *For $X_1L_{12} = L_{11}X_2 - X_2L_{22}$,*

1. *If the Kronecker product $I_{n-r} \otimes L_{11} - L_{22}^T \otimes I_r$ is nonsingular, there exists a unique solution for X_2;*
2. *If*
   $rank(cs(X_1L_{12})I_{n-r} \otimes L_{11} - L_{22}^T \otimes I_r = rank(I_{n-r} \otimes L_{11} - L_{22}^T \otimes I_r),$
   then there are solutions for X_2;
3. *If*
   $rank(cs(X_1L_{12})I_{n-r} \otimes L_{11} - L_{22}^T \otimes I_r) > rank(I_{n-r} \otimes L_{11} - L_{22}^T \otimes I_r),$
   then there are no solutions for X_2.

2.3 The Existence of Feedback Matrices in MIMO/SISO Control Systems

Consider the following multiple-input multiple-output (MIMO) linear control system

$$\begin{cases} \dot{x} = Ax + Bu, \\ y = Cx, \\ u = -Kx, \end{cases} \quad (2.9)$$

where $A \in \mathbb{R}^{n \times n}, B \in \mathbb{R}^{n \times m}, C \in \mathbb{R}^{l \times n}, m, l \leq n$ and $rank(B) = r$. Then, there exist nonsingular matrices P, V such that

$$PBV = \begin{pmatrix} I_r & 0 \\ 0 & 0 \end{pmatrix},$$

where I_r denotes the rank-r identity matrix.

2.3.1 Closed-Loop Control

We begin with the closed-loop system.

Theorem 2.9 *There exists a rank m, $m \times n$ matrix K such that*

$$ABK = BKA \iff PAP^{-1} = \begin{pmatrix} A_r & * \\ 0 & * \end{pmatrix}, \quad QAQ^{-1} = \begin{pmatrix} A_r & 0 \\ * & * \end{pmatrix}, \quad (2.10)$$

where A_r denotes the submatrix of PAP^{-1} of rank r and $*$ denotes the rest. Thus, $AP_r = P_r A_r$ where P_r denotes the forward r rows of P;

$$B = P^{-1} \begin{pmatrix} I_r & 0 \\ 0 & 0 \end{pmatrix}_{n \times m} V^{-1}; \quad (2.11)$$

I_r is the identity matrix of rank r; and P, V, Q are nonsingular matrices.

Proof For a rank m $m \times n$ matrix K, there exist nonsingular matrices V, Q such that $K = V \begin{pmatrix} I_m & 0 \end{pmatrix} Q$.

Substituting for B from (2.11) into $ABK = BKA$, we obtain

$$AP^{-1} \begin{pmatrix} I_r & 0 \\ 0 & 0 \end{pmatrix} V^{-1} V \begin{pmatrix} I_m & 0 \end{pmatrix} Q = P^{-1} \begin{pmatrix} I_r & 0 \\ 0 & 0 \end{pmatrix} V^{-1} V \begin{pmatrix} I_m & 0 \end{pmatrix} QA,$$

which simplifies to give

$$AP^{-1} \begin{pmatrix} I_r & 0 \\ 0 & 0 \end{pmatrix} \begin{pmatrix} I_m & 0 \end{pmatrix} Q = P^{-1} \begin{pmatrix} I_r & 0 \\ 0 & 0 \end{pmatrix} \begin{pmatrix} I_m & 0 \end{pmatrix} QA,$$

so that

$$PAP^{-1} \begin{pmatrix} I_r & 0 \\ 0 & 0 \end{pmatrix} = \begin{pmatrix} I_r & 0 \\ 0 & 0 \end{pmatrix} QAQ^{-1}.$$

Conversely, suppose $PAP^{-1} = \begin{pmatrix} A_r & * \\ 0 & * \end{pmatrix}$, where A_r denotes the forward r-order submatrix of PAP^{-1}. Let P_r denote the matrix whose r column vectors are made of the forward r columns of P^{-1}. Clearly, $AP_r = P_r A_r$.

Using the similarity of A with A^T, and the similarity of A_r with A_r^T, it follows that there exist a nonsingular rank-n matrix U and a nonsingular rank-m matrix V_1 such that $U^{-1} A^T U = A$ and $V_1^{-1} A_r^T V_1 = A_r$.

Since $AP_r = P_r A_r$, we obtain

$$U^{-1} A^T U P_r = P_r V_1^{-1} A_r^T V_1,$$

2.3 The Existence of Feedback Matrices in MIMO/SISO Control Systems

so that
$$A^T U P_r V_1^{-1} = U P_r V_1^{-1} A_r^T.$$

Let $Q_1 = U P_r V_1^{-1}$. Since U and V_1 are nonsingular, and Q_1 is an $(m \times r)$ matrix with rank r, we can construct the nonsingular matrix $Q = (Q_1 \; Q_2)$ to get

$$A^T Q^T = Q^T \begin{pmatrix} A_r^T & * \\ 0 & * \end{pmatrix}, \; QA = \begin{pmatrix} A_r & 0 \\ * & * \end{pmatrix} R_1, \; QAQ^{-1} = \begin{pmatrix} A_r & 0 \\ * & * \end{pmatrix}.$$

We now need some information on the feedback K. It is clear that $ABK = BKA$ if we define $K = V_1(I_m \; 0)Q$. So (2.10) holds.

In particular, suppose that $A \in \mathbb{R}^{n \times n}$, and $B = (b_1 \; b_2 \; \ldots \; b_n)^T \in \mathbb{R}^{n \times 1}$. Without loss of generality, we will assume that $b_1 \neq 0$. We now consider how $K = (k_1 \; k_2 \; \ldots \; k_n) \in \mathbb{R}^{1 \times n}$ satisfies $A(A - BK) = (A - BK)A$.

It is clear that $A(A - BK) = (A - BK)A$ is true when $K = 0$.

Theorem 2.10 *There exists a nonzero vector K such that*

$$ABK = BKA \iff A \begin{pmatrix} b_1 \\ 0 \\ \vdots \\ 0 \end{pmatrix} = \alpha \begin{pmatrix} b_1 \\ 0 \\ \vdots \\ 0 \end{pmatrix}, \quad (2.12)$$

where $\alpha = a_{11} + \frac{b_2}{b_1} a_{12} + \cdots + \frac{b_n}{b_1} a_{1n}$, i.e., α is an eigenvalue of A with corresponding eigenvector $(b_1 \; b_2 \; \ldots \; b_n)^T$. This follows from Theorem 2.9, and we omit the proof here.

Similarly, we obtain the following results.

Theorem 2.11 *Suppose (2.12) holds, A is Hurwitz (so that all its eigenvalues have strictly negative real parts) and there is at least one genuine eigenvalue of A for SISO control systems. Then, there is an eigenvector K such that $ABK = BKA$, and the trivial solution of $\dot{x} = (A - BK)x$ is asymptotically stable.*

Theorem 2.12 *Under the condition of (2.12), if no real eigenvalues exist for the matrix A, then for SISO control systems, there does not exist a real matrix K with $ABK = BKA$. In another words, one precondition for $ABK = BKA$ to be satisfied is the existence of at least one real eigenvalue for A.*

2.3.2 Open-Loop Control

We now turn to the open-loop systems case. Given matrices A, B, C, we investigate whether there exists a matrix K such that

$$(A - BK)C = C(A - BK). \tag{2.13}$$

If $AC = CA$, then (2.13) reduces to $BKC = CBK$. We already obtained some results for the case of the closed-loop control system (2.9). If matrices A, C do not commute, however, then (2.13) reduces to

$$AC - CA = BKC - CBK. \tag{2.14}$$

For MIMO systems, if $B \in \mathbb{R}^{n \times m}$ and B is of rank r, then there exist two transformation matrices P and Q such that

$$B = P \begin{pmatrix} I_r & 0 \\ 0 & 0 \end{pmatrix} Q,$$

so that (2.14) becomes

$$AC - CA = -CP \begin{pmatrix} I_r & 0 \\ 0 & 0 \end{pmatrix} QK + P \begin{pmatrix} I_r & 0 \\ 0 & 0 \end{pmatrix} QKC. \tag{2.15}$$

Therefore

$$P^{-1}(AC - CA) = -P^{-1}CP \begin{pmatrix} I_r & 0 \\ 0 & 0 \end{pmatrix} QK + \begin{pmatrix} I_r & 0 \\ 0 & 0 \end{pmatrix} QKC. \tag{2.16}$$

If we introduce

$$X = \begin{pmatrix} I_r & 0 \\ 0 & 0 \end{pmatrix} QK, \quad D = P^{-1}(AC - CA), \quad E = -P^{-1}CP,$$

then (2.16) becomes

$$D = EX + XC. \tag{2.17}$$

It is clear that the existence of X in (2.17) is equivalent to the existence of K in (2.14).

Let E_r be the forward r rows of E. By the Kronecker product $\otimes$ property and (2.17), it follows that there exists a matrix K such that $A - BK$ and C commute $\iff$ the solution of $cs(X_r)$ exists, where X_r denotes the forward r columns of X such that

$$\left([E_r \otimes I_n] + \begin{pmatrix} I_r \otimes C^T \\ 0 \\ \vdots \\ 0 \end{pmatrix} \right) cs(X_r) = cs(D). \tag{2.18}$$

From the above discussion, we readily obtain the following result.

2.3 The Existence of Feedback Matrices in MIMO/SISO Control Systems

Theorem 2.13 *The matrix K of (2.13) exists $\iff$ the solution $cs(X_r)$ of (2.18) exists.*

For (2.9), if $B \in \mathbb{R}^{n \times 1}$, $K \in \mathbb{R}^{1 \times n}$, then (2.18) becomes

$$\left([E_r \otimes I_n] + \begin{pmatrix} I_r \otimes C^T \\ 0 \\ \vdots \\ 0 \end{pmatrix}\right) K^T = cs(D). \tag{2.19}$$

Theorem 2.14 *For the matrix K in (2.13) and for $B \in \mathbb{R}^{n \times 1}$, $K \in \mathbb{R}^{1 \times n}$ exists $\iff$ $cs(X_r)$ of (2.19) exists.*

2.4 Equilibrium Points of SISO Control Systems with Single Input

In this section, we consider equilibrium points of single saturated input under the condition that $ABK = BKA$.

For convenience, let

$$B = Q_1 \begin{pmatrix} 1 \\ 0 \\ \vdots \\ 0 \end{pmatrix}, \quad K = \begin{pmatrix} 1 & 0 & \cdots & 0 \end{pmatrix} R_1, \tag{2.20}$$

where Q_1, R_1 are nonsingular matrices. Then, there exists a nonsingular matrix $C = (C_{ij})$ of rank n such that $Q_1^{-1} = CR_1$. A and $B_1 = BK$ satisfy the conditions of Theorems 2.1 and 2.2 $\iff$

$$ABK = BKA. \tag{2.21}$$

Theorem 2.15 *For system (2.1), if A, B, K satisfy assumptions (2.20) and (2.21), and*

$$\begin{cases} A \in Hurwitz, \\ A - BK \in Hurwitz, \end{cases} \tag{2.22}$$

then, apart from the stable solution at the origin, the remaining equilibrium solutions of (2.1) are all spurious stable equilibria.

Proof From the above

$$B_1 := BK = Q_1 \begin{pmatrix} c_{11} & c_{12} & \cdots & c_{1n} \\ 0 & 0 & \cdots & 0 \\ \vdots & \vdots & \ddots & \vdots \\ 0 & 0 & \cdots & 0 \end{pmatrix} Q_1^{-1},$$

where $C = (c_{ij})_{n \times n}$.

It is clear that

$$A - BK = Q_1 \begin{pmatrix} l_{11} - c_{11} & l_{12} - c_{12} & \cdots & l_{1n} - c_{1n} \\ 0 & l_{22} & \cdots & l_{2n} \\ \vdots & \vdots & \ddots & \vdots \\ 0 & l_{n2} & \cdots & l_{nn} \end{pmatrix} Q_1^{-1}, \qquad (2.23)$$

and

$$l_{11} \in \mathbb{R}^1, c_{11} \in \mathbb{R}^1. \qquad (2.24)$$

Since matrices $A - BK$ and A are stable and Q_1 is nonsingular, we have

$$l_{11} < 0, l_{11} - c_{11} < 0.$$

Otherwise, according to the Definition 1.1, we have

$$\begin{aligned} -K_{x_{eq},0,I} &= KA^{-1}Bu_{\lim} \\ &= (1\ 0\ \cdots\ 0)\, CQ_1^{-1} \left[Q_1 \begin{pmatrix} l_{11} & l_{12} & \cdots & l_{1n} \\ 0 & l_{22} & \cdots & l_{2n} \\ \vdots & \vdots & \ddots & \vdots \\ 0 & l_{n2} & \cdots & l_{nn} \end{pmatrix} Q_1^{-1} \right]^{-1} Q_1 \begin{pmatrix} 1 \\ 0 \\ \vdots \\ 0 \end{pmatrix} u_{\lim} \\ &\geq u_{\lim} \\ &= (1\ 0\ \cdots\ 0)\, I \begin{pmatrix} 1 \\ 0 \\ \vdots \\ 0 \end{pmatrix} u_{\lim}. \end{aligned} \qquad (2.25)$$

Since $u_{lim} > 0$, we have

$$(1\ 0\ \cdots\ 0)\, C \begin{pmatrix} l_{11} & l_{12} & \cdots & l_{1n} \\ 0 & l_{22} & \cdots & l_{2n} \\ \vdots & \vdots & \ddots & \vdots \\ 0 & l_{n2} & \cdots & l_{nn} \end{pmatrix}^{-1} \begin{pmatrix} 1 \\ 0 \\ \vdots \\ 0 \end{pmatrix} - (1\ 0\ \cdots\ 0)\, I \begin{pmatrix} 1 \\ 0 \\ \vdots \\ 0 \end{pmatrix} \geq 0. \qquad (2.26)$$

2.4 Equilibrium Points of SISO Control Systems with Single Input

Hence

$$(1\ 0\ \cdots\ 0) \left[C \begin{pmatrix} l_{11} & l_{12} & \cdots & l_{1n} \\ 0 & l_{22} & \cdots & l_{2n} \\ \vdots & \cdots & \ddots & \vdots \\ 0 & l_{n2} & \cdots & l_{nn} \end{pmatrix}^{-1} - I \right] \begin{pmatrix} 1 \\ 0 \\ \vdots \\ 0 \end{pmatrix} \geq 0, \qquad (2.27)$$

and

$$C \begin{pmatrix} l_{11} & l_{12} & \cdots & l_{1n} \\ 0 & l_{22} & \cdots & l_{2n} \\ \vdots & \cdots & \ddots & \vdots \\ 0 & l_{n2} & \cdots & l_{nn} \end{pmatrix}^{-1} - I = \begin{pmatrix} c_{11} \frac{1}{l_{11}} - 1 & E_{12} \\ E_{21} & E_{22} \end{pmatrix},$$

where $L_{22} = \begin{pmatrix} l_{22} & \cdots & l_{2n} \\ \vdots & \ddots & \vdots \\ l_{n2} & \cdots & l_{nn} \end{pmatrix}$, $\lambda_{A,1} = \begin{pmatrix} l_{11} & L_{12} \\ 0 & L_{22} \end{pmatrix}$, $C = \begin{pmatrix} c_{11} & C_{12} \\ C_{21} & C_{22} \end{pmatrix}$,

$E_{21} = \frac{1}{l_{11}} C_{21}, E_{12} = -\frac{c_{11}}{l_{11}} L_{12} L_{22}^{-1} + C_{12} L_{22}^{-1}, E_{22} = \frac{1}{l_{11}} C_{21} L_{12} L_{22}^{-1} + C_{22} L_{22}^{-1} - I.$

From (2.27), we have

$$c_{11} \frac{1}{l_{11}} - 1 \geq 0. \qquad (2.28)$$

By virtue of (2.24), (2.28) transforms to

$$c_{11} \frac{1}{l_{11}} - 1 \geq 0 \Leftrightarrow (c_{11} - l_{11}) l_{11} \geq 0.$$

This is in contradiction to $(c_{11} - l_{11}) > 0$, $l_{11} < 0$.
Thus, $x_{eq,0,1}$ is not a genuine equilibrium, and since $A \in Hurwitz$, $x_{eq,0,1}$ is also spurious stable.
Similarly, we can prove that $x_{eq,0,-1}$ is spurious stable, too. □

In the other words, we can rewrite Theorem 2.15 as follows.

Theorem 2.15 *: *For system (2.1), suppose A, B, K satisfy assumptions of (2.20) and (2.21), $l_{11} < 0, l_{11} - c_{11} < 0$, and $L_{22} \in Hurwitz$, then the nonzero equilibrium points of system (2.1) are all spurious stable ones, except for the genuine stable solution at the origin.*

We present several curious features of the equilibrium points below.

Theorem 2.16 *Under the assumptions of Theorem 2.16*, let $l_{11} > 0$, $l_{11} - c_{11} < 0$, then*

(I) The equilibrium points $x_{eq,0,1}$ and $x_{eq,0,-1}$ are both unstable genuine equilibria;
(II) Let $L_{22} \in Hurwitz$, then $x = 0$ is either stable or unstable.

Table 2.1 The equilibrium points under A and BK commuting

A,B,K	0	$x_{eq,0,1}$	$x_{eq,0,-1}$
$L_{22} \in$ Hurwitz, $l_{11} < 0, l_{11} - c_{11} < 0$	Ge,S	Sp,S	Sp,S
$L_{22} \in$ Hurwitz, $l_{11} > 0, l_{11} - c_{11} < 0$	Ge,S	Ge,US	Ge,US
$L_{22} \in$ Hurwitz, $l_{11} < 0, l_{11} - c_{11} > 0$	Ge,US	Ge,S	Ge,S
$L_{22} \in$ Hurwitz, $l_{11} > 0, l_{11} - c_{11} > 0$	Ge,US	Sp,US	Sp,US

Theorem 2.17 *Under the assumptions of Theorem 2.15, let $L_{22} \in Hurwitz$ and $l_{11} < 0$, $l_{11} - c_{11} > 0$, then $x = 0$ is unstable, while $x_{eq,0,1}$ and $x_{eq,0,-1}$ are both genuine stable equilibrium points.*

Theorem 2.18 *Under the assumptions of Theorem 2.15, let $L_{22} \in Hurwitz$, $l_{11} > 0$, $l_{11} - c_{11} > 0$, then $x = 0$ is unstable, $x_{eq,0,1}, x_{eq,0,-1}$ are spurious unstable equilibrium point.*

The above two results are obtained by the same method with Theorem 2.15. We omit the discussions.

We explain the above results by Table 2.1.

Here, Ge denotes genuine, Sp denotes spurious, S denotes stable, and US denotes unstable.

It is the fact from the above table that under the state matrices commuting, if $x = 0$ is genuine stable, then there may not exist any genuine stable nonzero equilibrium points under the certain control law. And if $x = 0$ is unstable, then there may exist genuine or spurious nonzero equilibrium points under the certain control law.

2.5 Example Analysis and Simulink

We give four examples to demonstrate Table 2.1.

Example 2.1 We demonstrate some strange points in the following saturating systems:
$$\dot{x}(t) = Ax(t) + Bu_s(t),$$

2.5 Example Analysis and Simulink

where $A = \begin{pmatrix} -1 & 1 & 0 \\ 0 & -0.5 & 1 \\ 0 & 0 & -1.5 \end{pmatrix}$, $B = \begin{pmatrix} 1 \\ 0 \\ 0 \end{pmatrix}$, and u_s denotes a saturated state feedback

$$u_s = \begin{cases} 5, & -[0.6\ -1.2\ -2.4]x \geq 5, \\ -[0.6\ -1.2\ -2.4]x, & \left|-[0.6\ -1.2\ -2.4]x\right| < 5, \\ -5, & -[0.6\ -1.2\ -2.4]x \leq -5, \end{cases}$$

where $K = [0.6\ -1.2\ -2.4]$.

It is clear that $ABK = BKA$, $-1 < 0$, $-1 - 0.6 < 0$, $\begin{pmatrix} -0.5 & 1 \\ 0 & -1.5 \end{pmatrix} \in Hurwitz$, then

$$\begin{cases} A \in Hurwitz, \\ A - BK \in Hurwitz. \end{cases}$$

While there exist the following strange equilibrium points:

$$x_{eq,1} = 0,\ x_{eq,2} = [5\ 0\ 0]^T,\ x_{eq,3} = [-5\ 0\ 0]^T,$$

which correspondingly satisfy

$$\dot{x}(t) = (A - BK)x(t),$$
$$\dot{x}(t) = Ax(t) + 5B,$$
$$\dot{x}(t) = Ax(t) - 5B.$$

Let

$$\mathbb{D}_1 = \{x\ |-5 < -[0.6\ -1.2\ -2.4]x < 5\},$$
$$\mathbb{D}_2 = \{x\ |-[0.6\ -1.2\ -2.4]x \geq 5\},$$
$$\mathbb{D}_3 = \{x\ |-[0.6\ -1.2\ -2.4]x \leq -5,\ \},$$

Clearly,

$$x_{eq,1} = 0 \in \mathbb{D}_1,$$
$$x_{eq,2} = [5\ 0\ 0]^T \notin \mathbb{D}_2,$$
$$x_{eq,3} = -[5\ 0\ 0]^T \notin \mathbb{D}_3.$$

From Definition 2.1, we know that $x_{eq,1}$ is a genuine stable equilibrium point, and $x_{eq,2}, x_{eq,3}$ are spurious stable equilibrium points.

Figures 2.2 and 2.3 show that the solution of Example 2.1 which starts from regions $\mathbb{D}_1, \mathbb{D}_2$, and $\mathbb{D}_3$ (including $x_{eq,2}, x_{eq,3}$) converges to $x_{eq,1}$. So, $x_{eq,1}$ is the genuine stable equilibrium point, and $x_{eq,2}$ and $x_{eq,3}$ are stable spurious equilibria.

Other equilibrium points except for the origin have to be spurious stable ones in analyzing the global asymptotic stability of the origin of control systems with saturated inputs.

Example 2.2 We demonstrate some special points in the following saturating systems:

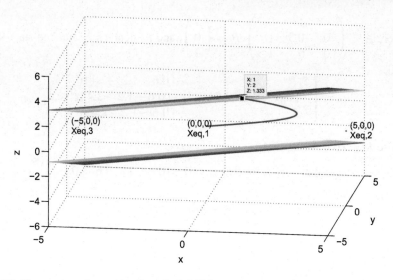

Fig. 2.2 Phase trajectories starting from (1, 2, 1.333)

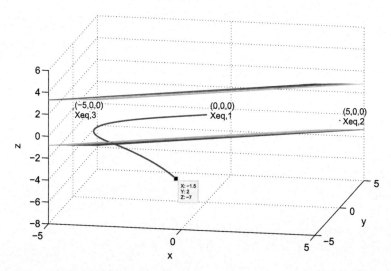

Fig. 2.3 Phase trajectories starting from (−1.5, 2, −7)

$$\dot{x}(t) = Ax(t) + Bu_s(t),$$

where $A = \begin{pmatrix} 1 & 1 & 0 \\ 0 & -0.5 & 1 \\ 0 & 0 & -1.5 \end{pmatrix}, B = \begin{pmatrix} 1 \\ 0 \\ 0 \end{pmatrix}$, and where u_s denotes a saturated state feedback

2.5 Example Analysis and Simulink

$$u_s = \begin{cases} 5, & -[\,1.5\ 1\ 0.4\,]x \geq 5, \\ -[\,1.5\ 1\ 0.4\,]x, & \left|-[\,1.5\ 1\ 0.4\,]x\right| < 5, \\ -5, & -[\,1.5\ 1\ 0.4\,]x \leq -5, \end{cases}$$

where $K = [\,1.5\ 1\ 0.4\,]$.

It is clear that $ABK = BKA, 1 > 0, 1 - 1.5 < 0, \begin{pmatrix} -0.5 & 1 \\ 0 & -1.5 \end{pmatrix} \in Hurwitz$, then

$$\begin{cases} A \notin Hurwitz, \\ A - BK \in Hurwitz. \end{cases}$$

While there exist the following strange equilibrium points:

$$x_{eq,1} = 0,\ x_{eq,2} = [-5\ 0\ 0]^T,\ x_{eq,3} = [5\ 0\ 0]^T,$$

which correspondingly satisfy

$$\dot{x}(t) = (A - BK)x(t),$$
$$\dot{x}(t) = Ax(t) + 5B,$$
$$\dot{x}(t) = Ax(t) - 5B.$$

Let

$$\mathbb{D}_1 = \{x \mid -5 < -[\,1.5\ 1\ 0.4\,]x < 5\},$$
$$\mathbb{D}_2 = \{x \mid -[\,1.5\ 1\ 0.4\,]x \geq 5\},$$
$$\mathbb{D}_3 = \{x \mid -[\,1.5\ 1\ 0.4\,]x \leq -5\}.$$

Clearly,

$$x_{eq,1} = 0 \in \mathbb{D}_1,$$
$$x_{eq,2} = [-5\ 0\ 0]^T \in \mathbb{D}_2,$$
$$x_{eq,3} = [5\ 0\ 0]^T \in \mathbb{D}_3.$$

From Definition 2.1, we know that $x_{eq,1}$ is a genuine stable equilibrium point, and $x_{eq,2}, x_{eq,3}$ are genuine unstable equilibrium points.

Figure 2.4 shows that the solutions of Example 2.2 which start near $x_{eq,1}$ will converge to $x_{eq,1}$. Figure 2.5 shows that some solutions of Example 2.2 which start near $x_{eq,2}$ or $x_{eq,3}$ will converge to $x_{eq,1}$, but others will diffuse to infinite.

Therefore, $x_{eq,1}$ is a genuine stable equilibrium point, and $x_{eq,2}, x_{eq,3}$ are unstable true equilibrium points. The maximum domain of convergence of $x_{eq,1}$ exists, but how to compute or describe it will be a challenging problem. It is very important to know whether there exist a pair or several pairs of nonzero stable true equilibrium points under specific feedback laws, when the state matrix is also unstable.

Example 2.3 We demonstrate some strange points in the following saturating systems:

$$\dot{x}(t) = Ax(t) + Bu_s(t),$$

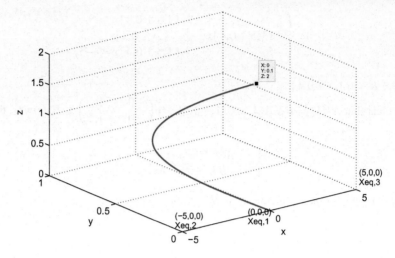

Fig. 2.4 Phase trajectories starting from (0, 0.1, 2)

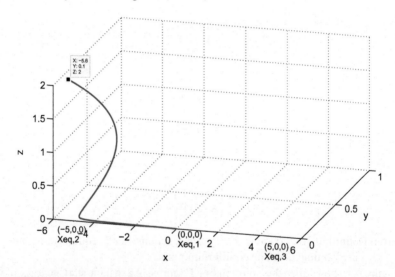

Fig. 2.5 Phase trajectories starting from (−5.6, 0.1, 2)

where $A = \begin{pmatrix} -1 & 1 & 0 \\ 0 & -0.5 & 1 \\ 0 & 0 & -1.5 \end{pmatrix}, B = \begin{pmatrix} 1 \\ 0 \\ 0 \end{pmatrix}$, and u_s denotes a saturated state feedback

$$u_s = \begin{cases} 5, & -[-1.5\ 3\ 6]x \geq 5, \\ -[-1.5\ 3\ 6]x, & |-[-1.5\ 3\ 6]x| < 5, \\ -5, & -[-1.5\ 3\ 6]x \leq -5, \end{cases}$$

where $K = [-1.5\ 3\ 6]$.

2.5 Example Analysis and Simulink

It is clear that $ABK = BKA$, $-1 < 0$, $-1 - (-1.5) > 0$, $\begin{pmatrix} -0.5 & 1 \\ 0 & -1.5 \end{pmatrix} \in Hurwitz$, then

$$\begin{cases} A \in Hurwitz, \\ A - BK \notin Hurwitz. \end{cases}$$

While there exist the following strange equilibrium points,

$$x_{eq,1} = 0, \; x_{eq,2} = [5 \; 0 \; 0]^T, \; x_{eq,3} = [-5 \; 0 \; 0]^T,$$

which correspondingly satisfy

$$\dot{x}(t) = (A - BK)x(t),$$
$$\dot{x}(t) = Ax(t) + 5B,$$
$$\dot{x}(t) = Ax(t) - 5B.$$

Let
$$\mathbb{D}_1 = \{x \mid -5 < -[-1.5 \; 3 \; 6]x < 5\},$$
$$\mathbb{D}_2 = \{x \mid -[-1.5 \; 3 \; 6]x \geq 5\},$$
$$\mathbb{D}_3 = \{x \mid -[-1.5 \; 3 \; 6]x \leq -5\}.$$

Clearly,
$$x_{eq,1} = 0 \in \mathbb{D}_1,$$
$$x_{eq,2} = [5 \; 0 \; 0]^T \in \mathbb{D}_2,$$
$$x_{eq,3} = [-5 \; 0 \; 0]^T \in \mathbb{D}_3.$$

From Definition 2.1, we know that $x_{eq,1}$ is a genuine unstable equilibrium point, $x_{eq,2}, x_{eq,3}$ are genuine stable equilibrium points.

Figures 2.6 and 2.7 show that the solutions of the Example 2.3, which start in $\mathbb{D}_1$, $\mathbb{D}_2$ and $\mathbb{D}_3$ will converge to $x_{eq,2}$ or $x_{eq,3}$. In conclusion, $x_{eq,1}$ is a genuine unstable equilibrium point, and $x_{eq,2}, x_{eq,3}$ are all genuine stable equilibrium points. From the above discussion, we know that the maximum domain of convergence of $x_{eq,2}$ or $x_{eq,3}$ will exist, but its computation or description will prove challenging.

Example 2.4 We demonstrate some strange points in the following saturating systems.
$$\dot{x}(t) = Ax(t) + Bu_s(t)$$

where
$$A = \begin{pmatrix} 1 & 1 & 0 \\ 0 & -0.5 & 1 \\ 0 & 0 & -1.5 \end{pmatrix}, \; B = \begin{pmatrix} 1 \\ 0 \\ 0 \end{pmatrix}$$

and where u_s denotes a saturated state feedback

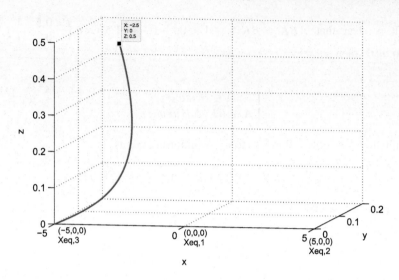

Fig. 2.6 Phase trajectories starting from $(-2.5, 0, 0.5)$

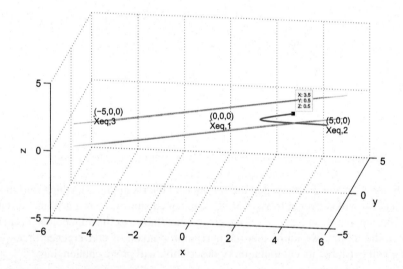

Fig. 2.7 Phase trajectories starting from $(3.5, 0.5, 0.5)$

$$u_s = \begin{cases} 5, & -[-1.5 \ -1 \ -0.4]x \geq 5, \\ -[-1.5 \ -1 \ -0.4]x, & \left|-[-1.5 \ -1 \ -0.4]x\right| < 5, \\ -5, & -[-1.5 \ -1 \ -0.4]x \leq -5, \end{cases}$$

and $K = [-1.5 \ -1 \ -0.4]$.

2.5 Example Analysis and Simulink

It is clear that $ABK = BKA, 1 > 0, 1 - (-1.5) > 0, \begin{pmatrix} -0.5 & 1 \\ 0 & -1.5 \end{pmatrix} \in Hurwitz$,

$$\begin{cases} A \notin Hurwitz, \\ A - BK \notin Hurwitz, \end{cases}$$

While there exist the following strange equilibrium points

$$x_{eq,1} = 0, \ x_{eq,2} = [-5\ 0\ 0]^T, \ x_{eq,3} = [5\ 0\ 0]^T$$

which correspondingly satisfy

$$\dot{x}(t) = (A - BK)x(t), \\ \dot{x}(t) = Ax(t) + 5B, \\ \dot{x}(t) = Ax(t) - 5B,$$

Let

$$\mathbb{D}_1 = \{x \mid -5 < -[-1.5\ -1\ -0.4]x < 5\}, \\ \mathbb{D}_2 = \{x \mid -[-1.5\ -1\ -0.4]x \geq 5\}, \\ \mathbb{D}_3 = \{x \mid -[-1.5\ -1\ -0.4]x \leq -5\}.$$

Clearly

$$x_{eq,1} = 0 \in \mathbb{D}_1, \\ x_{eq,2} = [-5\ 0\ 0]^T \notin \mathbb{D}_2, \\ x_{eq,3} = [5\ 0\ 0]^T \notin \mathbb{D}_3.$$

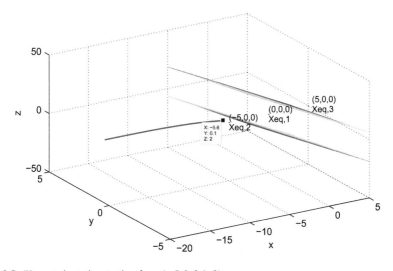

Fig. 2.8 Phase trajectories starting from $(-5.6, 0.1, 2)$

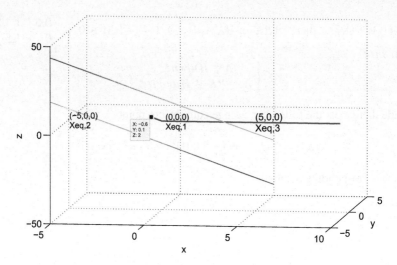

Fig. 2.9 Phase trajectories starting from $(-0.6, 0.1, 2)$

From Definition 2.1, we know that $x_{eq,1}$ is a genuine unstable equilibrium point and $x_{eq,2}$, $x_{eq,3}$ are spurious unstable equilibrium points.

Figures 2.8 and 2.9 show that the solutions of the Example 2.4 which start in $\mathbb{D}_1$, $\mathbb{D}_2$, $\mathbb{D}_3$ diffuse.

Therefore, $x_{eq,1}$ is a genuine unstable equilibrium point and $x_{eq,2}$, $x_{eq,3}$ are spurious unstable equilibrium points.

2.6 Conclusion

In this chapter we have introduced the spatial structure of control systems with saturated inputs and described its equations by a $0-1$ structure. Commutative matrices of MIMO linear systems are considered and the existence of the feedback matrices of a commutative state matrix set in MIMO closed-loops is reduced to an invariant subspace of a matrix A. The existence of feedback matrices in systems in open-loop is equivalent to the existence of the solution of a matrix equation denoted by the Kronecker product. The relationship between equilibrium points is discussed for linear systems with single saturated input under commutative matrices. We find that the other equilibrium points except for the origin, are all spurious stable ones in analyzing the global asymptotic stability of the origin of the control systems with saturated inputs. When the state matrix is unstable, it is possible for one or several pairs of nonzero stable genuine equilibrium points to exist under specific feedback laws.

Acknowledgements This work is Supported by National Key Research and Development Program of China (2017YFF0207400).

References

1. Dugard L, Verrist EI, editors. Stability and control of time-delay systems. Berlin: Springer; 1998. p. 303–31.
2. Henrion D, Tarbouriech S. LMI relaxations for robust stability of linear systems with saturating controls. Automatica. 1999;35:1599–604.
3. Horisberger HP, Belanger PR. Regulator for linear time invariant plants with uncertain parameters. IEEE Trans Autom Control. 1976;42:705–8.
4. Huang L. Stability theory. Beijing: Peking University Press; 1992. p. 235–83.
5. Kapoor N, Daoutidis P. An observer-based anti-windup scheme for nonlinear systems with input constraints. Int J Control. 1999;72(1):18–29.
6. Lee AW, Hedrick JK. Some new results on closed-loop stability in the presence of control saturation. Int J Control. 1995;62(3):619–51.
7. Kothare MV, Campo PJ, Morarior M, Nett CN. A unified framework for the study of anti-windup designs. Automatica. 1994;30(12):1869–83.
8. Edwards C, Postlethwaite I. Anti-windup and bumpless transfer schemes. Automatica. 1998;34(2):199–210.
9. Saeki M, Wada N. Synthesis of a static anti-windup compensator via linear matrix inequalities. Int J Robust Nonlinear Control. 2002;12:927–53.
10. Wu F, Soto M. Extended anti-windup control schemes for LTI and LFT systems with actuator saturations. Int J Robust Nonlinear Control. 2004;14:1255–81.
11. Molchanov AP, Pyatnitskiy YS. Criteria of asymptotic stability of differential and difference inclusions encountered in control theory. Syst Control Lett N-Holl. 1989;13:59–64.
12. Polanski K. On absolute stability analysis by polyhedric Lypunov functions. Automatica. 2000;36:573–8.
13. Scibile L, Kouvaritakis BIF. Stability region for a class of open-loop unstable linear systems: theory and application. Automatica. 2000;36(1):37–44.
14. Sontag ED, Sussmann HJ. Nonlinear output feedback design for linear systems with saturated control. In: Proceeding of 29th conference on decision and control, vol. 45(5); 1990. p. 719–721.
15. Tatsushi S, Yasuyuki F. Two conditions concerning common quadratic lyapunov functions for linear systems. IEEE Trans Autom Control. 1997;42(5):719–21.
16. Tarbouriech S, Gomes da JM. Synthesis of controllers for continuous-times delay systems with saturated control via LMIs. IEEE Trans Autom Control. 2000;45(1):105–11.
17. Favez JY, Mullhaupt PH, Srinivasan B, Bonvin D. Attraction region of planar linear systems with one unstable pole and saturated feedback. J Dyn Control Syst. 2006;12(3):331–55.
18. Guo S, Irene M, Han L, Xin WF, Feng XJ. Commuting matrices, equilibrium points for control systems with single saturated input. Appl Math Comput. 2015;259:987–1002.

Chapter 3
Stability and Closed Trajectory for Second-Order Control Systems with Single-Saturated Input

3.1 Introduction

The main contribution of this chapter is that a new method for analyzing nonlinear systems is presented by defining the new equilibrium points. The main results of this chapter are as follows: First, the new equilibrium points are defined. Second, the criteria of the new equilibrium points are presented too. Third, the sufficient conditions of GAS of the origin are presented for second-order control systems with a single-saturated input by qualitative analysis. Fourth, the existence of the closed trajectory is also considered for the same control systems. Finally, the structure of the commutative matrices is considered. And at the same time, we present the simple criterion equations of asymptotic stability of second-order control systems with single-saturated input when $ABK = BKA$.

The continuous-time system in paper [1, 6, 11], which is as follows:

$$\begin{cases} \dot{x}(t) = Ax(t) + Bu_s(t) \\ y(t) = C^T x(t) \\ u_s = sat(-Kx) \end{cases} \quad (3.1)$$

is considered, where $x(t) \in \mathbb{R}^2$ is the state vector, $y(t) \in \mathbb{R}^1$ is the output vector, $u_s(t) \in \mathbb{R}^1$ is the control input. A, B, C, K are matrices of appropriate dimensions and the system is minimal.

Saturated input $sat(-Kx)$ can be expressed as

$$sat(-Kx) = \begin{cases} u_{\lim}, & if -Kx \geq u_{\lim}, \\ -Kx, & if |-Kx| < u_{\lim}, \\ -u_{\lim}, & if -K_i x \leq -u_{\lim}, \end{cases} \quad (3.2)$$

where $u_{\lim}$ represents the saturation limit of the feedback system and $\|u_s\| \leq u_{\lim}$.

From the above saturating definition, it is reasonable that we transform the SISO system to the following control systems

© Springer Nature Singapore Pte Ltd. 2018
S. Guo and L. Han, *Stability and Control of Nonlinear Time-varying Systems*,
https://doi.org/10.1007/978-981-10-8908-4_3

$$\begin{cases} \dot{x}(t) = (A - \lambda BK)x(t) + \lambda^* Bu_{\lim}, \\ y = Cx(t), \end{cases} \quad (3.3)$$

where λ demonstrates 0 or 1, and λ^* is derived form according to

$$\begin{cases} (\lambda, \lambda^*) = (0, 1) \text{ iff } -Kx - u_{\lim} \geq 0, \\ (\lambda, \lambda^*) = (0, -1) \text{ iff } -Kx + u_{\lim} \leq 0, \\ (\lambda, \lambda^*) = (1, 0) \text{ iff } |-Kx| < u_{\lim}. \end{cases} \quad (3.4)$$

If $A - \lambda BK$ is nonsingular, it follows that

$$x_{eq,\lambda,\lambda^*} = -\lambda^*(A - \lambda BK)^{-1} Bu_{\lim}.$$

We explain its physical meaning as follows:

$$\begin{aligned} \mathbb{D}_{0,1} &= \{x \mid -Kx \geq u_{\lim}\}, \\ \mathbb{D}_{0,-1} &= \{x \mid -Kx \leq -u_{\lim}\}, \\ \mathbb{D}_{1,0} &= \{x \mid |-Kx| < u_{\lim}\}. \end{aligned}$$

For convenience, we give the following definitions.

For system (3.1), if matrix $A - \lambda BK$ is nonsingular, then

$$x_{eq,\lambda,\lambda^*} = -\lambda^*(A - \lambda BK)^{-1} Bu_{\lim}.$$

Definition 3.1 (I) For system (3.1), some pair (λ, λ^*) and its corresponding x_{eq,λ,λ^*} satisfying (3.3), if $x_{eq,\lambda,\lambda^*} \in \mathbb{D}_{\lambda,\lambda^*}$, we call x_{eq,λ,λ^*} a genuine equilibrium point. For such an equilibrium, if $(A - \lambda BK)$ is a stable (unstable) matrix, then x_{eq,λ,λ^*} is a genuine stable (unstable) equilibrium point.

(II) For system (3.1), some pair (λ, λ^*) and its corresponding x_{eq,λ,λ^*} satisfying (3.3), if $x_{eq,\lambda,\lambda^*} \notin \mathbb{D}_{\lambda,\lambda^*}$, then we call x_{eq,λ,λ^*} a spurious equilibrium point. For such spurious equilibria, if $(A - \lambda BK)$ is stable (unstable), then we say that x_{eq,λ,λ^*} is a spurious stable (unstable) equilibrium point.

Theorem 3.1 *If x_{eq,λ,λ^*} is a genuine stable (unstable) equilibrium point for system (3.1), then so is $x_{eq,\lambda,-\lambda^*}$. Similarly, if x_{eq,λ,λ^*} is a spurious stable (unstable) equilibrium point, then so is $x_{eq,\lambda,-\lambda^*}$. In particular, the origin $x = 0$ is always an equilibrium point for both the unsaturated and the saturated systems.*

The proof, which is readily obtained via Definition 3.1 and noting that $x_{eq,\lambda,-\lambda^*} = -x_{eq,\lambda,\lambda^*}$, is omitted here.

3.2 Criteria for Stability Analysis

We shall investigate the sufficient conditions on the globally asymptotical stability of the origin of the saturated control system (3.1) when $n = 2$, $m = 1$.

3.2 Criteria for Stability Analysis

Theorem 3.2 *Suppose that $n = 2$, $m = 1$ of system (3.1), let*

(a) The other equilibrium points are spurious stable equilibrium points except the stable origin;

(b) There does not exist any closed trajectory, then the origin of control system (3.1) with SISO-saturated input is globally asymptotical stable.

Proof That $x = 0$ is a genuine stable equilibrium point which implies that $A - BK$ is asymptotically stable. We deduce that there exists a positive definite symmetric matrix P in virtue of Lyapunov theorem such that

$$(A - BK)^T P + P(A - BK) < 0. \tag{3.5}$$

Let

$$\mathbb{S} = \{x \, | x^T Px \leq d\}.$$

The maximum d in $\mathbb{S} \subset \mathbb{D}_{I,0}$ is called the maximum invariant set and we define $\mathbb{S}_{I,0}$ as $\mathbb{S}_{I,0} := \{x \, | x^T Px \leq d_{\max}\}$. Suppose that $-Kx = u_{\lim}$ and $-Kx = -u_{\lim}$ be tangent lines, whose tangent points are $p'_{0,1}$ and $p''_{0,-1}$ (See Fig. 3.1).

Let $x_0 \in \mathbb{D}_{0,1}$, since $x_{eq,0,-1}$ and $x_{eq,0,1}$ are the spurious stable equilibrium points, we get the following trajectory

$$\begin{aligned}x_1(t) &= e^{A(t-T_0)}[x_0 + \int_{T_0}^{t} e^{A(t-s)} B\Lambda^* u_{\lim} dt] \\ &= e^{A(t-T_0)} x_0 + (e^{A(t-T_0)} - I) A^{-1} B u_{\lim}.\end{aligned} \tag{3.6}$$

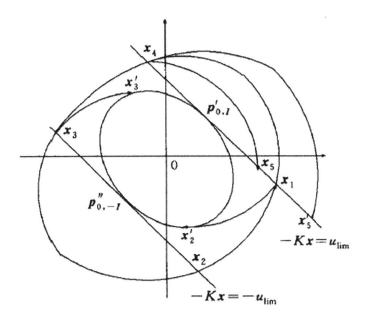

Fig. 3.1 GAS analysis about second-order linear system in a single saturation input

And that x_{eq,λ,λ^*} is a spurious stable equilibrium point implies that $x_{eq,\lambda,\lambda^*} \notin \mathbb{D}_{0,1}$ ($x_{eq,\lambda,\lambda^*} \notin \mathbb{D}_{0,-1}$). Here take the former for example. We deduce that there exists some time point T_1 such as $-Kx_1(T_1) = u_{\lim}$.

Let $x_1(T_1) = x(T_1, T_0, x_0)$ be as an initial value, its trajectory is

$$x_2(t) = e^{(A-BK)(t-T_1)} x_1(T_1).$$

If there exists a time point T_2' such as $x_2(T_2') \in \partial \mathbb{S}_{(1,0)}$, then we obtain the limit relation

$$\lim_{t \to \infty} x(t, T_2', x_2(T_2')) = 0.$$

If not, there exist a time point T_2 such that $-Kx_1(T_2) = -u_{\lim}$. Similarly, in the region $\mathbb{D}_{0,-1}$, there exists a time point T_3 such that

$$\begin{cases} x_3(T_3) = e^{A(T_3-T_2)}[x_2(T_2) - \int_{T_0}^{T_3} e^{A(T_3-s)} A^{-1} B u_{\lim} dt] \\ Kx_3(T_3) = -u_{\lim}. \end{cases}$$

Otherwise, there is a genuine stable origin equilibrium point in the region $\mathbb{D}_{1,0}$, if there exists a T_3' such that

$$x_4(T_3') = e^{(A-BK)(t-T_3)} x_{T_3} \in \partial \mathbb{S}_{I,0}.$$

We come to conclusion that

$$\lim_{t \to \infty} x(t, t_0, x_0) = 0.$$

If not, a sequence of x_k enters into the region $\mathbb{D}_{0,1}$, then there are three possibilities for x_k with initial value $x_4(T_4)$, which are as follows:

(a) $x(t, t_4, x_4)$ is superposing with $x(t, t_0, x_0)$ in the region $\mathbb{D}_{1,0}$;
(b) the joint point between $x(t, t_4, x_4)$ and $-Kx = u_{\lim}$ lies in the inner near x_1;
(c) the joint between $x(t, t_4, x_4)$ and $-Kx = u_{\lim}$ lies in the outer near x_1.

By the Theorem 3.2 (II), we learn that the case (a) is impossible, so we only need to check (b) and (c). If the case (c) holds, there exist x_6'' which lies outwards far from x_2, x_7'' which lies outwards far from x_3, and etc, we will obtain a sequence of x_i'' ($i = 6, 7, \ldots$) ($x_i''(t)$ denotes that the joint points between the trajectory $x(t, t_{i-4}, x_{i-4})$ and lines $-Kx = u_{\lim}$ or $-Kx = -u_{\lim}$) which will be far away from point $p_{0,1}'$ or from $p_{0,-1}$.

Let

$$\lim_{t \to \infty} \|x_i'' - p_{0,I}\| = \alpha < \infty, \alpha > 0.$$

It is a contradiction with the fact that there does not exist any closed trajectory. It follows from the uniqueness of the solution of (3.1) that

$$\lim_{t \to \infty} \|x_i'' - p_{0,I}\| = \infty \text{ or } \lim_{t \to \infty} \|x_i'' - p_{0,-1}\| = \infty.$$

3.2 Criteria for Stability Analysis

So
$$\lim_{t\to\infty} \|x_i^{''}\| \geq \lim_{t\to\infty} \|x_i^{''} - p_{0,1}\| - \|p_{0,1}\| = \infty.$$

Further
$$\lim_{t\to\infty} \|x_i^{''}\| = \infty.$$

With regard to the Theorem 3.2 (I), it follows that

$$\max Re(\lambda(A - BK)) < 0, \max Re(\lambda(A)) < 0.$$

where $\lambda(.)$ denotes the eigenvalue of the matrix.

It follows from Theorem 3.2 (II) that there exists $T(\varepsilon) > 0$ such that

$$\left\| e^{(A-\lambda BK)(t-t_0)} x_0 + \lambda^*[e^{(A-\lambda BK)(t-t_0)} - I](A - \lambda BK)^{-1} Bu_{\lim} \right\|$$
$$- \left\| \lambda^*(A - \lambda BK)^{-1} Bu_{\lim} \right\| < \varepsilon, t \geq T(\varepsilon).$$

So

$$\|x(t, t_0, x_0)\|$$
$$\leq \sup_{\forall x_0 \in \mathbb{R}^n, \forall t \in [0,\infty)} \{\|e^{(A-\lambda BK)(t-t_0)} x_0 + \lambda^* [e^{(A-\lambda BK)(t-t_0)} - I](A - \lambda BK)^{-1} Bu_{\lim}\|\}$$
$$= \max_{\forall x_0 \in \mathbb{R}^n, \forall \varepsilon(x_0)>0} \{ \sup_{\forall x_0 \in \mathbb{R}^n, t_0 \leq t \leq T(\varepsilon)} \{\|e^{(A-\lambda BK)(t-t_0)} x_0 + \lambda^*[e^{(A-\lambda BK)(t-t_0)} - I]$$
$$*(A - \lambda BK)^{-1} Bu_{\lim}\|, \max_{(\lambda,\lambda^*), \forall \varepsilon(x_0)>0} \{\|\lambda^*(A - \lambda BK)^{-1} Bu_{\lim}\| + \varepsilon\}\}$$
$$= \max_{\forall x_0 \in \mathbb{R}^n, \forall \varepsilon(x_0)>0, t_0 \leq t \leq T(\varepsilon)} \{\|e^{(A-BK)(t-t_0)} x_0\|,$$
$$\|e^{A(t-t_0)} x_0 \pm [e^{A(t-t_0)} - I] A^{-1} Bu_{\lim}\|, \|A^{-1} Bu_{\lim}\| + \varepsilon\}$$
$$< \infty.$$

It follows from the above discussions that the case (c) is impossible.

For the case (b), there are two possibilities below.

(I) In the region $\mathbb{D}_{1,0}$, the trajectory $x(t, t_4, x_4)$ will be intersect with $\partial \mathbb{S}_{1,0}$;

(II) Trajectory $x(t, t_4, x_4)$ will at last reach to the tangent points $p_{0,1}^{'}$ or $p_{0,-1}^{''}$.

If the case (I) holds, it is clear that $x(t, t_0, x_0) = 0$ is globally asymptotically stable. If not, the sequence of the joint points x_{4k+1} on the line $-Kx = u_{\lim}$ satisfies

$$0 < \overline{r(x_{4k+1} p_{1,0}^{'})} < \overline{r(x_1 p_{1,0}^{'})}, 0 > \overline{r(x_{4k} p_{1,0}^{'})} > \cdots > \overline{r(x_4 p_{1,0}^{'})}.$$

where $\overline{r}$ denotes the distance between two points.

Similarly, it follows the same result for a sequence of the joint points in the line $-Kx = -u_{\lim}$. The sequence $\overline{r(x_{4k+1} p_{1,0}^{'})}$ monotonically decreases as k increases, while the sequence $\overline{r(x_{4k} p_{1,0}^{'})}$ monotonously increases. Since the limit relation $\overline{r(x_{4k+1} p_{1,0}^{'})}$ is the existence and the uniqueness, it follows that

$$\lim_{t\to\infty}\overline{r(x_{4k+1}p'_{1,0})}=0,\ \lim_{t\to\infty}\overline{r(x_{4k}p'_{1,0})}=0.$$

That is to say that the trajectory $x(t,t_0,x_0)=0$ will pass through the point $p'_{0,1}$ or $p''_{0,-1}$.

Considering the positively invariant and contractive of the maximum ellipsoid $\mathbb{S}_{0,1}$, it follows that

$$\lim_{t\to\infty,\,x_0\in\mathbb{S}_{0,1}} x(t,t_0,x_0)=0.$$

Similarly, we obtain the same results with respect to $\forall x_0 \in \mathbb{D}_{0,-1}$ or $\forall x_0 \in \mathbb{D}_{0,1}$. In short, it follows that

$$\lim_{t\to\infty} x(t,t_0,x_0)=0,$$

with respect to $\forall x_0 \in \mathbb{R}^2$. □

3.3 Criteria to the Closed Trajectory

From the above discussions, we know that for Theorem 3.2 (II), that there does not exist any closed trajectory, is a precondition. More importantly, how to judge the closed trajectory is a meaningful problem. Now, we shall discuss some criteria about the existence of closed trajectories.

Suppose that there exists the feedback control law $u=-Kx$ such that matrices A, $A-BK$ are Hurwitz stable for (1). If there exists the closed trajectory in $\mathbb{R}^2$, then it must intersect with line $-Kx(t)=u_{\lim}$ and line $-Kx(t)=-u_{\lim}$. Since $x_{eq,0,1}$ and $x_{eq,0,-1}$ are the spurious equilibrium points, there exists some time point T_1 such that $-Kx(T_1)=u_{\lim}$ with respect to any initial point $x_0 \in \mathbb{D}_{1,0}$ and any initial time t_0. Let $\Delta T'_1 = T_1 - t_0$, if the trajectory $x(t,T_1,x_1)$ does not intersect with the maximum ellipsoid $\mathbb{S}_{1,0} \in \mathbb{D}_{1,0}$, then there exists time point T_2 where $x(t,T_1,x_1)$ intersects with $-Kx(t)=-u_{\lim}$, whose joint point is $x_2 = x(T_2,T_1,x_1)$. Let $\Delta T_2 = T_2 - T_1$. The trajectory $x(t,T_1,x_1)$ intersects with $-Kx(t)=-u_{\lim}$ in $\mathbb{D}_{1,0}$ whose passing time is ΔT_4 and whose joint point is x_4.

In the region $\mathbb{D}_{1,0}$, the trajectory $x(t,T_1,x_1)$ intersects with $-Kx(t)=u_{\lim}$ after time $\Delta T''_1$ whose joint point is x_1. From the above discussions, it follows that

$$\begin{cases} \Delta T_1 = \Delta T'_1 + \Delta T''_1, \\ \Delta T_2 = T_2 - T_1, \\ \Delta T_3 = T_3 - T_2, \\ \Delta T_4 = T_4 - T_3. \end{cases}$$

Suppose that the closed trajectory is $\overline{x_1x_2x_3x_4}$ where x_1, x_2, x_3, x_4 are joint points with lines $-Kx(t)=u_{\lim}$, and $-Kx(t)=-u_{\lim}$ (See Fig. 3.2).

3.3 Criteria to the Closed Trajectory

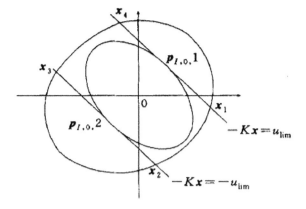

Fig. 3.2 The closed trajectory of second-order linear system with a single saturation input

The closed trajectory with the initial value x_0 is as follows:

$$\begin{cases} x_1(T_1) = e^{A\Delta T_1'}x_0 + (e^{A\Delta T_1} - I)A^{-1}Bu_{\lim}, \\ -Kx_1(T_1) = u_{\lim}. \\ x_2(T_2) = e^{(A-BK)\Delta T_2}x_1, \\ -Kx_2(T_2) = -u_{\lim}. \\ x_3(T_3) = e^{A\Delta T_3}x_2 - (e^{A\Delta T_3} - I)A^{-1}Bu_{\lim}, \\ -Kx_3(T_3) = -u_{\lim}. \\ x_4(T_4) = e^{(A-BK)\Delta T_4}x_3, \\ -Kx_4(T_4) = -u_{\lim}. \\ x_1(T_1) = e^{A\Delta T_1}x_4 + (e^{A\Delta T_1} - I)A^{-1}Bu_{\lim}, \\ -Kx_1(T_1) = u_{\lim}. \end{cases}$$

Since the closed trajectory $\overline{x_1 x_2 x_3 x_4}$ is periodic, the closed trajectories with initial values x_1, x_2, x_3, x_4 are defined as follows.

(1) The closed trajectories with an initial value x_1 is

$$\begin{cases} x_2(T_2) = e^{(A-BK)\Delta T_2}x_1, \\ -Kx_2(T_2) = u_{\lim}. \\ x_3(T_3) = e^{A\Delta T_3}x_2 - (e^{A\Delta T_3} - I)A^{-1}Bu_{\lim}, \\ -Kx_3(T_3) = -u_{\lim}. \\ x_4(T_4) = e^{(A-BK)\Delta T_4}x_3, \\ -Kx_4(T_4) = -u_{\lim}. \\ x_1(T_1) = e^{A\Delta T_1}x_4 + (e^{A\Delta T_1} - I)A^{-1}Bu_{\lim}, \\ -Kx_1(T_1) = u_{\lim}. \end{cases}$$

That is to say

$$\begin{cases} (e^{A\Delta T_1}e^{(A-BK)\Delta T_4}e^{A\Delta T_3}e^{(A-BK)\Delta T_2} - I)x_1 = e^{A\Delta T_1}e^{(A-BK)\Delta T_4}e^{A\Delta T_3}A^{-1}Bu_{\lim} \\ -e^{A\Delta T_1}e^{(A-BK)\Delta T_4}A^{-1}Bu_{\lim} - e^{A\Delta T_1}A^{-1}Bu_{\lim} + A^{-1}Bu_{\lim}, \\ -Kx_1 = u_{\lim}. \end{cases}$$

Let

$$\begin{cases} A_1 x_1 = B_1 A^{-1} B u_{\lim}, \\ -K x_1 = u_{\lim}, \end{cases} \tag{3.7}$$

where

$$\begin{cases} A_1 = e^{A\Delta T_1}e^{(A-BK)\Delta T_4}e^{A\Delta T_3}e^{(A-BK)\Delta T_2} - I, \\ B_1 = e^{A\Delta T_1}e^{(A-BK)\Delta T_4}e^{A\Delta T_3} - e^{A\Delta T_1}e^{(A-BK)\Delta T_4} - e^{A\Delta T_1} + I. \end{cases}$$

Similarly, we give the following closed trajectories equations with different initial values.

(2) The closed trajectories with an initial value x_2 is

$$\begin{cases} A_2 x_2 = B_2 A^{-1} B u_{\lim}, \\ -K x_2(T_2) = -u_{\lim}, \end{cases} \tag{3.8}$$

where

$$\begin{cases} A_2 = e^{(A-BK)\Delta T_2}e^{A\Delta T_1}e^{(A-BK)\Delta T_4}e^{A\Delta T_3} - I, \\ B_2 = e^{(A-BK)\Delta T_2}e^{A\Delta T_1}e^{(A-BK)\Delta T_4}e^{A\Delta T_3} - e^{(A-BK)\Delta T_2}e^{A\Delta T_1}e^{(A-BK)\Delta T_4} \\ -e^{(A-BK)\Delta T_2}e^{A\Delta T_1} + e^{(A-BK)\Delta T_2}. \end{cases}$$

(3) The closed trajectories with an initial value x_3 is

$$\begin{cases} A_3 x_3 = B_3 A^{-1} B u_{\lim}, \\ -K x_3 = -u_{\lim}, \end{cases} \tag{3.9}$$

where

$$\begin{cases} A_3 = e^{A\Delta T_3}e^{(A-BK)\Delta T_2}e^{A\Delta T_1}e^{(A-BK)\Delta T_4} - I, \\ B_3 = -e^{A\Delta T_3}e^{(A-BK)\Delta T_2}e^{A\Delta T_1} + e^{A\Delta T_3}e^{(A-BK)\Delta T_2} + e^{A\Delta T_3} - I. \end{cases}$$

(4) The closed trajectories with an initial value x_4 is

$$\begin{cases} A_4 x_4 = B_4 A^{-1} B u_{\lim}, \\ -K x_4 = u_{\lim}, \end{cases} \tag{3.10}$$

3.3 Criteria to the Closed Trajectory

where

$$\begin{cases} A_4 = e^{(A-BK)\Delta T_4}e^{A\Delta T_3}e^{(A-BK)\Delta T_2}e^{A\Delta T_1} - I, \\ B_4 = -e^{(A-BK)\Delta T_4}e^{A\Delta T_3}e^{(A-BK)\Delta T_2}e^{A\Delta T_1} + e^{(A-BK)\Delta T_4}e^{A\Delta T_3}e^{(A-BK)\Delta T_2} \\ -e^{(A-BK)\Delta T_4}e^{A\Delta T_3} + e^{(A-BK)\Delta T_4}. \end{cases}$$

If the trajectory $\overline{x_1 x_2 x_3 x_4}$ is closed, then (3.7)–(3.10) simultaneously hold. On the other hand, if points x_1, x_2, x_3, x_4 which satisfy (3.7) as initial values must simultaneously satisfy (3.7)–(3.10) with regard to uniqueness of the solution.

We again consider the fact of uniqueness of the solution and nonsingular matrices $e^{(A-BK)\Delta T}$, $e^{A\Delta T}$, it follows that the above four trajectories are superposing.

Hence,

$$\begin{cases} x_1(t) &= e^{A(t-T_0)}x_0 + (e^{A(t-T_0)} - I)A^{-1}Bu_{\lim}, \\ -Kx_1(t) &\geq u_{lim}, \\ t &\in [t_0, T_1], \\ x_1(T_1) &= e^{A\Delta T_1'}x_0 + (e^{A\Delta T_1} - I)A^{-1}Bu_{\lim}, \\ -Kx_1(T_1) &= u_{lim}. \\ x_2(t) &= e^{(A-BK)(t-T_1)}x_1, \\ \|-Kx_2(T_2)\| &< u_{lim}, \\ t &\in [T_1, T_2), \\ x_2(T_2) &= e^{(A-BK)\Delta T_2}x_1, \\ -Kx_2(T_2) &= -u_{lim}. \\ x_3(t) &= e^{A(t-T_2)}x_2 - (e^{A(t-T_2)} - I)A^{-1}Bu_{\lim}, \\ -Kx_3(t) &\leq -u_{lim}, \\ t &\in [T_2, T_3], \\ x_3(T_3) &= e^{A\Delta T_3}x_2 - (e^{A\Delta T_3} - I)A^{-1}Bu_{\lim}, \\ -Kx_3(T_3) &= -u_{lim}. \\ x_4(t) &= e^{(A-BK)(t-T_3)}x_3 \\ \|-Kx_4(t)\| &< u_{lim} \\ t &\in [T_3, T_4] \\ x_4(T_4) &= e^{(A-BK)\Delta T_4}x_3 \\ -Kx_4(T_4) &= u_{lim} \\ x_1(t) &= e^{A(t-T_4)}x_4 + (e^{A\Delta T_1} - I)A^{-1}Bu_{\lim}, \\ -Kx_1(T_1) &\geq u_{lim}, \\ t &\in [T_4, \Delta T_1 + T_4], \\ x_1(\Delta T_1 + T_4) &= e^{A\Delta T_1}x_4 + (e^{A\Delta T_1} - I)A^{-1}Bu_{\lim}, \\ -Kx_1(\Delta T_1 + T_4) &= u_{lim}. \end{cases} \quad (3.11)$$

We consider the simplicities of (3.7)–(3.11) below. If A_1, A_2, A_3, A_4 are invertible, (3.7)–(3.11) transforms to

$$\begin{cases} -KA_1^{-1}B_1A^{-1}Bu_{\lim} = u_{\lim}, \\ -KA_2^{-1}B_2A^{-1}Bu_{\lim} = -u_{\lim}, \\ -KA_3^{-1}B_3A^{-1}Bu_{\lim} = -u_{\lim}, \\ -KA_4^{-1}B_4A^{-1}Bu_{\lim} = u_{\lim}. \end{cases} \quad (3.12)$$

Because of the constant $u_{\lim} > 0$, (3.12) transforms to

$$\begin{cases} -KA_1^{-1}B_1A^{-1}B = 1, \\ -KA_2^{-1}B_2A^{-1}B = -1, \\ -KA_3^{-1}B_3A^{-1}B = -1, \\ -KA_4^{-1}B_4A^{-1}B = 1. \end{cases} \quad (3.13)$$

For (3.13), if there exists a group of positive constants ΔT_1, ΔT_2, ΔT_3, ΔT_4 such that (3.13) holds, then there exists closed trajectory. If for every group of positive constants ΔT_1, ΔT_2, ΔT_3, ΔT_4, at least one equality of (3.13) does not hold, then there does not exist any closed trajectory.

If matrices A_1, A_2, A_3, A_4 are singular, we consider inequalities

$$\begin{cases} Rank(A_1, B_1A^{-1}B) > RankA_1, \\ Rank(A_2, B_2A^{-1}B) > RankA_2, \\ Rank(A_3, B_3A^{-1}B) > RankA_3, \\ Rank(A_4, B_4A^{-1}B) > RankA_4. \end{cases} \quad (3.14)$$

For any group of positive constants ΔT_1, ΔT_2, ΔT_3, ΔT_4, if at least one inequalities of (3.14) does not hold, then there does not exist any solution for algebraic inequalities (3.14) and any closed trajectory of (3.1). □

From the above discussions, we can easily obtain the following results.

Theorem 3.3 *For system (3.1), if*

(I) There exist a group of positive constants ΔT_1, ΔT_2, ΔT_3, ΔT_4 such that (3.14) holds, then there exists a closed trajectory.

(II) For any group of positive constants ΔT_1, ΔT_2, ΔT_3, ΔT_4, at least one equalities of (3.13) does not hold or satisfies at least one of (3.14), then there does exist any closed trajectory.

3.4 The Commutative Case

In this section, we analyze that when state matrices are commuting, the more simplified sufficient conditions on the GAS of the zero solution and the more simplified existence of the closed trajectory for the second-order control systems with single input compared to the above section. In particular, we transform the determination condition of the closed trajectory into the existence of the common solution for a set of higher order equations satisfying certain conditions. The above proposed

3.4 The Commutative Case

conclusions provide us a more comprehensive understanding of the GAS of the zero solution for the second-order control systems with single input.

First, we give the following results.

Theorem 3.4 $AB = BA$, $rank(A) = n$, $rank(B) = 1$ is only and if only that there exist nonsingular matrices Q_1, C such that

$$A = Q_1 \begin{pmatrix} l_{11} & L_{12} \\ 0 & L_{22} \end{pmatrix} Q_1^{-1}, \quad B = Q_1 \begin{pmatrix} c_{11} & C_{12} \\ 0 & 0 \end{pmatrix} Q_1^{-1},$$

where l_{11}, c_{11} are constants, $L_{12}, C_{12} \in \mathbb{R}^{1\times(n-1)}$.

The above results are easily obtained, and we omit the above proof process.

Consider the system (3.1) or (3.3), we can simplify the identification process using Theorem 3.3 when $ABK = BKA$.

$$\begin{aligned} A_1 &= A_2 \\ &= A_3 \\ &= A_4 \\ &= e^{A(\Delta T_1 + \Delta T_3)} e^{(A-BK)(\Delta T_2 + \Delta T_4)} - I. \end{aligned} \quad (3.15)$$

Similar to the above conclusions, (3.12) can be replaced into

$$\begin{cases} -KA_1B_1A^{-1}B = 1, \\ -KA_1B_2A^{-1}B = -1, \\ -KA_1B_3A^{-1}B = -1, \\ -KA_1B_4A^{-1}B = 1. \end{cases} \quad (3.16)$$

Combining with Theorem 3.3, we can simplify the form of A_1, B_1, K, $A^{-1}B$ when $ABK = BKA$,

$$\begin{aligned} &(e^{A(\Delta T_1+\Delta T_3)} e^{(A-BK)(\Delta T_2+\Delta T_4)} \\ &= Q_1 e^{\begin{pmatrix} l_{11} & L_{12} \\ 0 & L_{22} \end{pmatrix}(\Delta T_1+\Delta T_3)} e^{\begin{pmatrix} l_{11}-c_{11} & L_{12}-C_{12} \\ 0 & L_{22} \end{pmatrix}(\Delta T_2+\Delta T_4)} Q_1^{-1}. \end{aligned}$$

Now considering the particularity of the $(1, 1)$ element of an upper triangular matrix which is similar to A, $A - BK$, we easily obtain

$$e^{A(\Delta T_1+\Delta T_3)} e^{(A-BK)(\Delta T_2+\Delta T_4)} = Q_1 \begin{pmatrix} e^{l_{11}(\Delta T_1+\Delta T_3)+(l_{11}-c_{11})(\Delta T_2+\Delta T_4)} & * \\ 0 & * \end{pmatrix} Q_1^{-1}.$$

So

$$\begin{aligned} A_1 &= Q_1 (\begin{pmatrix} e^{l_{11}(\Delta T_1+\Delta T_3)+(l_{11}-c_{11})(\Delta T_2+\Delta T_4)} & * \\ 0 & * \end{pmatrix} - I) Q_1^{-1} \\ &= Q_1 \begin{pmatrix} e^{l_{11}(\Delta T_1+\Delta T_3)+(l_{11}-c_{11})(\Delta T_2+\Delta T_4)} - 1 & * \\ 0 & * \end{pmatrix} Q_1^{-1}. \end{aligned} \quad (3.17)$$

Similarly, we can get

$$B_1 = Q_1 \begin{pmatrix} e^{l_{11}(\Delta T_1+\Delta T_3)+(l_{11}-c_{11})(\Delta T_2+\Delta T_4)} \\ -e^{l_{11}\Delta T_1+(l_{11}-c_{11})\Delta T_4} & * \\ -e^{l_{11}\Delta T_1} + 1 \\ 0 & * \end{pmatrix} Q_1^{-1},$$

$$B_2 = Q_1 \begin{pmatrix} e^{l_{11}(\Delta T_1+\Delta T_3)+(l_{11}-c_{11})(\Delta T_2+\Delta T_4)} \\ -e^{l_{11}\Delta T_1+(l_{11}-c_{11})(\Delta T_2+\Delta T_4)} \\ -e^{l_{11}\Delta T_1+(l_{11}-c_{11})\Delta T_2} & * \\ +e^{(l_{11}-c_{11})\Delta T_2} \\ 0 & * \end{pmatrix} Q_1^{-1},$$

$$B_3 = Q_1 \begin{pmatrix} -e^{l_{11}(\Delta T_1+\Delta T_3)+(l_{11}-c_{11})\Delta T_2} \\ +e^{l_{11}\Delta T_3+(l_{11}-c_{11})\Delta T_2} & * \\ +e^{l_{11}\Delta T_3} - 1 \\ 0 & * \end{pmatrix} Q_1^{-1},$$

$$B_4 = Q_1 \begin{pmatrix} -e^{l_{11}(\Delta T_1+\Delta T_3)+(l_{11}-c_{11})(\Delta T_2+\Delta T_4)} \\ +e^{l_{11}\Delta T_3+(l_{11}-c_{11})(\Delta T_2+\Delta T_4)} \\ +e^{l_{11}\Delta T_3+(l_{11}-c_{11})\Delta T_4} & * \\ -e^{(l_{11}-c_{11})\Delta T_4} \\ 0 & * \end{pmatrix} Q_1^{-1}.$$

Therefore, (3.16) may be equivalent to

$$-c_{11}[e^{l_{11}(\Delta T_1+\Delta T_3)+(l_{11}-c_{11})(\Delta T_2+\Delta T_4)} - 1]^{-1}(e^{l_{11}(\Delta T_1+\Delta T_3)+(l_{11}-c_{11})\Delta T_4}$$
$$-e^{l_{11}\Delta T_1+(l_{11}-c_{11})\Delta T_4} - e^{l_{11}\Delta T_1} + 1)]l_{11}^{-1} = 1,$$
$$-c_{11}[e^{l_{11}(\Delta T_1+\Delta T_3)+(l_{11}-c_{11})(\Delta T_2+\Delta T_4)} - 1]^{-1}(e^{l_{11}(\Delta T_1+\Delta T_3)+(l_{11}-c_{11})(\Delta T_4+\Delta T_2)}$$
$$-e^{l_{11}\Delta T_1+(l_{11}-c_{11})(\Delta T_4+\Delta T_2)} - e^{l_{11}\Delta T_1+(l_{11}-c_{11})\Delta T_2} + e^{(l_{11}-c_{11})\Delta T_2})]l_{11}^{-1} = -1.$$

$$-c_{11}[e^{l_{11}(\Delta T_1+\Delta T_3)+(l_{11}-c_{11})(\Delta T_2+\Delta T_4)} - 1]^{-1}(-e^{l_{11}(\Delta T_1+\Delta T_3)+(l_{11}-c_{11})\Delta T_2}$$
$$+e^{l_{11}\Delta T_3+(l_{11}-c_{11})\Delta T_2} + e^{l_{11}\Delta T_3} - 1)]l_{11}^{-1} = -1,$$
$$-c_{11}[e^{l_{11}(\Delta T_1+\Delta T_3)+(l_{11}-c_{11})(\Delta T_2+\Delta T_4)} - 1]^{-1}(-e^{l_{11}(\Delta T_1+\Delta T_3)+(l_{11}-c_{11})\Delta T_4}$$
$$+e^{l_{11}\Delta T_3+(l_{11}-c_{11})(\Delta T_4+\Delta T_2)} + e^{l_{11}\Delta T_3+(l_{11}-c_{11})\Delta T_4} - e^{(l_{11}-c_{11})\Delta T_4})]l_{11}^{-1} = 1. \quad (3.18)$$

We know that (3.16) may be equivalent with the following question if

$$e^{l_{11}(\Delta T_1+\Delta T_3)+(l_{11}-c_{11})(\Delta T_2+\Delta T_4)} - 1 \quad (3.19)$$

is equal to 0.

3.4 The Commutative Case

Since A, $A - BK$ are Hurwitz matrix, l_{11}, $l_{11} - c_{11}$ are obviously negative. Therefore, for any positive number ΔT_1, ΔT_2, ΔT_3, ΔT_1, there must be

$$e^{l_{11}(\Delta T_1 + \Delta T_3) + (l_{11} - c_{11})(\Delta T_2 + \Delta T_4)} - 1 < 0. \tag{3.20}$$

Therefore, we only need to study the case of (3.18).
Also, we know that $c_{11}, l_{11} \in R^1$ from the above results, and suppose

$$\begin{cases} e^{l_{11} \Delta T_1} := x_1, \\ e^{l_{11} \Delta T_3} := x_3, \\ e^{(l_{11} - c_{11}) \Delta T_2} := x_2, \\ e^{(l_{11} - c_{11}) \Delta T_4} := x_4. \end{cases} \tag{3.21}$$

Then, (3.18) may be changed into the following equations

$$\begin{cases} \frac{l_{11}}{c_{11}} x_1 x_2 x_3 x_4 + x_1 x_3 x_4 - x_1 x_4 + (1 - \frac{l_{11}}{c_{11}}) = 0, \\ (1 - \frac{l_{11}}{c_{11}}) x_1 x_2 x_3 x_4 - x_1 x_2 x_4 - x_1 x_2 + x_2 = 0, \\ -\frac{l_{11}}{c_{11}} x_1 x_2 x_3 x_4 + x_1 x_3 x_4 - x_1 x_2 x_3 + x_2 x_3 + x_3 - (1 - \frac{l_{11}}{c_{11}}) = 0, \\ \frac{l_{11}}{c_{11}} x_1 x_2 x_3 x_4 + x_1 x_3 x_4 - x_1 x_3 x_4 + x_2 x_3 x_4 + x_3 x_4 - x_4 = 0. \end{cases} \tag{3.22}$$

whether there exist a group of solutions (x_1, x_2, x_3, x_4), $0 < x_i < 1$, $i = 1, 2, 3, 4$.
Further, (3.22) can also be transformed into the following higher order equations

$$\begin{cases} \frac{l_{11}}{c_{11}} x^4 + x^3 - x^2 + (1 - \frac{l_{11}}{c_{11}}) = 0, \\ (1 - \frac{l_{11}}{c_{11}}) x^4 - x^3 - x^2 + x = 0, \\ -\frac{l_{11}}{c_{11}} x^4 + x^2 + x - (1 - \frac{l_{11}}{c_{11}}) = 0, \\ \frac{l_{11}}{c_{11}} x^4 + x^3 + x^2 - x = 0, \end{cases} \tag{3.23}$$

whether there exists a common positive number solution x whose less than 1.
If we suppose that

$$x = e^{l_{11} \Delta T} = 0,$$

or

$$x = e^{(l_{11} - c_{11}) \Delta T} = 0.$$

And taking into account the following facts

$$\max Re(\lambda(A)) < 0, \max Re(\lambda(A - BK)) < 0,$$

then there must be

$$\Delta T = \infty.$$

While the closed trajectory requires

$$\Delta T < \infty,$$

so we conclude that

$$x = e^{l_{11}\Delta T} \neq 0, x = e^{(l_{11}-c_{11})\Delta T} \neq 0.$$

Therefore, (3.23) is equivalent to the following problem

$$\begin{cases} \frac{l_{11}}{c_{11}}x^4 + x^3 - x^2 + (1 - \frac{l_{11}}{c_{11}}) = 0, \\ (1 - \frac{l_{11}}{c_{11}})x^3 - x^2 - x + 1 = 0, \\ -\frac{l_{11}}{c_{11}}x^4 + x^2 + x - (1 - \frac{l_{11}}{c_{11}}) = 0, \\ \frac{l_{11}}{c_{11}}x^3 + x^2 + x - 1 = 0. \end{cases} \quad (3.24)$$

whether there exists a common root which is a positive number less than 1.

(By contradiction) If there exists a positive root which is less than 1, then, for (3.24), the first equation plus the third equation, we obtain

$$x^3 + x = 0.$$

Similarly, the second equation plus the fourth equation, we obtain

$$x^3 = 0.$$

Obviously, it is impossible. So, (3.24) does not exist any positive root which is less than 1. □

To sum up the above discussions, we outline the following results.

Theorem 3.5 *For (3.1), suppose A, $A - BK$ are Hurwitz, and $ABK = BKA$, then*
(1) There do not exist any closed trajectory for (3.1);
(2) The zero solution of the system (3.1) is globally asymptotically stable.
Based on Theorems 3.1, 3.2 and 3.3, we can be obtain the above conclusions.

3.5 Example and Simulation

Consider the following second-order single-saturated input control system

$$\begin{cases} \dot{y}_1(t) = -y_1(t) + 2y_2(t) + u_s, \\ \dot{y}_2(t) = -3y_2(t), \\ u_s = sat(-0.5y_1(t) - y_2(t)), \end{cases} \quad (3.25)$$

where saturated input u_s is defined as

3.5 Example and Simulation

Fig. 3.3 The movement phase trajectories starting from (5, 5)

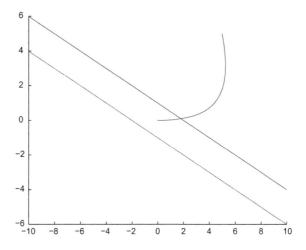

Fig. 3.4 The movement phase trajectories starting from (5, 5)

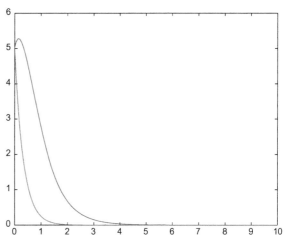

$$sat(-0.5y_1(t) - y_2(t)) = \begin{cases} 1, & if \ -0.5y_1(t) - y_2(t) \geq 1, \\ -0.5y_1(t) - y_2(t), & if \ |-0.5y_1(t) - y_2(t)| < 1, \\ -1, & if \ -0.5y_1(t) - y_2(t) \leq -1. \end{cases}$$

Clearly, A,B,K satisfy the commutative condition and A, $A - BK$ are Hurwitz stable. According to the conditions of Theorem 3.4, the zero solution of (3.25) is GAS. Through the following simulation, we illustrate the effectiveness of the above conclusions.

(1) Figures 3.3 and 3.4 show phase portrait $x_2(x_1)$ and trajectories $x_1(t)$, $x_2(t)$ passing initial the points$(5, 5)(-0.5 * 5 - 5) = -7.5 \leq -1$), which denote that the zero solution of the system (3.25) is asymptotically stable.

Fig. 3.5 The movement phase trajectories starting from $(-5, -5)$

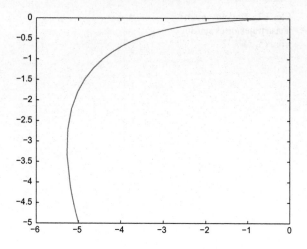

Fig. 3.6 The movement phase trajectories starting from $(-5, -5)$

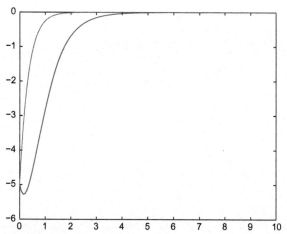

(2) Figures 3.5 and 3.6 show phase portrait $x_2(x_1)$ and trajectories $x_1(t)$, $x_2(t)$ passing initial the points $(-5, -5)(-0.5 * 5 - (-5)) = 2.5 \geq 1)$, which denote that the zero solution of the system (3.25) is asymptotically stable.

(3) Figures 3.7 and 3.8 show phase portrait $x_2(x_1)$ and trajectories $x_1(t)$, $x_2(t)$ passing initial the points $(5, -2)(-0.5 * 5 - (-2)) = -0.5 \leq 1)$, which denote the zero solution of the system (3.25) is asymptotically stable.

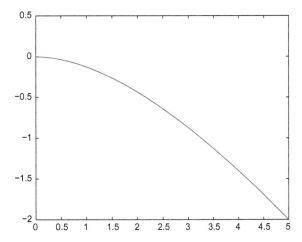

Fig. 3.7 The movement phase trajectories starting from $(5, -2)$

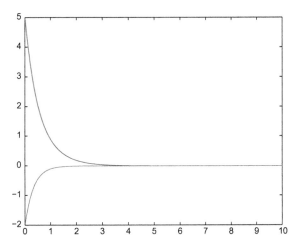

Fig. 3.8 The movement phase trajectories starting from $(5, -2)$

3.6 Conclusion

We have introduced the spatial structure of control systems with saturated inputs and described its equations by $0 - 1$ structure. Commutative matrices of MIMO linear systems are considered, and the existence of the feedback matrices of commutative state matrix set in the MIMO closed loops is reduced to invariance subspace of matrix A, and the existence of feedback matrices in the open-loop systems is equivalent to the existence of the solution of Lyapunov matrix equation. The relationship among equilibrium points is discussed for linear system with single-saturated input under commutative matrices. The new method for control system with saturation inputs is presented. The sufficient conditions on the asymptotic stability of the origin of the linear system in the presence of a single saturation input is presented, and the existence of closed trajectory is for the same control systems also converted into the existence of solution for the algebraic equations.

Acknowledgements This work is Supported by National Key Research and Development Program of China (2017YFF0207400).

References

1. Dugard L, Verrist EI, editors. Stability and control of time-delay systems. Berlin: Springer; 1998. p. 303–10.
2. Henrion D, Tarbouriech S. LMI relaxations for robust stability of linear systems with saturating controls. Automatica. 1999;35:1599–604.
3. Horisberger HP, Belanger PR. Regulator for linear, time invariant plants with uncertain parameters. IEEE Trans Autom Control. 1976;42:705–8.
4. Huang L. Stability theory. Beijing: Peking University Press; 1992. p. 235–83.
5. Kapoor N, Daoutidis P. An observer-based anti-windup scheme for nonlinear systems with input constraints. Int J Control. 1999;72(1):18–29.
6. Lee AW, Hedrick JK. Some new results on closed-loop stability in the presence of control saturation. Int J Control. 1995;62(3):619–51.
7. Liberzon D, Hespanta JP, Morse AS. Stability of switched systems: a Lie-algebraic condition. Syst Control Lett North-Holland. 1999;37:117–22.
8. Mareada KS, Balakrishnan J. A common Lyapunov function for stable LTI systems with commuting A-martices. IEEE Trans Autom Control. 1994;39(12):2469–71.
9. Mancilla-Aguilar JL, Garcia RA. A converse Lyapunov theorem for nonlinear switched systems. Syst Control Lett North-Holland. 2000;41(1):69–71.
10. Molchanov AP, Pyatnitskiy YS. Criteria of asymptotic stability of differential and difference inclusions encountered in control theory. Syst Control Lett North-Holland. 1989;13:59–64.
11. Polanski K. On absolute stability analysis by polyhedric Lypunov functions. Automatica. 2000;36:573–8.
12. Scibile L, Kouvaritakis BIF. Stability region for a class of open-loop unstable linear systems: theory and application. Automatica. 2000;36(1):37–44.
13. Sontag ED, Sussmann HJ. Nonlinear output feedback design for linear systems with saturated control. In: Proceeding of 29th conference on decision and control; 1990. vol. 45(5). p. 719–721.
14. Tatsushi, Yasuyuki F. Two conditions concerning common quadratic Lyapunov functions for linear systems. IEEE Trans Autom Control. 1997;42(5):750–761.
15. Tarbouriech S, Gomes da JM. Synthesis of controllers for continuous-times delay systems with saturated control via LMIs. IEEE Trans Autom Control. 2000;45(1):105–11.

Chapter 4
Equilibrium Points of Second-Order Linear Systems with Single Saturated Input

4.1 Introduction

Input saturation is encountered in all control system implementations because of the physical limits of actual actuators. The analysis and synthesis of controllers for dynamic systems subject to actuator saturation have been attracting increasingly more attention. Common examples of such limits are the deflection limits in aircraft actuators, the voltage limits in electrical actuators, the limits on flow volume or rate in hydraulic actuators, and so on. Although such limits obviously restrict the performance achievable in the systems of which they are part, if these limits are not treated carefully and if the relevant controllers do not account for these limits appropriately, peculiar and pernicious behaviors may be observed. In particular, actuator saturation or rate limits have been implicated in various aircraft crashes and so on.

Lee and Hedrick [1] used the hypercube in $\mathbb{R}^n$ to obtain GAS of the origin. The same method is used in paper [1–18] and some results about GAS of the origin are obtained. It is fact that the control is allowed to saturate when the resulting control signal exceeds the limit. We find that there exist nonzero equilibrium points when saturation happens. It is necessary to learn how the other equilibrium points except the origin function in analyzing the control systems with saturated inputs. In [17], commutative matrices of multiple input multiple output (MIMO) linear systems are considered. The existence of the feedback matrices of a commutative state matrix set in the MIMO closed loops is reduced to the existence of an invariant subspace of a matrix A. The existence of feedback matrices in systems in open loop is equivalent to the existence of the solution of matrix equations denoted by Kronecker products. By defining new equilibrium points, the relationship between equilibrium points is discussed for a linear system with a single saturated input under $ABK = BKA$. Four criteria for equilibrium points are outlined for such linear systems.

In this chapter, the main contribution is to present many interested results about equilibrium points for $2nd$ differential systems with single saturated input. First, we present general results about equilibrium points relationship based on classical qualitative methods and equilibrium point definitions. Second, we present some discussions on many cases about $A, A - BK$ and their corresponding numerical simulations.

4.2 Problem Statements, Equilibrium Points

Consider the following system:

$$\dot{x}(t) = Ax(t) + Bu_s, \qquad (4.1)$$

where

$$u_s = sat(-Kx) = \begin{cases} u_{\lim}, & \text{if } -Kx \geq u_{\lim}, \\ -Kx, & \text{if } |-Kx| < u_{\lim}, \\ -u_{\lim}, & \text{if } -Kx \leq -u_{\lim}. \end{cases}$$

and $A = \begin{bmatrix} a_{11} & a_{12} \\ a_{21} & a_{22} \end{bmatrix}$, $B = \begin{bmatrix} b_1 \\ b_2 \end{bmatrix}$, $K = \begin{bmatrix} k_1 & k_2 \end{bmatrix}$.

When A is nonsingular (i.e., $det(A) \neq 0$), there exist the following strange equilibrium points

$$x_{eq,1,0} = [0, 0]^T,$$

$$x_{eq,0,1} = -A^{-1}Bu_{\lim} = \left[\frac{a_{12}b_2 - a_{22}b_1}{a_{11}a_{22} - a_{12}a_{21}} u_{\lim} \quad \frac{a_{21}b_1 - a_{11}b_2}{a_{11}a_{22} - a_{12}a_{21}} u_{\lim} \right]^T = -\frac{1}{det(A)} A^* Bu_{\lim},$$

$$x_{eq,0,-1} = A^{-1}Bu_{\lim} = \left[-\frac{a_{12}b_2 - a_{22}b_1}{a_{11}a_{22} - a_{12}a_{21}} u_{\lim} \quad -\frac{a_{21}b_1 - a_{11}b_2}{a_{11}a_{22} - a_{12}a_{21}} u_{\lim} \right]^T = \frac{1}{det(A)} A^* Bu_{\lim},$$

$$A^{-1} = \begin{pmatrix} \dfrac{a_{22}}{a_{11}a_{22} - a_{12}a_{21}} & -\dfrac{a_{12}}{a_{11}a_{22} - a_{12}a_{21}} \\ -\dfrac{a_{21}}{a_{11}a_{22} - a_{12}a_{21}} & \dfrac{a_{11}}{a_{11}a_{22} - a_{12}a_{21}} \end{pmatrix} = \frac{1}{det(A)} \begin{pmatrix} a_{22} & -a_{12} \\ -a_{21} & a_{11} \end{pmatrix} = \frac{1}{det(A)} A^*,$$

which correspondingly satisfy the following equations:

$$\dot{x}(t) = (A - BK)x(t),$$
$$\dot{x}(t) = Ax(t) + Bu_{\lim},$$
$$\dot{x}(t) = Ax(t) - Bu_{\lim},$$

or satisfy the following equations:

$$\dot{x}(t) = (A - \lambda BK)x(t) + \lambda^* Bu_{\lim},$$

4.2 Problem Statements, Equilibrium Points

where (λ, λ^*) is defined as

$$\begin{cases} (\lambda, \lambda^*) = (0, 1) \Leftrightarrow \text{if } -Kx \geq u_{\lim}, \\ (\lambda, \lambda^*) = (1, 0) \Leftrightarrow \text{if } \|-Kx\| \leq u_{\lim}, \\ (\lambda, \lambda^*) = (0, -1) \Leftrightarrow \text{if } -Kx \leq -u_{\lim}. \end{cases}$$

Here, we define

$$\begin{cases} \mathbb{D}_{1,0} = \{x| -u_{\lim} < -Kx < u_{\lim}\}, \\ \mathbb{D}_{0,1} = \{x| -Kx \geq u_{\lim}\}, \\ \mathbb{D}_{0,-1} = \{x| -Kx \leq -u_{\lim}\}. \end{cases}$$

Then, we have the following definitions.

Definition 4.1 (I) For system (4.1), some pair (λ, λ^*) and its corresponding x_{eq,λ,λ^*} satisfying (4.1), if $x_{eq,\lambda,\lambda^*} \in \mathbb{D}_{\lambda,\lambda^*}$, we call x_{eq,λ,λ^*} a genuine equilibrium point. For such an equilibrium, if $(A - \lambda BK)$ is a stable (unstable) matrix, then x_{eq,λ,λ^*} is a genuine stable (unstable) equilibrium point.

(II) For system (4.1), some pair (λ, λ^*) and its corresponding x_{eq,λ,λ^*} satisfying (4.1), if $x_{eq,\lambda,\lambda^*} \notin \mathbb{D}_{\lambda,\lambda^*}$, then we call x_{eq,λ,λ^*} a spurious equilibrium point. For such spurious equilibria, if $(A - \lambda BK)$ is stable (unstable), then we say that x_{eq,λ,λ^*} is a spurious stable (unstable) equilibrium point.

Note 4.1 Clearly, $x_{eq,1,0} \in \mathbb{D}_{1,0}$, we say that $x_{eq,1,0}$ is always genuine equilibrium point; we can know that $x_{eq,0,1} \in \mathbb{D}_{0,1}$ is if and only if (iff) $x_{eq,0,-1} \in \mathbb{D}_{0,-1}$. If not, $x_{eq,0,\pm 1}$ are both spurious equilibrium points.

Note 4.2 $x = 0$ is an equilibrium of (4.1) provided $det(A - BK) \neq 0$, while $det(A - BK) = det(A) - prod, det(A) = a_{11}a_{22} - a_{12}a_{21}$, $prod = k_1(a_{22}b_1 - a_{12}b_2) + k_2(a_{11}b_2 - a_{21}b_1)$.

Note 4.3 The stability of $x = 0$ is determined from

$$det(A - BK - \lambda I_2) = 0,$$

i.e.,

$$\lambda^2 - Tr(A - BK)\lambda + det(A - BK) = 0,$$

where

$$Tr(A - BK) = Tr\begin{pmatrix} a_{11} - b_1k_1 & a_{12} - b_1k_2 \\ a_{21} - b_2k_1 & a_{22} - b_2k_2 \end{pmatrix}$$
$$= a_{11} + a_{22} - (b_1k_1 + b_2k_2)$$
$$= tr(A) - <B, K>,$$

So,

$$\lambda^2 - (Tr(A) - <B, K>)\lambda + det(A) - prod = 0.$$

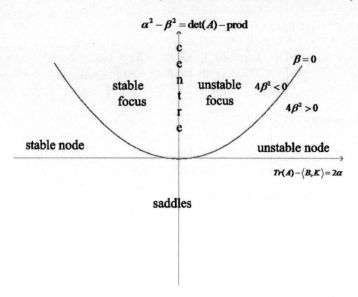

Fig. 4.1 $x_{eq,0,1}, x_{eq,0,-1}$ classification

Further, we have

$$\lambda = \tfrac{1}{2}(tr(A) - <B,K>) \pm \tfrac{1}{2}((tr(A) - <B,K>)^2 - 4(det(A) - prod))^{\tfrac{1}{2}} = \alpha \pm \beta.$$

where $\alpha = \tfrac{1}{2}(tr(A) - <B,K>)$, $\beta = \tfrac{1}{2}((tr(A) - <B,K>)^2 - 4(det(A) - prod))^{\tfrac{1}{2}}$.

Then, the stability of $x = 0$ is classified through Fig. 4.1.

Note 4.4 Based on Definition 4.1, we consider whether $x_{eq,0,1}$ is genuine or not.

$$x_{eq,0,1} = \begin{pmatrix} x_1^+ \\ x_2^+ \end{pmatrix} = -A^{-1}Bu_{lim} = -det(A)^{-1}\begin{pmatrix} a_{22}b_1 - a_{12}b_2 \\ a_{21}b_1 + a_{11}b_2 \end{pmatrix} u_{lim}.$$

Further, we have
$$x_1^+ k_1 + x_2^+ k_2 = -det(A)^{-1} prod \cdot u_{lim}.$$

If $x_{eq,0,1}$ is a genuine equilibrium point iff

$$-Kx_{eq,0,1} = -(x_1^+ k_1 + x_2^+ k_2) = det(A)^{-1} prod \cdot u_{lim} \geq u_{lim}.$$

Since $u_{lim} > 0$, $x_{eq,0,1}$ is genuine iff $det(A)^{-1} prod \geq 1$, and $x_{eq,0,1}$ is spurious iff $det(A)^{-1} prod < 1$.

The same conditions for $x_{eq,0,-1}$ are outlined too, and we omit their discussions.

4.2 Problem Statements, Equilibrium Points

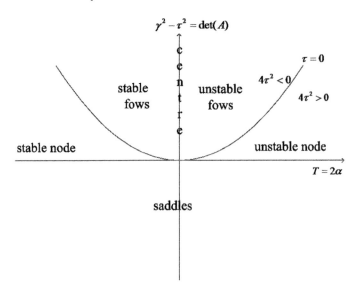

Fig. 4.2 $x_{eq,0,1}, x_{eq,0,-1}$ classification

Note 4.5 The stability of $x_{eq,0,\pm 1}$ is given by eigenvalue of A, i.e.,

$$\Lambda^2 - Tr(A)\Lambda + det(A) = 0.$$

Further, we have

$$\Lambda = \tfrac{1}{2}Tr(A) \pm \tfrac{1}{2}(Tr(A)^2 - 4det(A))^{\frac{1}{2}} = \gamma \pm \tau,$$

where $\gamma = \frac{Tr(A)}{2}, \tau = \tfrac{1}{2}(Tr(A)^2 - 4det(A))^{\frac{1}{2}}$.

Then, the stability of $x_{eq,0,\pm 1}$ is classified through the following Fig. 4.2. Combining the above facts, we easily obtain the following results:

$$\begin{aligned}\alpha^2 - \beta^2 &= a_{11}a_{22} - a_{12}a_{21} + a_{12}b_2k_1 - a_{22}b_1k_1 - a_{11}b_2k_2 + a_{21}b_1k_2\\ &= det(A - BK)\\ &= det(A) - prod,\\ \gamma^2 - \tau^2 &= a_{11}a_{22} - a_{12}a_{21}\\ &= det(A).\end{aligned}$$

Theorem 4.1 *If $det(A) \neq 0$, then $x_{eq,0,1} \in \mathbb{D}_{0,1}$ (or $x_{eq,0,-1} \in \mathbb{D}_{0,-1}$) for the system (4.1) iff*

$$det(A)^{-1} prod \geq 1;$$

Similarly, $x_{eq,0,1} \notin \mathbb{D}_{0,1}$ (or $x_{eq,0,-1} \notin \mathbb{D}_{0,-1}$) for the system (4.1) iff

$$det(A)^{-1} prod < 1.$$

Further, it is easy to get the following results from Theorem 4.1.

Theorem 4.2 *If* $det(A) \neq 0$, *then* $x_{eq,0,1} \in \mathbb{D}_{0,1}$ (or $x_{eq,0,-1} \in \mathbb{D}_{0,-1}$) *for the system (4.1) iff*

$$\begin{cases} det(A) > 0, \\ det(A) - prod \leq 0; \end{cases}$$

or

$$\begin{cases} det(A) < 0, \\ det(A) - prod \geq 0; \end{cases}$$

Similarly, $x_{eq,0,1} \notin \mathbb{D}_{0,1}$ (or $x_{eq,0,-1} \notin \mathbb{D}_{0,-1}$) for the system (4.1) iff

$$\begin{cases} det(A) > 0, \\ det(A) - prod \geq 0; \end{cases}$$

or

$$\begin{cases} det(A) < 0, \\ det(A) - prod \leq 0. \end{cases}$$

Combining Theorem 4.1 with Theorem 4.2, we easily obtain the following results.

Theorem 4.3 *If* $det(A) \neq 0$, *then* $x_{eq,0,1} \in \mathbb{D}_{0,1}$ (or $x_{eq,0,-1} \in \mathbb{D}_{0,-1}$) *for the system (4.1) iff*

$$det(A) \cdot det(A - BK) \leq 0;$$

Similarly, $x_{eq,0,1} \notin \mathbb{D}_{0,1}$ (or $x_{eq,0,-1} \notin \mathbb{D}_{0,-1}$) for the system (4.1) iff

$$det(A) \cdot det(A - BK) > 0.$$

4.3 Some Discussions and Numerical Simulation

In this section, we shall discussion equilibrium points for many cases and their corresponding numerical simulation. In the following numerical simulation, green points denote the initial points, and yellow lines (or red lines) and blue lines denote saturated lines.

4.3.1 $x_{eq,1,0}$ Being Stable Focus

Since $x_{eq,1,0} = 0$ is stable focus, it is clear that $\alpha < 0$ and $\beta^2 < 0$ and $\alpha^2 - \beta^2 = det(A - BK) = det(A) - prod > 0$.

Theorem 4.4 *If A is $\gamma < 0$ and $\tau^2 < 0$, then $x_{eq,0,\pm1}$ for the system (4.1) are both spurious stable focuses.*

Proof From $\gamma < 0$ and $\tau^2 < 0$, we have $\gamma^2 - \tau^2 = det(A) > 0$, which means

$$det(A - BK) \cdot det(A) > 0.$$

Combining with Theorem 4.3, we obtain $x_{eq,0,\pm1}$ are spurious stable focus. □

Example 4.1 In the system (4.1), if

$$A = \begin{bmatrix} -1 & 2 \\ -2 & -1 \end{bmatrix}, B = \begin{bmatrix} 2 \\ 1 \end{bmatrix}, K = \begin{bmatrix} 1 & -2 \end{bmatrix}, u_{\lim} = 1,$$

then the eigenvalues of A are $-1 + 2i, -1 - 2i$, the eigenvalues of $A - BK$, $-1 + 3.742i$ and $-1 - 3.742i$, $x_{eq,1,0} = \begin{bmatrix} 0 & 0 \end{bmatrix}^T$ is genuine stable focus, $x_{eq,0,1} = \begin{bmatrix} 0.8 & -0.6 \end{bmatrix}^T$, $x_{eq,0,-1} = \begin{bmatrix} -0.8 & 0.6 \end{bmatrix}^T$ are both spurious stable focus (Fig. 4.3).

Theorem 4.5 *If A with $\gamma > 0 (\gamma = 0)$ and $\tau^2 < 0$, then $x_{eq,0,\pm1}$ for the system (4.1) are both spurious unstable focuses (or spurious center).*

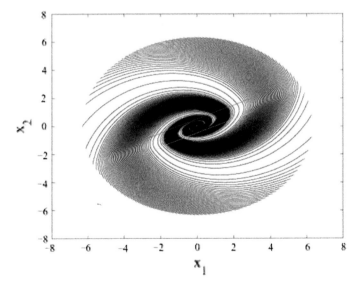

Fig. 4.3 Example 4.1

Proof Since A with $\gamma \geq 0$ and $\tau^2 < 0$, which means

$$det(A - BK) \cdot det(A) > 0.$$

Combining with Theorem 4.3, we obtain $x_{eq,0,\pm 1}$ are spurious unstable focuses (or spurious unstable center). □

Example 4.2 In the system (4.1), if

$$A = \begin{bmatrix} 1 & 3 \\ -3 & 1 \end{bmatrix}, \ B = \begin{bmatrix} 1 \\ 2 \end{bmatrix}, \ K = \begin{bmatrix} 2 & 2 \end{bmatrix}, \ u_{\lim} = 1,$$

then the eigenvalues of A are $1 + 3i, 1 - 3i$, the eigenvalues of $A - BK$ are $-2 + 2.449i, -2 - 2.449i$, $x_{eq,1,0} = \begin{bmatrix} 0 & 0 \end{bmatrix}^T$ is a stable genuine equilibrium point, $x_{eq,0,1} = \begin{bmatrix} 0.5 & -0.5 \end{bmatrix}^T$, $x_{eq,0,-1} = \begin{bmatrix} -0.5 & 0.5 \end{bmatrix}^T$ are both spurious unstable focus (Fig. 4.4).

Theorem 4.6 *If A with $\gamma \pm \tau < 0$, then $x_{eq,0,\pm 1}$ for the system (4.1) are both spurious stable nodes.*

Similarly with Theorem 4.2 or Theorem 4.3, from $\gamma \pm \tau < 0$, we have $(\gamma + \tau)(\gamma - \tau) = det(A) > 0$, which means

$$det(A - BK) \cdot det(A) > 0.$$

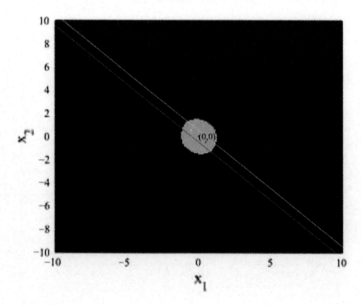

Fig. 4.4 Example 4.2

4.3 Some Discussions and Numerical Simulation

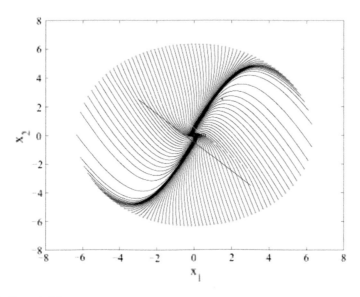

Fig. 4.5 Example 4.3

Combining with Theorem 4.3, we obtain $x_{eq,0,\pm 1}$ are spurious stable nodes. □

Example 4.3 In the system (4.1), if

$$A = \begin{bmatrix} -3 & 1 \\ 6 & -4 \end{bmatrix}, B = \begin{bmatrix} 2 \\ 1 \end{bmatrix}, K = \begin{bmatrix} 2 & 2 \end{bmatrix}, u_{lim} = 1,$$

then the eigenvalues of A are $-1, -6$, the eigenvalues of $A - BK$ are $-6.5 + 3.428i, -6.5 - 3.428i$, $x_{eq,1,0} = \begin{bmatrix} 0 & 0 \end{bmatrix}^T$ is a stable genuine focus, $x_{eq,0,1} = \begin{bmatrix} 1.5 & 2.5 \end{bmatrix}^T$, $x_{eq,0,-1} = \begin{bmatrix} -1.5 & -2.5 \end{bmatrix}^T$ are both spurious stable nodes (Fig. 4.5).

Theorem 4.7 *If A with $\gamma \pm \tau > 0$, then $x_{eq,0,\pm 1}$ are both spurious unstable nodes for the system (4.1).*

It is clear that $\gamma \pm \tau > 0$, we have $(\gamma + \tau)(\gamma - \tau) = det(A) > 0$, which means

$$det(A - BK) \cdot det(A) > 0.$$

Combining with Theorem 4.3, we obtain $x_{eq,0,\pm 1}$ are spurious unstable nodes. □

Example 4.4 In the system (4.1), if

$$A = \begin{bmatrix} 3 & 1 \\ 2 & 2 \end{bmatrix}, B = \begin{bmatrix} 2 \\ 1 \end{bmatrix}, K = \begin{bmatrix} 1 & 4 \end{bmatrix}, u_{lim} = 1,$$

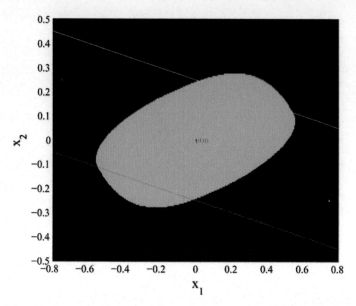

Fig. 4.6 Example 4.4

then the eigenvalues of A are $4, 1$, the eigenvalues of $A - BK$ are $-0.5 + 2.179i$, $-0.5 - 2.179i$, $x_{eq,1,0} = \begin{bmatrix} 0 & 0 \end{bmatrix}^T$ is a stable genuine focus, $x_{eq,0,1} = \begin{bmatrix} -0.75 & 0.25 \end{bmatrix}^T$, $x_{eq,0,-1} = \begin{bmatrix} 0.75 & -0.25 \end{bmatrix}^T$ are both spurious unstable nodes (Fig. 4.6).

Theorem 4.8 *If A with $(\gamma + \tau)(\gamma - \tau) < 0$, then $x_{eq,0,\pm 1}$ for the system (4.1) are both genuine saddles.*

It is clear that $(\gamma + \tau)(\gamma - \tau) < 0$, we have $\gamma^2 - \tau^2 = \det(A) < 0$, which means

$$\det(A - BK) \cdot \det(A) < 0.$$

Combining with Theorem 4.3, we obtain $x_{eq,0,\pm 1}$ are genuine saddles. □

Example 4.5 In the system (4.1), if

$$A = \begin{bmatrix} 5 & -1 \\ 6 & -2 \end{bmatrix}, B = \begin{bmatrix} 4 \\ 2 \end{bmatrix}, K = \begin{bmatrix} 2 & 2 \end{bmatrix}, u_{lim} = 1,$$

then the eigenvalues of A are $4, -1$, the eigenvalues of $A - BK$ are $-4.5 + 3.969i$, $-4.5 - 3.969i$, $x_{eq,1,0} = \begin{bmatrix} 0 & 0 \end{bmatrix}^T$ is genuine stable focus, $x_{eq,0,1} = \begin{bmatrix} -1.5 & -3.5 \end{bmatrix}^T$, $x_{eq,0,-1} = \begin{bmatrix} 1.5 & 3.5 \end{bmatrix}^T$ are both genuine saddles (Fig. 4.7).

4.3 Some Discussions and Numerical Simulation

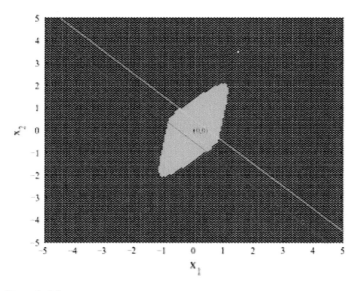

Fig. 4.7 Example 4.5

4.3.2 $x_{eq,1,0}$ Being Unstable Focus (or Center)

Since $x_{eq,1,0}$ is unstable focus (or center), we easily obtain $\alpha \geq 0$ and $\beta^2 < 0$. It is obviously that $\alpha^2 - \beta^2 = det(A - BK) > 0$.

Similarly, we obtain the following results.

Theorem 4.9 *If A with $\gamma < 0, \tau^2 < 0$, then $x_{eq,0,\pm 1}$ for the system (4.1) are both spurious stable focus.*

Example 4.6 In the system (4.1), if

$$A = \begin{bmatrix} -1 & 2 \\ -2 & -1 \end{bmatrix}, B = \begin{bmatrix} 2 \\ 1 \end{bmatrix}, K = \begin{bmatrix} 2 & -8 \end{bmatrix}, u_{\lim} = 1$$

then the eigenvalues of A are $-1 + 2i, -1 - 2i$, the eigenvalues of $A - BK$ are $1 + 6i, 1 - 6i$, $x_{eq,1,0} = \begin{bmatrix} 0 & 0 \end{bmatrix}^T$ is a genuine unstable focus, $x_{eq,0,1} = \begin{bmatrix} 0.8 & -0.6 \end{bmatrix}^T$, $x_{eq,0,-1} = \begin{bmatrix} -0.8 & 0.6 \end{bmatrix}^T$ are both spurious stable focus (Figs. 4.8 and 4.9).

Theorem 4.10 *If A with $\gamma > 0 (\gamma = 0), \tau^2 < 0$, then $x_{eq,0,\pm 1}$ for the system (4.1) are both spurious unstable focuses (spurious center).*

Example 4.7 In the system (4.1), if

$$A = \begin{bmatrix} 4 & 3 \\ -3 & 4 \end{bmatrix}, B = \begin{bmatrix} 1 \\ 2 \end{bmatrix}, K = \begin{bmatrix} 2 & 2 \end{bmatrix}, u_{\lim} = 1,$$

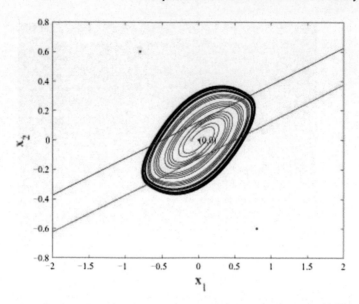

Fig. 4.8 Example 4.6(a)

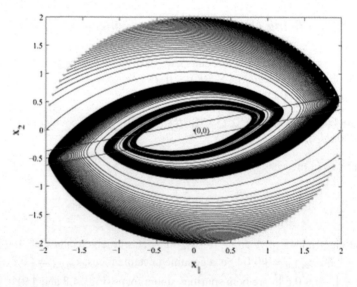

Fig. 4.9 Example 4.6(b)

4.3 Some Discussions and Numerical Simulation

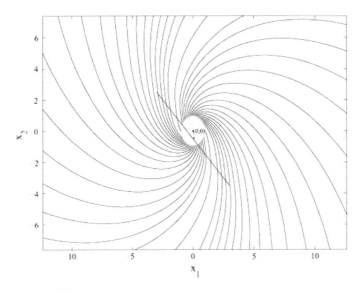

Fig. 4.10 Example 4.7

then the eigenvalues of A are $4 + 3i, 4 - 3i$, the eigenvalues of $A - BK$ are $1 + 2.449i, 1 - 2.449i$, $x_{eq,1,0} = \begin{bmatrix} 0 & 0 \end{bmatrix}^T$ is a genuine unstable focus, $x_{eq,0,1} = \begin{bmatrix} 0.08 & -0.44 \end{bmatrix}^T$, $x_{eq,0,-1} = \begin{bmatrix} -0.08 & 0.44 \end{bmatrix}^T$ are both spurious unstable focuses (Fig. 4.10).

Theorem 4.11 *If A with $\gamma \pm \tau < 0$, then $x_{eq,0,\pm 1}$ for the system (4.1) are both spurious stable nodes.*

Example 4.8 In the system (4.1), if

$$A = \begin{bmatrix} -3 & 1 \\ 6 & -4 \end{bmatrix}, \quad B = \begin{bmatrix} 2 \\ 1 \end{bmatrix}, \quad K = \begin{bmatrix} -8 & 6 \end{bmatrix}, \quad u_{\lim} = 1,$$

then the eigenvalues of A are $-1, -6$, the eigenvalues of $A - BK$ are $1.5 + 4.664i$, $1.5 - 4.664i$, $x_{eq,1,0} = \begin{bmatrix} 0 & 0 \end{bmatrix}^T$ is a unstable genuine focus, $x_{eq,0,1} = \begin{bmatrix} 1.5 & 2.5 \end{bmatrix}^T$, $x_{eq,0,-1} = \begin{bmatrix} -1.5 & -2.5 \end{bmatrix}^T$ are both spurious stable nodes (Figs. 4.11 and 4.12).

Theorem 4.12 *If A with $\gamma \pm \tau > 0$, then $x_{eq,0,\pm 1}$ for the system (4.1) are both spurious unstable nodes.*

Example 4.9 In the system (4.1), if

$$A = \begin{bmatrix} 3 & 1 \\ 2 & 2 \end{bmatrix}, \quad B = \begin{bmatrix} 2 \\ 1 \end{bmatrix}, \quad K = \begin{bmatrix} -1 & 2 \end{bmatrix}, \quad u = 1,$$

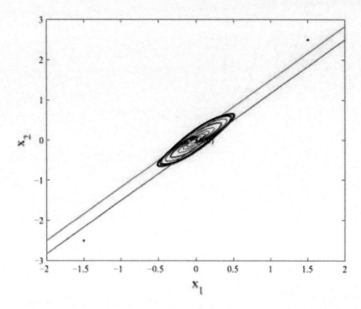

Fig. 4.11 Example 4.8(a)

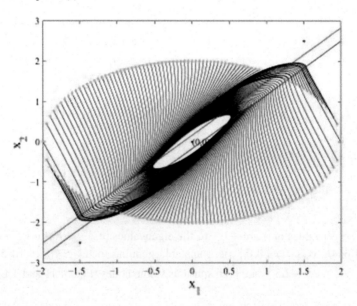

Fig. 4.12 Example 4.8(b)

4.3 Some Discussions and Numerical Simulation

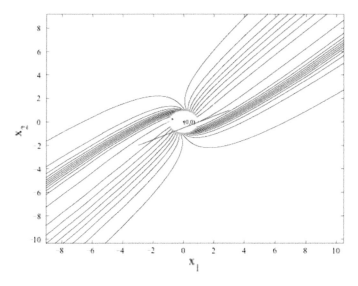

Fig. 4.13 Example 4.9

then the eigenvalues of A are 4, 1, the eigenvalues of $A - BK$ are $2.5 + 1.658i, 2.5 - 1.658$, $x_{eq,1,0} = [0\ 0]^T$ is a genuine unstable focus, $x_{eq,0,1} = [-0.75\ 0.25]^T$, $x_{eq,0,-1} = [0.75\ -0.25]^T$ are both spurious unstable nodes (Fig. 4.13).

Theorem 4.13 *If A with $(\gamma + \tau)(\gamma - \tau) < 0$, then $x_{eq,0,\pm1}$ for the system (4.1) are both genuine saddles.*

Example 4.10 In the system (4.1), if

$$A = \begin{bmatrix} 5 & -1 \\ 6 & -2 \end{bmatrix}, B = \begin{bmatrix} 1 \\ 2 \end{bmatrix}, K = [0.25\ 0.25], u_{lim} = 1,$$

then the eigenvalues of A are $4, -1$, the eigenvalues of $A - BK$ are $0.75 + 0.661i$, $0.75 - 0.661i$, $x_{eq,1,0} = [0\ 0]^T$ is a unstable genuine focus, $x_{eq,0,1} = [-1.5\ -3.5]^T$, $x_{eq,0,-1} = [1.5\ 3.5]^T$ are both genuine saddles (Fig. 4.14).

4.3.3 $x_{eq,1,0}$ Being Genuine Stable Node

Since $x_{eq,1,0}$ is genuine stable node, we know that $\alpha \pm \beta < 0$, and $(\alpha + \beta)(\alpha - \beta) = det(A - BK) > 0$.

Similarly, we obtain the following results.

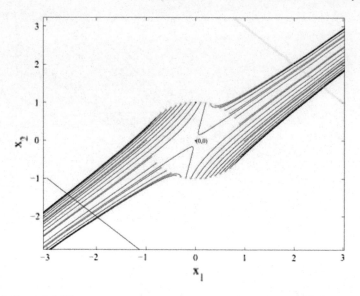

Fig. 4.14 Example 4.10

Theorem 4.14 *If A with $\gamma < 0$ and $\tau^2 < 0$, then $x_{eq,0,\pm 1}$ for the system (4.1) are both spurious stable focuses.*

Example 4.11 In the system (4.1), if

$$A = \begin{bmatrix} -1 & 2 \\ -2 & -1 \end{bmatrix}, B = \begin{bmatrix} 2 \\ 1 \end{bmatrix}, K = \begin{bmatrix} 6 & -2 \end{bmatrix}, u_{lim} = 1,$$

then the eigenvalues of A are $-1 + 2i, -1 - 2i$, the eigenvalues of $A - BK$ are $-7, -5$, $x_{eq,1,0} = \begin{bmatrix} 0 & 0 \end{bmatrix}^T$ is a stable genuine node, $x_{eq,0,1} = \begin{bmatrix} 0.8 & -0.6 \end{bmatrix}^T$, $x_{eq,0,-1} = \begin{bmatrix} -0.8 & 0.6 \end{bmatrix}^T$ are both spurious stable focuses (Fig. 4.15).

Theorem 4.15 *If A with $\gamma > 0 (\gamma = 0)$ and $\tau^2 < 0$, then $x_{eq,0,\pm 1}$ for the system (4.1) are both spurious unstable focuses (spurious centers).*

Example 4.12 In the system (4.1), if

$$A = \begin{bmatrix} 1 & 2 \\ -2 & 1 \end{bmatrix}, B = \begin{bmatrix} 1 \\ 2 \end{bmatrix}, K = \begin{bmatrix} 2 & 2 \end{bmatrix}, u_{lim} = 1,$$

then the eigenvalues of A are $1 + 2i, 1 - 2i$, the eigenvalues of $A - BK$ are $-3, -1$, $x_{eq,1,0} = \begin{bmatrix} 0 & 0 \end{bmatrix}^T$ is a genuine stable node, $x_{eq,0,1} = \begin{bmatrix} 0.6 & -0.8 \end{bmatrix}^T$, $x_{eq,0,-1} = \begin{bmatrix} -0.6 & 0.8 \end{bmatrix}^T$ are both spurious unstable focuses (Fig. 4.16).

4.3 Some Discussions and Numerical Simulation

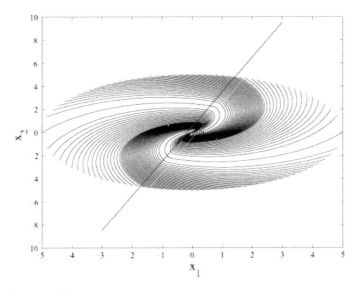

Fig. 4.15 Example 4.11

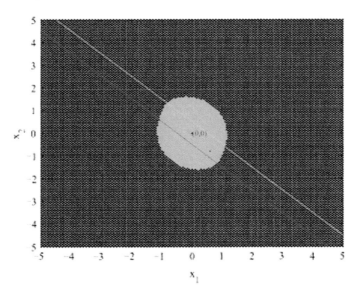

Fig. 4.16 Example 4.12

Theorem 4.16 *If A with $\gamma \pm \tau < 0$, then $x_{eq,0,\pm 1}$ for the system (4.1) are both spurious stable nodes.*

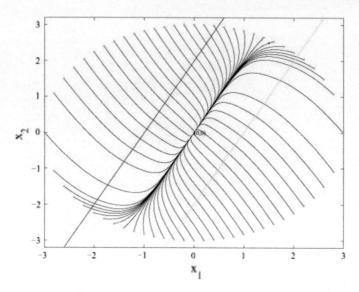

Fig. 4.17 Example 4.13

Example 4.13 In the system (4.1), if

$$A = \begin{bmatrix} -3 & 1 \\ 6 & -4 \end{bmatrix}, B = \begin{bmatrix} 2 \\ 1 \end{bmatrix}, K = \begin{bmatrix} 1 & -0.5 \end{bmatrix}, u_{\lim} = 1,$$

then the eigenvalues of A are $-1, -6$ the eigenvalues of $A - BK$ are $-7.5, -1$, $x_{eq,1,0} = \begin{bmatrix} 0 & 0 \end{bmatrix}^T$ is a genuine stable node, $x_{eq,0,1} = \begin{bmatrix} 1.5 & 2.5 \end{bmatrix}^T$, $x_{eq,0,-1} = \begin{bmatrix} -1.5 & -2.5 \end{bmatrix}^T$ are both spurious stable nodes (Fig. 4.17).

Theorem 4.17 *If A with $\gamma \pm \tau > 0$, then $x_{eq,0,\pm 1}$ for the system (4.1) are both spurious unstable nodes.*

Example 4.14 In the system (4.1), if

$$A = \begin{bmatrix} 3 & 1 \\ 2 & 2 \end{bmatrix}, B = \begin{bmatrix} 2 \\ 1 \end{bmatrix}, K = \begin{bmatrix} 1 & 11 \end{bmatrix}, u_{\lim} = 1,$$

then the eigenvalues of A are $4, 1$, the eigenvalues of $A - BK$ are $-2, -6$, $x_{eq,1,0} = \begin{bmatrix} 0 & 0 \end{bmatrix}^T$ is a genuine stable node, $x_{eq,0,1} = \begin{bmatrix} -0.75 & 0.25 \end{bmatrix}^T$, $x_{eq,0,-1} = \begin{bmatrix} 0.75 & -0.25 \end{bmatrix}^T$ are both spurious unstable nodes (Fig. 4.18).

Theorem 4.18 *If A with $(\gamma + \tau)(\gamma - \tau) < 0$, then $x_{eq,0,\pm 1}$ for the system (4.1) are both genuine saddles.*

4.3 Some Discussions and Numerical Simulation

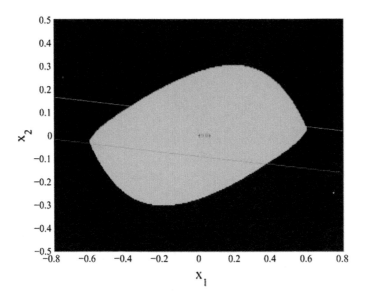

Fig. 4.18 Example 4.14

Example 4.15 In the system (4.1), if

$$A = \begin{bmatrix} 5 & -1 \\ 6 & -2 \end{bmatrix}, \ B = \begin{bmatrix} 4 \\ 2 \end{bmatrix}, \ K = \begin{bmatrix} 3 & 2 \end{bmatrix}, \ u_{\lim} = 1,$$

then the eigenvalues of A are $4, -1$, the eigenvalues of $A - BK$ are $-7, -6$, $x_{eq,1,0} = \begin{bmatrix} 0 & 0 \end{bmatrix}^T$ is a genuine stable node, $x_{eq,0,1} = \begin{bmatrix} -1.5 & -3.5 \end{bmatrix}^T$, $x_{eq,0,-1} = \begin{bmatrix} 1.5 & 3.5 \end{bmatrix}^T$ are both genuine saddles (Fig. 4.19).

4.3.4 $x_{eq,1,0}$ Being Unstable Node

Since $x_{eq,1,0}$ is unstable node, we know that $\alpha \pm \beta > 0$ and $(\alpha + \beta)(\alpha - \beta) = det(A - BK) > 0$.

Theorem 4.19 *If A with $\gamma < 0$ and $\tau^2 < 0$, then $x_{eq,0,\pm 1}$ for the system (4.1) are both spurious stable focuses.*

Example 4.16 In the system (4.1), if

$$A = \begin{bmatrix} -1 & 2 \\ -2 & -1 \end{bmatrix}, \ B = \begin{bmatrix} 2 \\ 1 \end{bmatrix}, \ K = \begin{bmatrix} -2 & -6 \end{bmatrix}, \ u_{\lim} = 1,$$

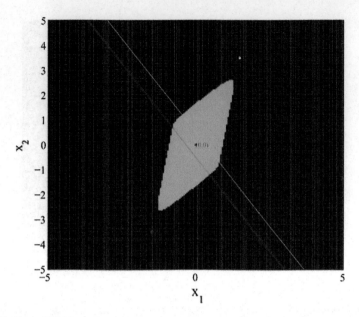

Fig. 4.19 Example 4.15

then the eigenvalues of A are $-1+4i, -1-4i$, the eigenvalues of $A - BK$ are $3, 5$, $x_{eq,1,0} = \begin{bmatrix} 0 & 0 \end{bmatrix}^T$ is a genuine unstable node, $x_{eq,0,1} = \begin{bmatrix} 0.8 & -0.6 \end{bmatrix}^T$, $x_{eq,0,-1} = \begin{bmatrix} -0.8 & 0.6 \end{bmatrix}^T$ are both spurious stable focuses (Figs. 4.20 and 4.21).

Theorem 4.20 *If A with $\gamma > 0 (\gamma = 0)$ and $\tau^2 < 0$, then $x_{eq,0,\pm 1}$ for the system (4.1) are both spurious unstable focuses (spurious centers).*

Example 4.17 In the system (4.1), if

$$A = \begin{bmatrix} 1 & 2 \\ -2 & 1 \end{bmatrix}, \quad B = \begin{bmatrix} 1 \\ 2 \end{bmatrix}, \quad K = \begin{bmatrix} -2 & -2 \end{bmatrix}, \quad u_{\lim} = 1,$$

then the eigenvalues of A are $1+2i, 1-2i$, the eigenvalues of $A - BK$ are $1, 7$, $x_{eq,1,0} = \begin{bmatrix} 0 & 0 \end{bmatrix}^T$ is a genuine unstable node, $x_{eq,0,1} = \begin{bmatrix} 0.6 & -0.8 \end{bmatrix}^T$, $x_{eq,0,-1} = \begin{bmatrix} -0.6 & 0.8 \end{bmatrix}^T$ are both spurious unstable focuses (Fig. 4.22).

Theorem 4.21 *If A with $\gamma \pm \tau < 0$, then $x_{eq,0,\pm 1}$ for the system (4.1) are both spurious stable nodes.*

Example 4.18 In the system (4.1), if

$$A = \begin{bmatrix} -3 & 1 \\ 6 & -4 \end{bmatrix}, \quad B = \begin{bmatrix} 2 \\ 1 \end{bmatrix}, \quad K = \begin{bmatrix} -10 & 6 \end{bmatrix}, \quad u_{\lim} = 1,$$

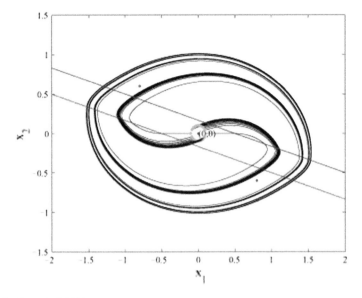

Fig. 4.20 Example 4.16(a)

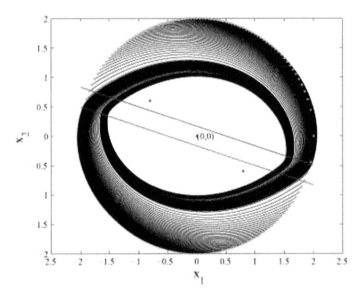

Fig. 4.21 Example 4.16(b)

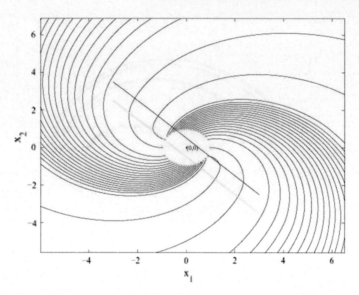

Fig. 4.22 Example 4.17

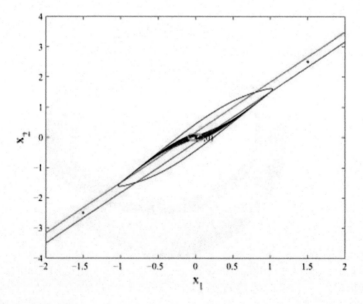

Fig. 4.23 Example 4.18(a)

then the eigenvalues of A are $-1, -6$, the eigenvalues of $A - BK$ are $6, 1$, $x_{eq,1,0} = \begin{bmatrix} 0 & 0 \end{bmatrix}^T$ is a genuine unstable node, $x_{eq,0,1} = \begin{bmatrix} 1.5 & 2.5 \end{bmatrix}^T$, $x_{eq,0,-1} = \begin{bmatrix} -1.5 & -2.5 \end{bmatrix}^T$ are both spurious stable nodes (Figs. 4.23 and 4.24).

4.3 Some Discussions and Numerical Simulation

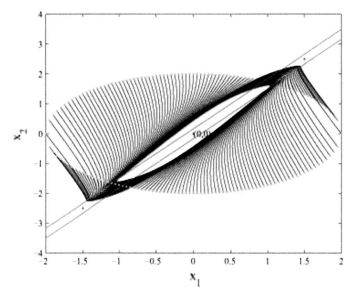

Fig. 4.24 Example 4.18(b)

Theorem 4.22 *If A with $\gamma \pm \tau > 0$, then $x_{eq,0,\pm 1}$ for the system (4.1) are both spurious unstable nodes.*

Example 4.19 In the system (4.1), if

$$A = \begin{bmatrix} 3 & 1 \\ 2 & 2 \end{bmatrix}, B = \begin{bmatrix} 2 \\ 1 \end{bmatrix}, K = \begin{bmatrix} -2 & 2 \end{bmatrix}, u_{\lim} = 1,$$

then the eigenvalues of A are $4, 1$, the eigenvalues of $A - BK$ are $4, 3$, $x_{eq,1,0} = \begin{bmatrix} 0 & 0 \end{bmatrix}^T$ is a genuine unstable node, $x_{eq,0,1} = \begin{bmatrix} -0.75 & 0.25 \end{bmatrix}^T$, $x_{eq,0,-1} = \begin{bmatrix} 0.75 & -0.25 \end{bmatrix}^T$ are both spurious unstable nodes (Fig. 4.25).

Theorem 4.23 *If A with $(\gamma + \tau)(\gamma - \tau) < 0$, then $x_{eq,0,\pm 1}$ for the system (4.1) are both genuine saddles.*

Example 4.20 In the system (4.1), if

$$A = \begin{bmatrix} 5 & -1 \\ 6 & -2 \end{bmatrix}, B = \begin{bmatrix} 4 \\ 2 \end{bmatrix}, K = \begin{bmatrix} -2 & 2 \end{bmatrix}, u_{\lim} = 1,$$

then the eigenvalues of A are $4, -1$, the eigenvalues of $A - BK$ are $4, 3$, $x_{eq,1,0} = \begin{bmatrix} 0 & 0 \end{bmatrix}^T$ is a genuine unstable node, $x_{eq,0,1} = \begin{bmatrix} -1.5 & -3.5 \end{bmatrix}^T$, $x_{eq,0,-1} = \begin{bmatrix} 1.5 & 3.5 \end{bmatrix}^T$ are both genuine saddles (Fig. 4.26).

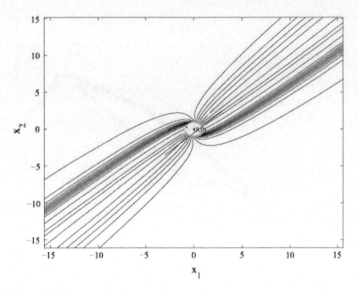

Fig. 4.25 Example 4.19

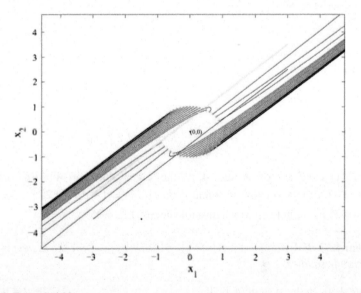

Fig. 4.26 Example 4.20

4.3.5 $x_{eq,1,0}$ Being Saddle

Since $x_{eq,1,0}$ is genuine saddle, we know that $(\alpha + \beta)(\alpha - \beta) < 0$ and $\alpha^2 - \beta^2 = det(A - BK) < 0$.

4.3 Some Discussions and Numerical Simulation

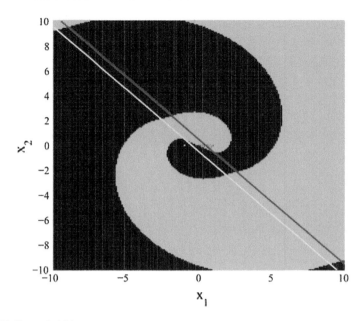

Fig. 4.27 Example 4.21

Theorem 4.24 *If A with $\gamma < 0$ and $\tau^2 < 0$, then $x_{eq,0,\pm 1}$ for the system (4.1) are both genuine stable focuses.*

Example 4.21 In the system (4.1), if

$$A = \begin{bmatrix} -1 & 2 \\ -2 & -1 \end{bmatrix}, \ B = \begin{bmatrix} 1 \\ 2 \end{bmatrix}, \ K = \begin{bmatrix} -2 & -2 \end{bmatrix}, \ u = 1,$$

then the eigenvalues of A are $-1+2i, -1-2i$, the eigenvalues of $A - BK$ are $-1, 5$, $x_{eq,1,0} = \begin{bmatrix} 0 & 0 \end{bmatrix}^T$ is a genuine saddle, $x_{eq,0,1} = \begin{bmatrix} 1 & 0 \end{bmatrix}^T$, $x_{eq,0,-1} = \begin{bmatrix} -1 & 0 \end{bmatrix}^T$ are both genuine stable focuses (Fig. 4.27).

Theorem 4.25 *If A with $\gamma > 0 (\gamma = 0)$ and $\tau^2 < 0$, then $x_{eq,0,\pm 1}$ for the system (4.1) are both genuine unstable focuses (genuine centers).*

Example 4.22 In the system (4.1), if

$$A = \begin{bmatrix} 1 & 2 \\ -2 & 1 \end{bmatrix}, \ B = \begin{bmatrix} 1 \\ 2 \end{bmatrix}, \ K = \begin{bmatrix} -2 & 2 \end{bmatrix}, \ u_{\lim} = 1,$$

then the eigenvalues of A are $1+2i, 1-2i$, the eigenvalues of $A - BK$ are $-3, 3$, $x_{eq,1,0} = \begin{bmatrix} 0 & 0 \end{bmatrix}^T$ is a genuine saddle, $x_{eq,0,1} = \begin{bmatrix} 0.6 & -0.8 \end{bmatrix}^T$, $x_{eq,0,-1} = \begin{bmatrix} -0.6 & 0.8 \end{bmatrix}^T$ are both genuine unstable focuses (Fig. 4.28).

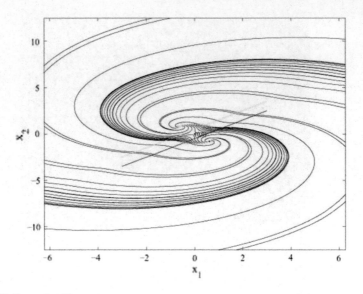

Fig. 4.28 Example 4.22

Theorem 4.26 *If A with $\gamma \pm \tau < 0$, then $x_{eq,0,\pm 1}$ for the system (4.1) are both genuine stable nodes.*

Example 4.23 In the system (4.1), if

$$A = \begin{bmatrix} -3 & 1 \\ 6 & -4 \end{bmatrix}, B = \begin{bmatrix} 2 \\ 1 \end{bmatrix}, K = \begin{bmatrix} -6 & 2 \end{bmatrix}, u = 1,$$

then the eigenvalues of A are $-1, -6$, the eigenvalues of $A - BK$ are $6, -3$, $x_{eq,1,0} = \begin{bmatrix} 0 & 0 \end{bmatrix}^T$ is a genuine saddle, $x_{eq,0,1} = \begin{bmatrix} 1.5 & 2.5 \end{bmatrix}^T$, $x_{eq,0,-1} = \begin{bmatrix} -1.5 & -2.5 \end{bmatrix}^T$ are both genuine stable nodes (Fig. 4.29).

Theorem 4.27 *If A with $\gamma \pm \tau > 0$, then $x_{eq,0,\pm 1}$ for the system (4.1) are both genuine unstable nodes.*

Example 4.24 In the system (4.1), if

$$A = \begin{bmatrix} 3 & 1 \\ 2 & 2 \end{bmatrix}, B = \begin{bmatrix} 2 \\ 1 \end{bmatrix}, K = \begin{bmatrix} 2 & -2 \end{bmatrix}, u_{\lim} = 1,$$

then the eigenvalues of A are $4, 1$, the eigenvalues of $A - BK$ are $-1, 4$, $x_{eq,1,0} = \begin{bmatrix} 0 & 0 \end{bmatrix}^T$ is a genuine saddle, $x_{eq,0,1} = \begin{bmatrix} -0.75 & 0.25 \end{bmatrix}^T$, $x_{eq,0,-1} = \begin{bmatrix} 0.75 & -0.25 \end{bmatrix}^T$ are both genuine unstable nodes (Fig. 4.30).

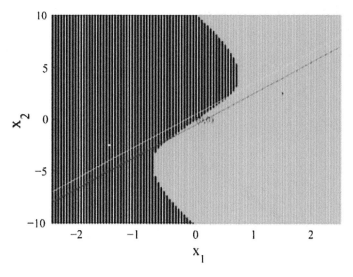

Fig. 4.29 Example 4.23

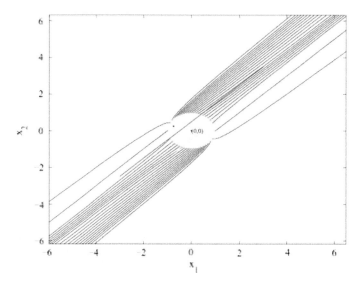

Fig. 4.30 Example 4.24

Theorem 4.28 *If A with $(\gamma + \tau)(\gamma - \tau) < 0$, then $x_{eq,0,\pm 1}$ for the system (4.1) are both spurious saddles.*

Example 4.25 In the system (4.1), if

$$A = \begin{bmatrix} 5 & -1 \\ 6 & -2 \end{bmatrix}, \ B = \begin{bmatrix} 4 \\ 2 \end{bmatrix}, \ K = \begin{bmatrix} 2 & -2 \end{bmatrix}, \ u_{\lim} = 1,$$

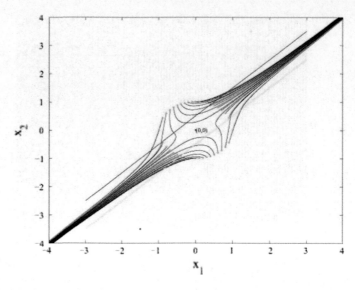

Fig. 4.31 Example 4.25

then the eigenvalues of A are $4, -1$, the eigenvalues of $A - BK$ are $-5, 4$, $x_{eq,1,0} = \begin{bmatrix} 0 & 0 \end{bmatrix}^T$ is a genuine saddle, $x_{eq,0,1} = \begin{bmatrix} -1.5 & -3.5 \end{bmatrix}^T$, $x_{eq,0,-1} = \begin{bmatrix} 1.5 & 3.5 \end{bmatrix}^T$ are both spurious saddles (Fig. 4.31).

Note 4.6 If $x_{eq,0,\pm 1}$ for the system (4.1) are both spurious stable, then $x_{eq,1,0}$ may realize the globally asymptotical stability under some additional conditions. See Chap. 3

Note 4.7 If $x_{eq,1,0}$ is unstable focus, and $x_{eq,0,\pm 1}$ are both spurious stable for the system (4.1), then there may exist at least a limited cycle. See Examples 4.6, 4.9, 4.18 and 4.19. That how many limited cycles for the system (4.1) or MIMO differential systems with many saturated inputs is an unresolved problem.

Note 4.8 If $x_{eq,1,0}$ is unstable, and $x_{eq,0,\pm 1}$ are both genuine stable nodes for the system (4.1), then $x_{eq,0,\pm 1}$ are local asymptotical stability. See Examples 4.21 and 4.23. Here, we do not discuss their local attraction region problems in the chapter.

4.4 Conclusion

This chapter is dedicated to presenting equilibrium points, stability, and limited circles for second differential systems with single saturated input. Based on new systematic techniques which can be formally presented in the chapter, we show equilibrium points, stability, and limited cycles of second differential systems with single saturated input. The focus of this paper is on the so-called 0–1 algebra-geometry type structure equations and their equilibrium points relationship under many cases. The chapter attempts to provide constructive qualitative or equilibrium points conditions for second linear systems with saturated inputs. By defining new equilibrium points, the relationship between equilibrium points is discussed for second linear systems with a single saturated input. Moreover, many enough interesting examples, including their corresponding Simulink plots, are shown to illustrate the above results.

Acknowledgements This work is Supported by National Key Research and Development Program of China (2017YFF0207400).

References

1. Lee AW, Hedrick JK. Some new results on closed-loop stability in the presence of control saturation. Int J Control. 1995;62(3):619–51.
2. Henrion D, Tarbouriech S. LMI relaxations for robust stability of linear systems with saturating controls. Automatica. 1999;35:1599–604.
3. Horisberger HP, Belanger PR. Regulator for linear, time invariant plants with uncertain parameters. IEEE Trans Autom Control. 1976;42:705–8.
4. Huang L. Stability theory. Beijing: Peking University Press; 1992. p. 235–83.
5. Kapoor N, Daoutidis P. An observer-based anti-windup scheme for nonlinear systems with input constraints. Int J Control. 1999;72(1):18–29.
6. Dugard L, Verrist EI, editors. Stability and control of time-delay systems. Berlin: Springer; 1998. p. 303–31.
7. Kothare MV, Campo PJ, Morarior M, Nett CN. A unified framework for the study of anti-windup designs. Automatica. 1994;30(12):1869–83.
8. Edwards C, Postlethwaite I. Anti-windup and bumpless transfer schemes. Automatica. 1998;34(2):199–210.
9. Saeki M, Wada N. Synthesis of a static anti-windup compensator via linear matrix inequalities. Int J Robust Nonlinear Control. 2002;12:927–53.
10. Wu F, Soto M. Extended anti-windup control schemes for LTI and LFT systems with actuator saturations. Int J Robust Nonlinear Control. 2004;14:1255–81.
11. Molchanov AP, Pyatnitskiy YS. Criteria of asymptotic stability of differential and difference inclusions encountered in control theory. In: Systems and control letters, vol. 13. Amsterdam: North-Holland; 1989. p. 59–64.
12. Polanski K. On absolute stability analysis by polyhedric Lypunov functions. Automatica. 2000;36:573–8.
13. Scibile L, Kouvaritakis BIF. Stability region for a class of open-loop unstable linear systems: theory and application. Automatica. 2000;36(1):37–44.
14. Sontag ED, Sussmann HJ. Nonlinear output feedback design for linear systems with saturated control. In: Proceeding of 29th conference on decision and control, vol. 45(5);1990. p. 719–721.
15. Tatsushi S, Yasuyuki F. Two conditions concerning common quadratic Lyapunov functions for linear systems. IEEE Trans Autom Control. 1997;42(5):719–21.
16. Tarbouriech S, Gomes da JM. Synthesis of controllers for continuous-times delay systems with saturated control via LMIs. IEEE Trans Autom Control. 2000;45(1):105–11.
17. Guo S, Irene M, Han L, Xin WF, Feng XJ. Commuting matrices, equilibrium points for control systems with single saturated input. Appl Math Comput. 2015;259:987–1002.
18. Favez JY, Mullhaupt Ph, Srinivasan B, Bonvin D. Attraction region of planar linear systems with one unstable pole and saturated feedback. J Dyn Control Syst. 2006;12(3):331–55.

Part II
Stability Analysis for Three Types of Nonlinear Time-Varying Systems

Chapter 5
Fuzzy Observer, Fuzzy Controller Design, and Common Hurwitz Matrices for a Class of Uncertain Nonlinear Systems

5.1 Introduction

It is well known that observers and controllers play an important role in modern control theory and practice. To design a state observer and a controller for a system, it is usually necessary to know the model structure and its parameters. But in the real time, because of the attrition of the system or the variation of the environment, we cannot get the structure and the parameters exactly. So it is necessary to develop a method to design the observer and controller for the uncertain systems, especially the nonlinear systems.

Actually, the adaptive observers for linear systems have been studied since 1970s and the first nonlinear adaptive observer was proposed in 1988, where the SISO nonlinear systems were considered. In recent years, with the development of intelligent control technologies, these technologies led to a great success and provided effective approaches to handle nonlinear systems. Especially, there were many classical work in designing the state observer, controller or stability. For example, Marino [1] proposed an adaptive observer which presented an arbitrary fast exponential rate of convergence for both parameters and state estimates. In [2, 3], a design methodology for stabilization of a class of nonlinear systems with a Takagi–Sugeno fuzzy model was presented, and this problem was reduced to linear matrix inequalities (LMIs), and at the same time LMIs were efficiently solved by convex programming techniques. Ma [4] presented a fuzzy controller and a fuzzy observer on the basis of the T-S fuzzy model, which proved the separation property. Chen [5] proposed a disturbance observer based on the variable structure system, but it is only used for the minimum-phase dynamical systems. Precup [6] presented a development method for fuzzy controllers based on a linearized model of the controlled plant. The stability conditions, including the relative degree of stability, were expressed by a set of indices that verify the character of the equilibrium point and its uniqueness. This proposed method was applicable to the control of a class of nonlinear plants of arbitrary order (mainly second-order systems), with the right-hand term of a continuous and partially differentiable function. Hamdy [7] developed an observer based on dynamic fuzzy logic system scheme for a class of uncertain nonlinear SISO systems. Park [8] presented a design and an analysis of the online parameter estimator for the plant model whose structure was represented by the general T-S fuzzy model. However, for this approach, the state of the system must be known, but its realization was difficult. Lin [9] considered the problem of observer-based H_∞ control for TCS fuzzy systems with time delay. Gao [10] proposed a T-S fuzzy observer for a class of nonlinear system with unknown output disturbances. Kamel [11] gave an unknown inputs observer to estimate the states and the unknown inputs for a class of uncertain systems by using an LMI approach, but this method ignored the uncertainty of the input matrix. Tong [12] presented a sufficient condition for robust stabilization of the nonlinear system with uncertain parameters, and he also tried to consider state estimation as well as system identification, such as the fuzzy adaptive observer design and fuzzy adaptive output feedback control design by using fuzzy logic systems to approximate the unknown nonlinear functions under the framework of the backstepping design, seeing [13, 15–17]. Liu [18] gave an adaptive fuzzy-neural observer design algorithm for a class of nonlinear SISO systems with both a completely unknown function and an unknown dead-zone input. Here, he used the fuzzy-neural networks to approximate the unknown function. Mehdi [19] considered a robust adaptive fuzzy sliding mode control scheme that is proposed to overcome the synchronization problem for a class of unknown nonlinear chaotic. His main contribution was to attempt to solve the problem of dead zone in the synchronization of chaotic gyros under the unknown structure of the system, uncertainties, and unknown external disturbances. Boulkroune [20] investigated a fuzzy approximation-based indirect adaptive control for a class of unknown non-affine nonlinear systems with unknown control direction, and gave equivalent affine-like model in replacing the original non-affine systems by using a Taylor series expansion. Tanaka [21] presented a sum-of-squares (SOS) approach to polynomial fuzzy observer designs, which improved the LMI-based approach to some degree. In practice of quantum systems, it is difficult to implement continuous weak measurement, which limits the applications of the feedback stabilization strategy based on continuous measurement. The incoherent control strategy is easy to be implemented. However, it cannot deal with other uncertainties in the control process. Chen [22] observed a corresponding relationship between a mixed quantum state and probabilistic fuzzy logic and proposed an approach of control design using a PFE for quantum systems with uncertainties and initial mixed states.

In this chapter, a fuzzy state observer and a fuzzy controller are developed for a class of uncertain nonlinear systems. The systems are represented by more general fuzzy modeling. Many interested results are obtained as follows: First, by constructing a special Lyapunov function approach, the adaptive observer laws including many Riccati equations, two differentiators and many solvable conditions about the above Riccati equations are presented. Second, based on a general Lyapunov function approach, the proposed controllers are designed to guarantee the stability of the overall closed-loop system, and many solvability conditions on the proposed controllers are analyzed too. More importantly, we give the structure of the common stable matrixes for this kind of control problem, including their disturbance structures. In addition, we extend Daniel's work on switched system and propose

5.1 Introduction

a sufficient condition to find common Hurwitz matrices based on the Lie-algebraic theory. Finally, numerical simulations on the magnetic levitation system show the effectiveness of our approaches.

This chapter is organized as follows. In Sect. 5.2, the problem statement and the general nonlinear systems are given. In Sect. 5.3, we design an adaptive observer and analyze its solvable conditions. In Sect. 5.4, we design a fuzzy controller and analyze its solvable conditions. In Sect. 5.5, we discuss the structures of the common Hurwitz matrixes and Lie-algebraic sufficient conditions. In addition, many simulations on some Magnetic levitation systems are presented. Finally, the conclusions are collected in Sect. 5.7.

5.2 Problem Statement

Consider the nonlinear system as follows:

$$\begin{aligned}\dot{x}(t) &= f(x(t),u(t)) + \Delta f(x(t),u(t)), \\ y(t) &= Cx(t),\end{aligned} \quad (5.1)$$

where $x(t) \in \mathbb{R}^n$, $u(t) \in \mathbb{R}$, $y(t) \in \mathbb{R}$ are, respectively, the state, input, and output of the system. $f(x(t), u(t)) \in \mathbb{R}^n$ is known nonlinear function with $x(t)$, $u(t)$, and $\Delta f(x(t), u(t)) \in \mathbb{R}^n$ is uncertain bounded disturbance term of the system (5.1). We define some fuzzy IF-THEN rules to represent the nonlinear systems (5.1) as follows:

Rule i : If $z_1(t)$ is M_1^i, and ... and $z_k(t)$ is M_k^i, then

$$\begin{aligned}\dot{x}(t) &= (A_i x(t) + B_i u(t)) + (E_{1i} \Delta A_i F_{1i} x(t) + E_{2i} \Delta B_i F_{2i} u(t)), \\ y_i(t) &= C_i x(t),\end{aligned}$$

where Rule i ($i = 1, 2, \ldots, r$) denotes the ith rule, r is the number of the IF-THEN rules, M_j^i ($j = 1, 2, \ldots, p$; $i = 1, 2, \ldots, r$) are the fuzzy sets, $A_i \in \mathbb{R}^{n \times n}$, $B_i \in \mathbb{R}^n$, ($i = 1, 2, \ldots, r$) (A_i, B_i is controllable) are the known parameter matrixes of the system $f(x(t), u(t))$, $E_{1i} \Delta A_i F_{1i} \in \mathbb{R}^{n \times n}$, $E_{2i} \Delta B_i F_{2i} \in \mathbb{R}^{n \times n}$ ($i = 1, 2, \ldots, r$) are unknown parameter matrixes which reflect the uncertainty $\Delta f(x(t), u(t))$, $C_i = C \in \mathbb{R}^{m \times n}$ ($i = 1, 2, \ldots, r$), E_{1i}, E_{2i} are full column rank matrices, F_{1i} and F_{2i} ($i = 1, 2, \ldots, r$) are full row rank matrices, $z_1(t), \ldots, z_n(t)$ are premise variables which do not depend on the input variables, and unimodular matrices ΔA_i, ΔB_i such that $\Delta A^T{}_i \Delta A_i \leq I_i$, $\Delta B^T{}_j \Delta B_j \leq I_j$, $I_i \in \mathbb{R}^{r_i \times r_i}$, $I_j \in \mathbb{R}^{r_j \times r_j}$, $(i, j = 1, 2, \ldots, r)$, $r_i, r_j \leq m \leq n$. Here, we always assume that the systems represented by the rule i is minimal. For notation convenience, we do not write the t of the variables.

If given a pair of (x, u), the above fuzzy system can be rephrased as

$$\dot{x} = \frac{\sum_{i=1}^{r} \omega_i(z)[A_i x + B_i u + E_{1i} \Delta A_i F_{1i} x + E_{2i} \Delta B_i F_{2i} u]}{\sum_{i=1}^{r} \omega_i(z)},$$

$$y = \frac{\sum_{i=1}^{r} \omega_i(z) C_i(x)}{\sum_{i=1}^{r} \omega_i(z)},$$

(5.2)

where $\omega_i(z) = \prod_{j=1}^{n} M_j^i(z_j)$, and $M_j^i(z_j)$ is grade of membership of z_j in M_j^i.

We suppose that $\omega_i(z) \geq 0$, $for\ i = 1, 2 \ldots r$ and make

$$\mu_i(z) = \frac{\omega_i(z)}{\sum_{j=1}^{r} \omega_j(z)}, \sum_{i=1}^{r} \mu_i(z) = 1, \mu_i(z) \geq 0, i = 1, 2, \ldots, r.$$

Then system (5.2) can be rephrased as

$$\dot{x} = \sum_{i=1}^{r} \mu_i(z)[A_i x + B_i u + E_{1i} \Delta A_i F_{1i} x + E_{2i} \Delta B_i F_{2i} u], \quad (5.3a)$$

$$y = \sum_{i=1}^{r} \mu_i(z) C_i x. \quad (5.3b)$$

Finally, we outline many important results and definitions, which are useful in upcoming discussion. Since their proofs can be easily obtained, we omit their discussions.

Lemma 5.1 *For any positive* $Q = Q^T > 0, F^T(t) F(t) \leq I$, *then*

$$F^T(t) Q F(t) - Q < 0$$

always holds.

Lemma 5.2 *For any positive real number ε, the following inequality*

$$2uv \leq \varepsilon u^T u + \frac{1}{\varepsilon} v^T v, \forall u, v \in \mathbb{R}^n,$$

always holds.

For classical Lyapunov equation, we introduce the following lemma.

5.2 Problem Statement

Lemma 5.3 *Let A_0 be Hurwitz stable, then for a given matrix $Q = Q^T > 0$, there always exists a $P = P^T > 0$ and $L = L^T > 0$ such that*

$$A_0^T P + P A_0 = -Q = -LL < 0,$$

where $L^T = L$, $\Lambda = diag[\lambda_1^{\frac{1}{2}}, \lambda_2^{\frac{1}{2}}, \ldots, \lambda_n^{\frac{1}{2}}]$, λ_i, $(i = 1, 2, \ldots, n)$ is the eigenvalue of Q, $S^T S = I$.

It is easily obtained by matrix theory, and we omit its proof procedure. □

In this chapter, we use the following definition.

Definition 5.1 For any matrix $A \in \mathbb{R}^{n \times m}$, and for any vector $\gamma \in \mathbb{R}^n$, we define some norms as follows:

$$\|A\| = \sqrt{\lambda_{\max}(A^T A)}, \quad \|\gamma\| = \sqrt{(\gamma^T \gamma)}, \quad \|A\|_F = (\sum_{i=1}^n \sum_{j=1}^m \|a_{ij}\|^2)^{\frac{1}{2}}.$$

Using Lemma 5.3, we easily obtain the following results.

Theorem 5.1 *For any $Q^T = Q > 0$, $P^T = P > 0$, $P, Q \in \mathbb{R}^{n \times n}$, then*

$$\lambda_{min}(Q - \varepsilon P^2) \geq \lambda_{min}(Q)[I - \varepsilon \|L^{-1} P\|^2],$$

where $\lambda_{min}(Q)$ denotes the minimum eigenvalue of Q, ε denotes any positive number, and others are defined as the above in Lemma 5.3.

Combining classical Lyapunov theory and matrices theory, the desired results are easily obtained, and we omit its discussion. □

5.3 Design the Fuzzy Observer

In practice, because all of the states are not fully measurable, it is necessary to design a fuzzy observer in order to estimate the states of the system (5.3).

For (5.3), we define the ith rule for the fuzzy observer as follows:

Rule i : If $z_1(t)$ is M_1^i, and ... and $z_k(t)$ is M_k^i,
Then
$$\dot{\hat{x}} = A_i \hat{x} + B_i u + E_{1i} \Delta A_i F_{1i} \hat{x} + E_{2i} \Delta B_i F_{2i} u + L_i (y - \hat{y}),$$
$$\hat{y}_i = C_i \hat{x},$$

where $L_i (i = 1, 2, \ldots, r)$ are observation error matrixes, $\hat{x}$ is the estimate of x, and y and $\hat{y}$, respectively, are the final output of the fuzzy system and the fuzzy observer.

So, we can get the final estimated state and output of the fuzzy observer as follows:

$$\dot{\hat{x}} = \sum_{i=1}^{r} \mu_i(z)[A_i\hat{x} + B_i u + E_{1i}\Delta A_i F_{1i}\hat{x} + E_{2i}\Delta B_i F_{2i}u + L_i(y - \hat{y})], \quad (5.4a)$$

$$\hat{y} = \sum_{i=1}^{r} \mu_i(z) C_i \hat{x}, \quad (5.4b)$$

where we use the same weight $\mu_i(z)$ as the weight of ith rule of the fuzzy system (5.3).

By substituting (5.4b) and (5.3) into (5.4a), we have

$$\dot{\hat{x}} = \sum_{i=1}^{r} \mu_i [A_i\hat{x} + B_i u + E_{1i}\Delta \hat{A}_i F_{1i} x + E_{2i}\Delta \hat{B}_i F_{2i}u + L_i(y - \sum_{j=1}^{r} \mu_j C_j \hat{x})]$$

$$= \sum_{i=1}^{r}\sum_{j=1}^{r} \mu_i \mu_j [(A_i - L_i C_j)\hat{x} + B_i u + E_{1i}\Delta \hat{A}_i F_{1i}\hat{x} + E_{2i}\Delta \hat{B}_i F_{2i} + L_i y]. \quad (5.5)$$

Let $e = \hat{x} - x$. Then from (5.3) and (5.5), we can obtain

$$\dot{e} = \dot{\hat{x}} - \dot{x}$$

$$= \sum_{i=1}^{r}\sum_{j=1}^{r} \mu_i \mu_j [(A_i - L_i C_j)e + E_{1i}\Delta \hat{A}_i F_{1i}\hat{x} - E_{1i}\Delta \hat{A}_i F_{1i} x + E_{2i}\Delta \hat{B}_i F_{2i}u]$$

$$= \sum_{i=1}^{r}\sum_{j=1}^{r} \mu_i \mu_j [(A_i - L_i C_j)e + E_{1i}\Delta A_i F_{1i} e + E_{1i}\Delta \tilde{A}_i F_{1i}\hat{x} + E_{2i}\Delta \tilde{B}_i F_{2i}u], \quad (5.6)$$

where $\Delta \tilde{A}_{i,eq} = \Delta \hat{A}_i$, and $\Delta \tilde{B}_{i,eq} = \Delta \hat{B}_i$, $\hat{x}_{eq} = 0$.

The objective is to find the adaptive law which makes

$$\lim_{t \to \infty} e(t) = 0.$$

By choosing the following function as the Lyapunov function:

$$V(e, \Delta \tilde{A}_i, \Delta \tilde{B}_i) = e^T P e + \sum_{i=1}^{r} tr(\frac{\Delta \tilde{A}_i^T \Delta \tilde{A}_i}{\gamma_1}) + \sum_{i=1}^{r} tr(\frac{\Delta \tilde{B}_i^T \Delta \tilde{B}_i}{\gamma_2}), \quad (5.7)$$

where $\gamma_1, \gamma_2 > 0$ are constant scalars, and P is a common symmetric positive definite matrix.

Then

$$\dot{V} = \dot{e}^T P e + e^T P \dot{e} + 2\sum_{i=1}^{r} tr(\frac{\dot{\tilde{A}}_i^T \Delta \tilde{A}_i}{\gamma_1}) + 2\sum_{i=1}^{r} tr(\frac{\dot{\tilde{B}}_i^T \Delta \tilde{B}_i}{\gamma_2}), \quad (5.8)$$

5.3 Design the Fuzzy Observer

where

$$
\begin{aligned}
(e^T Pe)' &= \dot{e}^T Pe + e^T P\dot{e} \\
&= \sum_{i=1}^{r}\sum_{i=1}^{r} \mu_i \mu_j [(A_i - L_i C_j)e + E_{1i}\Delta A_i F_{1i}e + E_{1i}\Delta\tilde{A}_i F_{1i}\hat{x} + E_{2i}\Delta\tilde{B}_i F_{2i}u]^T Pe \\
&\quad + e^T P \sum_{i=1}^{r}\sum_{i=1}^{r} \mu_i \mu_j [(A_i - L_i C_j)e + E_{1i}\Delta A_i F_{1i}e + E_{1i}\Delta\tilde{A}_i F_{1i}\hat{x} + E_{2i}\Delta\tilde{B}_i F_{2i}u] \\
&\leq \sum_{i=1}^{r}\sum_{j=1^r} \mu_i \mu_j \{e^T [(A_i - L_i C_j)^T P + P(A_i - L_i C_j)]e\} \\
&\quad + \sum_{i=1}^{r} \mu_i [\varepsilon e^T P^2 e + \varepsilon^{-1} e^T F_{1i}^T E_{1i}^T E_{1i} F_{1i} e + 2e^T PE\Delta\tilde{A}_i F_a \hat{x} + 2e^T PE\Delta\tilde{B}_i F_b u].
\end{aligned}
\tag{5.9}
$$

By substituting (5.9) into (5.8), we obtain

$$
\begin{aligned}
\dot{V} &\leq \sum_{i=1}^{r}\sum_{j=1}^{r} \mu_i \mu_j \{e^T [(A_i - L_i C_j)^T P + P(A_i - L_i C_j)]e\} \\
&\quad + \sum_{i=1}^{r} \mu_i [\varepsilon e^T P^2 e + \varepsilon^{-1} e^T F_{1i}^T E_{1i}^T E_{1i} F_{1i} e + 2e^T PE_{1i}\Delta\tilde{A}_i F_{1i}\hat{x} + 2e^T PE_{2i}\Delta\tilde{B}_i F_{2i}u] \\
&\quad + 2\sum_{i=1}^{r} tr(\frac{\Delta\tilde{A}_i \Delta\dot{\tilde{A}}_i}{\gamma_1}) + 2\sum_{i=1}^{r} tr(\frac{\Delta\tilde{B}_i \Delta\dot{\tilde{B}}_i}{\gamma_2}) \\
&= \sum_{i=1}^{r}\sum_{j=1}^{r} \mu_i \mu_j \{e^T [(A_i - L_i C_j)^T P + P(A_i - L_i C_j) + \varepsilon P^2 + \varepsilon^{-1} F_{1i}^T E_{1i}^T E_{1i} F_{1i}]e\} \\
&\quad + 2\sum_{i=1}^{r} tr[\frac{\Delta\tilde{A}_i \Delta\dot{\tilde{A}}_i}{\gamma_1} + \mu_i(\Delta\tilde{A}_i F_{1i}\hat{x})(e^T P E_{1i})] + 2\sum_{i=1}^{r} tr[\frac{\Delta\tilde{B}_i \Delta\dot{\tilde{B}}_i}{\gamma_2} \\
&\quad + \mu_i(\Delta\tilde{B}_i F_{2i}u)(e^T P E_{2i})].
\end{aligned}
$$

It is clear if there exists a common symmetric positive definite matrix P and $\varepsilon > 0$ such that

$$
(A_i - L_i C_j)^T P + P(A_i - L_i C_j) + \varepsilon P^2 + \varepsilon^{-1} F_{1i}^T E_{1i}^T E_{1i} F_{1i} < 0.
$$

and

$$
\begin{cases}
\dfrac{\Delta\tilde{A}_i \Delta\dot{\tilde{A}}_i^T}{\gamma_1} + \mu_i(\Delta\tilde{A}_i F_{1i}\hat{x})(e^T P E_{1i}) = 0, \\[2mm]
\dfrac{\Delta\tilde{B}_i \Delta\dot{\tilde{B}}_i^T}{\gamma_2} + \mu_i(\Delta\tilde{B}_i F_{2i}u)(e^T P E_{2i}) = 0.
\end{cases}
$$

That is

$$\begin{cases} (A_i - L_i C_j)^T P + P(A_i - L_i C_j) + \varepsilon P^2 + \varepsilon^{-1} F_{1i}^T E_{1i}^T E_{1i} F_{1i} < 0, & (5.10a) \\ \Delta \dot{\hat{A}}_i = \Delta \dot{\tilde{A}}_i = \gamma_1 \mu_i E_{1i}^T P e \hat{x}^T F_{1i}^T, & (5.10b) \\ \Delta \dot{\hat{B}}_i = \Delta \dot{\tilde{B}}_i = \gamma_2 \mu_i E_{2i}^T P e u^T F_{2i}^T. & (5.10c) \end{cases}$$

then $\dot{V} < 0$.

According to the above analysis, we can get the following theorem:

Theorem 5.2 *The fuzzy observer system (5.4) can estimate the state of plant system (5.3) exactly ($|e(t)| \to 0$) as $t \to \infty$ if there exists a common symmetric positive definite matrix P and a constant scalar $\varepsilon > 0$ such (5.10a), (5.10) and (5.10c).*

Proof If there exists a common symmetric positive definite matrix P and a constant scalar $\varepsilon > 0$ such (5.10a), (5.10b), and (5.10c), suppose that $\Delta \hat{A}_i$ and $\Delta \hat{B}_i$ satisfy (5.10b) and (5.10c). Then by choosing the Lyapunov function V as (5.7), it directly follows that the time derivative $\dot{V}$ of V satisfies

$$\dot{V} \le \sum_{i=1}^{r} \sum_{j=1}^{r} \mu_1 \mu_j e^T O_{ij} e \le 0,$$

where $O_{ij} = (A_i - L_i C_j)^T P + P(A_i - L_i C_j) + \varepsilon P^2 + \varepsilon^{-1} F_{1i}^T E_{1i}^T E_{1i} F_{1i}$ are common symmetric negative definite matrices. It is known that the function V is a Lyapunov function for the system (5.6) where x and u are independent bounded function of time, and $\dot{V} \le 0$, and for all $e(t) \ne 0$, $\dot{V} < 0$. Based on the Lyapunov stability theory, the equilibrium of e, $\Delta \tilde{A}$ and $\Delta \tilde{B}$ are stable. It is obviously that $\dot{V} = 0$ if and only if $e = 0$. As we know that the Lyapunov function $V > 0$ and it is bounded, then it is clear that $\dot{V} \to 0$ as $t \to \infty$, which means $|e| \to 0$ as $t \to \infty$. From adaptive laws (5.10b) and (5.10c), we can get that $\Delta \dot{\tilde{A}} \to 0$ and $\Delta \dot{\tilde{B}} \to 0$ when $|e| \to 0$. Furthermore, we have

$$\lim_{t \to \infty} \sum_{i=1}^{r} \sum_{j=1}^{r} \mu_i \mu_j [E \Delta \tilde{A}_i F_a \hat{x}(t) + E \Delta \tilde{B}_i F_b u(t)] = 0.$$

Then, it is obvious that the fuzzy observer system (5.4) can estimate the state of plant system (5.3) exactly. □

Note 5.1 For (5.7), if we choose $V(e, \Delta \tilde{A}_i, \Delta \tilde{B}_i) = e^T P e + \sum_{i=1}^{r} \sum_{k=1}^{r_i} \lambda_k (\frac{\Delta \tilde{A}_i^T \Delta \tilde{A}_i}{\gamma_1}) + \sum_{j=1}^{r} \sum_{k=1}^{r_j} \lambda_k (\frac{\Delta \tilde{B}_j^T \Delta \tilde{B}_j}{\gamma_2})$, or $V(e, \Delta \tilde{A}_i, \Delta \tilde{B}_i) = e^T P e + \sum_{i=1}^{r} (\frac{\|\Delta \tilde{A}_i\|_F^2}{\gamma_1}) + \sum_{i=1}^{r} (\frac{\|\Delta \tilde{B}_i\|_F^2}{\gamma_2})$, we obtain the same result as Theorem 5.2. This is because $\|\Delta \tilde{A}_i\|_F^2 = \sum_{k=1}^{r_i} \lambda_k (\Delta \tilde{A}_i^T \Delta \tilde{A}_i) = tr(\Delta \tilde{A}_i^T \Delta \tilde{A}_i)$.

5.3 Design the Fuzzy Observer

From Theorem 5.2, we find that how to realize (5.10a) is an important problem about observer design, and discuss the relation among ε and Q of the (5.10a) as follows.

Defining
$$-Q_{ij}^0 = (A_i - L_i C_j)^T P + P(A_i - L_i C_j), \qquad (5.11)$$
$$P_{1i}^0 = F_{1i}^T E_{1i}^T E_{1i} F_{1i},$$

then, we easily obtain

$$((A_i - L_i C_j)^T P + P(A_i - L_i C_j))P^{-2}((A_i - L_i C_j)^T P + P(A_i - L_i C_j))$$
$$-4F_{1i}^T E_{1i}^T E_{1i} F_{1i} > 0,$$
$$(\varepsilon I - \tfrac{1}{2} P^{-1}((A_i - L_i C_j)^T P + P(A_i - L_i C_j))P^{-1})^2$$
$$< (\tfrac{1}{2} P^{-1}((A_i - L_i C_j)^T P + P(A_i - L_i C_j))P^{-1})^2 - P^{-1} F_{1i}^T E_{1i}^T E_{1i} F_{1i} P^{-1},$$
$$\forall i, j = 1, 2, \ldots, r.$$
$$(5.12)$$

Conversely, if (5.12) holds, then we easily obtain the fact (5.10a). □

From the above discussion, we convert Theorem 5.2 into the following result.

Theorem 5.3 *The fuzzy observer system (5.4) with (5.10b), (5.10c) can estimate the state of plant system (5.3) exactly ($|e(t)| \to 0$) as $t \to \infty$, if there exists a common symmetric positive definite matrix P and a constant scalar $\varepsilon > 0$ such with (5.12).*

For (5.12), if the dimension of $A_i - L_i C_j$ is very large, it is difficult to compute (5.12). So, by virtue of Definition 5.1 and (5.11), we convert (5.12) into simpler two possible scalar inequalities below:

$$\begin{cases} -\lambda_{min}(Q_{ij}^0 - \varepsilon P^2) + \varepsilon^{-1} \|E_{1i} F_{1i}\|^2 < 0, \forall i, j = 1, 2, \ldots, r, & (5.13a) \\ -\lambda_{min}(Q_{ij}^0) + \varepsilon \|P\|^2 + \varepsilon^{-1} \|E_{1i} F_{1i}\|^2 < 0, \forall i, j = 1, 2, \ldots, r. & (5.13b) \end{cases}$$

For (5.13a), let

$$\lambda_{min}(Q_{ij}^0)[1 - \varepsilon \|L^{-1} P\|^2] - \tfrac{1}{\varepsilon} \|E_{1i} F_{1i}\|^2 > 0. \qquad (5.14)$$

By virtue of Theorems 5.1 and 5.3, we give the following solvable results of (5.10a).

Theorem 5.4 *If*

$$\Delta_{10} = \lambda_{min}(Q_{ij}^0)^2 - 4\|L^{-1} P\|^2 \lambda_{min}(Q_{ij}^0) \|E_{1i} F_{1i}\|^2 > 0,$$
$$\varepsilon_{11}^0 = max(0, \frac{\lambda_{min}(Q_{ij}^0) - \Delta_{10}^{\frac{1}{2}}}{2\|L^{-1} P\|^2 \lambda_{min}(Q_{ij}^0)}), \varepsilon_{12}^0 = \frac{\lambda_{min}(Q_{ij}^0) + \Delta_{10}^{\frac{1}{2}}}{2\|L^{-1} P\|^2 \lambda_{min}(Q_{ij}^0)}, \qquad (5.15)$$

(1) *For any* $\varepsilon \in D_1(Q_{ij}^0) = \{\varepsilon | \varepsilon \in (\varepsilon_{11}^0, \varepsilon_{12}^0), (Q_{ij}^0)^T = Q_{ij}^0 > 0\}$, *then (5.10a) holds.*
(2) *For any* $0 < \varepsilon \leq \varepsilon_{11}^0$ *or* $\varepsilon \geq \varepsilon_{12}^0$, $\Delta_{10} > 0$, *(5.10a) does not hold.*

(3) If $\triangle_1 \leq 0$, that is to say, there does not exist any ε to satisfy (5.10a).

The proof procedure is easily obtained by Theorem 5.3, and we omit its discussion. □

For (5.131b), we have similar results as in Theorem 5.4.

Theorem 5.5 *If*

$$\triangle_{11} = \lambda_{min}^2(Q_{ij}^0) - 4\|P\|^2\|E_{1i}F_{1i}\|^2 > 0,$$
$$\varepsilon_{11}^1 = max(0, \frac{\lambda_{min}(Q_{ij}^0) - \triangle_{11}^{\frac{1}{2}}}{2\|P\|^2}), \varepsilon_{12}^1 = \frac{\lambda_{min}(Q_{ij}^0) + \triangle_{11}^{\frac{1}{2}}}{2\|P\|^2}, \quad (5.16)$$

(1) *For any $\varepsilon \in D_2(Q_{ij}^0) = \{\varepsilon | (\varepsilon_{11}^1, \varepsilon_{12}^1), (Q_{ij}^0)^T = Q_{ij}^0 > 0\}$, then (5.10a) holds.*
(2) *For any $0 < \varepsilon \leq \varepsilon_{11}^1$ or $\varepsilon \geq \varepsilon_{12}^1$, $\triangle_{11} > 0$, (5.10a) does not hold.*
(3) *If $\triangle_{11} \leq 0$, that is to say, there does not exist any ε to satisfy (5.10a).*

5.4 Design the Fuzzy Controller

In this section, we will design the fuzzy controller. First, it is supposed that the fuzzy system (5.3) is locally controllable, and its local state feedback controllers are designed as follows:

Rule i: If $z_1(t)$ is M_1^i, and ... and $z_k(t)$ is M_k^i,
Then $u(t) = -K_i\hat{x}(t)$.
So the final output of the fuzzy controller is

$$u(t) = -\sum_{i=1}^{r}\mu_i(z)K_i\hat{x}(t), \quad (5.17)$$

where $\mu_i(z)$ is defined as the same as (5.3).

By substituting (5.17) into (5.3a), we obtain

$$\dot{x} = \sum_{i=1}^{r}\sum_{j=1}^{r}\mu_i\mu_j[A_i + E_{1i}\triangle A_i F_{1i} - (B_i + E_{2i}\triangle B_i F_{2i})K_j]x. \quad (5.18)$$

Theorem 5.6 *The fuzzy control system described by (5.18) is globally asymptotically stable if there exists a common positive definite matrix P such that*

$$(A_i - B_iK_j)^T P + P(A_i - B_iK_j) + 2\varepsilon P^2 + \varepsilon^{-1}F_{1i}^T E_{1i}^T E_{1i} F_{1i} + \varepsilon^{-1}K_j^T F_{2i}^T E_{2i}^T E_{2i} F_{2i} K_j < 0. \quad (5.19)$$

$\forall i$ and j.

5.4 Design the Fuzzy Controller

Proof By choosing the following function as the Lyapunov function:

$$V(x) = x^T P x,$$

where P is a common positive definite matrix.
Then

$$\dot{V}(x) = \dot{x}^T P x + x^T P \dot{x}$$

$$= \sum_{i=1}^{r}\sum_{j=1}^{r} \mu_i \mu_j \{x^T[(A_i - B_i K_j)^T P + P(A_i - B_i K_j)]x\}$$

$$+ 2\sum_{i=1}^{r}\sum_{j=1}^{r} \mu_i \mu_j [x^T P E \Delta A_i F_a x - x^T P E \Delta B_i F_b K_j x]$$

$$\leq \sum_{i=1}^{r}\sum_{j=1}^{r} \mu_i \mu_j \{x^T[(A_i - B_i K_j)^T P + P(A_i - B_i K_j)]x\}$$

$$+ \sum_{i=1}^{r}\sum_{j=1}^{r} \mu_i \mu_j [\varepsilon x^T 2P^2 x + \varepsilon^{-1} x^T F_{1i}^T E_{1i}^T E_{1i} F_{1i} x + \varepsilon^{-1} x^T K_j^T F_{2i}^T E_{2i}^T E_{2i} F_{2i} K_j x],$$

so if there exists a common positive definite matrix P such that

$$(A_i - B_i K_j)^T P + P(A_i - B_i K_j) + 2\varepsilon P^2 + \varepsilon^{-1} F_{1i}^T E_{1i}^T E_{1i} F_{1i} + \varepsilon^{-1} K_j^T F_{2i}^T E_{2i}^T E_{2i} F_{2i} K_j < 0. \quad (5.20)$$

which means

$$V(x) = x^T P x \leq 0.$$

Then, the fuzzy control system described by (5.18) is globally asymptotically stable. □

Based on similar reasons, we find that how to realize (5.20) is an important question in controller design, and discuss the relation among ε and Q of the (5.20) as follows.

Defining

$$\begin{aligned} -Q_{ij}^1 &= (A_i - B_i K_j)^T P + P(A_i - B_i K_j), \\ P_{ij}^1 &= F_{1i}^T E_{1i}^T E_{1i} F_{1i} + K_j^T F_{2i}^T E_{2i}^T E_{2i} F_{2i} K_j, \end{aligned} \quad (5.21)$$

then, we have

$$((A_i - B_i K_j)^T P + P(A_i - B_i K_j))P^{-2}((A_i - B_i K_j)^T P + P(A_i - B_i K_j))$$
$$-4(F_{1i}^T E_{1i}^T E_{1i} F_{1i} + K_j^T F_{2i}^T E_{2i}^T E_{2i} F_{2i} K_j) > 0,$$
$$(\varepsilon I - \tfrac{1}{4} P^{-1}((A_i - B_i K_j)^T P + P(A_i - B_i K_j))P^{-1})^2$$
$$< (\tfrac{1}{4} P^{-1}((A_i - B_i K_j)^T P + P(A_i - B_i K_j))P^{-1})^2 - \tfrac{1}{2} P^{-1}(F_{1i}^T E_{1i}^T E_{1i} F_{1i} + K_j^T F_{2i}^T E_{2i}^T E_{2i} F_{2i} K_j)P^{-1},$$
$$\forall i, j = 1, 2, \ldots, r.$$

(5.22)

Conversely, if (5.22) holds, then we easily obtain (5.19). □

From the above discussions, we convert Theorem 5.6 into the following result.

Theorem 5.7 *The fuzzy nonlinear system (5.4) can estimate the state of plant system (5.3) exactly ($|e(t)| \to 0$) as $t \to \infty$ if there exists a common symmetric positive definite matrix P and a constant scalar $\varepsilon > 0$ such with (5.22).*

Similarly with (5.10a), for (5.19), if the dimension of $A_i - B_i K_j$ is very large, it is difficult to compute (5.19). So, by virtue of Definition 5.1 and (5.21), we convert (5.19) into simpler two possible scalar inequalities below:

$$\begin{cases} -\lambda_{min}(Q^1_{ij} - 2\varepsilon P^2) + \varepsilon^{-1}\|P^1_{ij}\| < 0, \forall i, j = 1, 2, \ldots, r, & (5.23a) \\ -\lambda_{min}(Q^1_{ij}) + \varepsilon\|P\|^2 + \varepsilon^{-1}\|P^1_{ij}\| < 0, \forall i, j = 1, 2, \ldots, r. & (5.23b) \end{cases}$$

let

$$\lambda_{min}(Q^1_{ij})[1 - \varepsilon\|L^{-1}P\|^2] - \tfrac{1}{\varepsilon}\|P^1_{1i}\| > 0. \tag{5.24}$$

For (5.23a), we give the following solvable results.

Theorem 5.8 *If*

$$\Delta_{21} = \lambda_{min}(Q^1_{ij})^2 - 4\|L^{-1}P\|^2 \lambda_{min}(Q^1_{ij})(\|E_{1i}F_{1i}\|^2 + \|E_{2i}F_{2i}K_j\|^2) > 0,$$
$$\varepsilon^0_{21} = max(0, \tfrac{\lambda_{min}(Q^1_{ij}) - \Delta^{\frac{1}{2}}_{21}}{2\|L^{-1}P\|^2 \lambda_{min}(Q^1_{ij})}), \varepsilon^0_{22} = \tfrac{\lambda_{min}(Q^1_{ij}) + \Delta^{\frac{1}{2}}_{21}}{2\|L^{-1}P\|^2 \lambda_{min}(Q^1_{ij})}. \tag{5.25}$$

(1) *For any $\varepsilon \in D_{21}(Q^1_{ij}) = \{\varepsilon | \varepsilon \in (\varepsilon^0_{21}, \varepsilon^0_{22}), (Q^1_{ij})^T = Q^1_{ij} > 0\}$, then (5.20) holds.*
(2) *For any $0 < \varepsilon \leq \varepsilon^0_{21}$ or $\varepsilon \geq \varepsilon^0_{22}$, $\Delta_{21} > 0$, (5.20) does not hold.*
(3) *If $\Delta_{21} \leq 0$, that is to say, there does not exist any ε to satisfy (5.20).*

Similarly, for (5.23b), we have other solvability results as follows.

Theorem 5.9 *If*

$$\Delta_{22} = \lambda^2_{min}(Q^1_{ij}) - 4\|P\|^2(\|E_{1i}F_{1i}\|^2 + \|E_{2i}F_{2i}K_j\|^2) > 0,$$
$$\varepsilon^1_{21} = max(0, \tfrac{\lambda_{min}(Q^1_{ij}) - \Delta^{\frac{1}{2}}_{22}}{2\|P\|^2}), \varepsilon^1_{22} = \tfrac{\lambda_{min}(Q^1_{ij}) + \Delta^{\frac{1}{2}}_{22}}{2\|P\|^2}, \tag{5.26}$$

(1) *For any $\varepsilon \in D_{22}(Q) = \{\varepsilon | (\varepsilon^1_{21}, \varepsilon^1_{22}), (Q^1_{ij})^T = Q^1_{ij} > 0\}$, then (5.20) holds.*
(2) *For any $0 < \varepsilon \leq \varepsilon^1_{21}$ or $\varepsilon \geq \varepsilon^1_{22}$, $\Delta_{22} > 0$, (5.20) does not hold.*
(3) *If $\Delta_{22} \leq 0$, that is to say, there does not exist any ε to satisfy (5.20).*

If there does not exist the uncertain parameters, we can get the following corollary which was presented in [3]:

Corollary 5.1 *[3] Assume that $\Delta A_i = 0$ and $\Delta B_i = 0$, the fuzzy control system is globally asymptotically stable if there exists a common positive definite matrix P such that*

$$(A_i - B_i K_j)^T P + P(A_i - B_i K_j) < 0. \forall i, j.$$

It follows directly from Theorem 5.3. □

5.5 Structures of the Common Hurwitz Matrices

From the above discussions, it is necessary to consider the structures of $A_i - B_i K_j$. In this section, we can outline some conclusions on these problems.

First, we give common Hurwitz matrices definition.

Definition 5.2 If A_i $(i = 1, 2, \ldots, r)$ are common Hurwitz matrices, then there exists a $P = P^T > 0$ such as $A_i^T P + P A_i < 0$ $(i = 1, 2, \ldots, r)$.

Theorem 5.10 The matrices $A_i, (i = 1, 2, \ldots, r)$ are common Hurwitz matrices if and only if there exists symmetric positive definite matrices P, Q_i and skew-Hermitian matrices S_i such that

$$A_i = P^{-1}(S_i - \frac{1}{2} Q_i), (i = 1, 2, \ldots, r). \tag{5.27}$$

Proof On the one hand, we suppose there exists symmetric positive definite matrices P, Q_i and skew-Hermitian matrices S_i such (5.27). So we have

$$P A_i = S_i - \frac{1}{2} Q_i, \ A_i^T P = S_i^T - \frac{1}{2} Q_i^T = -S_i - \frac{1}{2} Q_i.$$

Then, we can get

$$P A_i + A_i^T P = -Q_i,$$

where P and Q_i are symmetric positive definite matrices, so the A_i are the common Hurwitz matrices.

On the other hand, if A_i are common Hurwitz matrices, from Theorem 5.3 we can get a symmetric positive definite matrix P which satisfies

$$P A_i + A_i^T P = -Q_i,$$

where Q_i are symmetric positive definite matrices.

Let $S_i = \frac{1}{2} Q_i + P A_i$, then

$$S_i^T = \frac{1}{2} Q_i^T + A_i^T Q = -\frac{1}{2} Q_i - P A_i = -S_i.$$

So, S_i are skew-Hermitian matrices; further, we obtain

$$Q^{-1}(S_i - \frac{1}{2} Q_i) = Q^{-1}(\frac{1}{2} Q_i + P A_i - \frac{1}{2} Q_i) = A_i.$$

□

Furthermore, we discuss the structure of $\Delta A_{ij} = B_i K_j$ which makes $A_i + B_i K_j$ are stable.

Theorem 5.11 *When $A_i, A_i + B_i K_j$ are common Hurwitz matrices, i.e., there exists a common symmetric positive definite matrix P such that $PA_i + A_i^T P = -Q_i < 0$, then $A_i + B_i K_j$ are also common Hurwitz matrices if and only if there exists symmetric matrix Q_{ij} and skew-Hermitian matrices S_{ij} such that*

$$B_i K_j = (S_{ij} - \frac{1}{2} Q_{ij}) P^{-1}, \tag{5.28}$$

where $\lambda_{min}(P^{-1} Q_{ij} P^{-1} + Q_i) > 0$.

Proof On one hand, we suppose that there exists symmetric matrix Q_i, skew-Hermitian matrices S_i satisfies (5.29) and $\lambda_{min}(P^{-1} Q_{ij} P^{-1} + Q_i) > 0$.

From $PA_i + A_i^T P = -Q_i < 0$, we can get

$$A_i P^{-1} + P^{-1} A_i^T = -P^{-1} Q_i P^{-1} < 0.$$

From (5.29), we have

$$\begin{aligned} B_i K_j P^{-1} + P^{-1} (B_i K_j)^T &= (S_{ij} - \tfrac{1}{2} Q_i) + (S_{ij} - \tfrac{1}{2} Q_i)^T \\ &= S_{ij} + S_{ij}^T - \tfrac{1}{2} Q_i - \tfrac{1}{2} Q_i \\ &= -Q_{ij}. \end{aligned}$$

Further

$$(A_i + B_i K_j) P^{-1} + P^{-1} (A_i + B_i K_j)^T = -(P^{-1} Q_i P^{-1} + Q_{ij}).$$

Combining with $\lambda_{min}(P^{-1} Q_i P^{-1} + Q_{ij}) > 0$, it means $A_i + B_i K_j$ are common Hurwitz matrices.

On the other hand, if $A_i + B_i K_j$ are common Hurwitz matrices, then

$$P(A_i + B_i K_j) + (A_i + B_i K_j)^T P = -X_{ij} < 0,$$

where X_{ij} are symmetric positive definite matrices.

Then, we have

$$P(A_i + B_i K_j) + (A_i + B_i K_j)^T P = -X_{ij}$$
$$P B_i K_j + (B_i K_j)^T P = -(X_{ij} - Q_{ij}).$$

Let $Q_{ij} = P^{-1}(X_{ij} - Q_i) P^{-1}$ be symmetric matrices, which means

$$Q_{ij} = Q_{ij}^T = -P B_i K_j - (B_i K_j)^T P.$$

Due to $X_{ij} > 0$, it is obvious that

5.5 Structures of the Common Hurwitz Matrices

$$Q_{ij} + P^{-1}Q_i P^{-1} > 0,$$

which implies $\lambda_{min}(Q_{ij} + P^{-1}Q_i P^{-1}) > 0$.

Defining

$$S_{ij} = \frac{1}{2}Q_{ij} + (B_i K_j)^T P = \frac{1}{2}P^{-1}(X_{ij} - Q_i)P^{-1} + (B_i K_j)^T P,$$

then

$$S_{ij}^T = \frac{1}{2}P^{-1}(X_{ij} - Q_i)P^{-1} + PB_i K_j = -\frac{1}{2}P^{-1}(X_{ij} - Q_i)P^{-1} - (B_i K_j)^T P = -S_{ij}.$$

So S_{ij} are skew-Hermitian matrices.

Further, we have

$$(S_{ij} - \tfrac{1}{2}Q_{ij})P^{-1} = \tfrac{1}{2}P^{-1}(X_{ij} - Q_i)P^{-1} + (B_i K_j)^T P - \tfrac{1}{2}P^{-1}(X_{ij} - Q_i)P^{-1})P^{-1} = B_i K_j.$$

The proof is similar with Theorem 5.10, and we omit its discussion. □

Note 5.2 It means that if a disturbance happens in the system and the system is demanded to keep asymptotically stable if and only if the disturbed matrix has the structure given by Theorem 5.11.

Theorem 5.11 is given by another form as follows.

Theorem 5.12 *When A_i, $A_i + B_i K_j$ are common Hurwitz matrices, i.e., there exists a common symmetric positive definite matrix P such $PA_i + A_i^T P = -Q_i < 0$, then $A_i + B_i K_j$ are common Hurwitz matrices if and only if there exists symmetric matrix Q_{ij} and skew-Hermitian matrices S_{ij} such that*

$$B_i K_j = P^{-1}(S_{ij} - \frac{1}{2}Q_{ij}), \tag{5.29}$$

where $Q_i - Q_{ij} > 0$.

Similarly, we present the structures of $E_{2i} \Delta B_i(t) F_{2i}$, $(i = 1, 2, \ldots, r)$ as follows.

Theorem 5.13 *If A_i are common Hurwitz matrices, that is to say, there exists a common symmetric positive definite matrix P such that $PA_i + A_i^T P = -Q_i < 0$, and $A_i + E_{2i} \Delta B_i(t) F_{2i} K_j$ are too common Hurwitz matrices if and only if there exist symmetric matrix $Q_{ij}(t)$ and skew-Hermitian matrices $S_{ij}(t)$ such that*

$$E_{2i} \Delta B_i(t) F_{2i} K_j = P^{-1}(S_{ij}(t) - \frac{1}{2}Q_{ij}(t)), \tag{5.30}$$

where $\lambda_{min}(Q_i - Q_{ij}(t)) > 0$.

It is easily obtained from Theorem 5.10 or Theorem 5.11, and we omit its proof procedure. □

In [23], Daniel presented a sufficient condition to find a common symmetric positive definite matrix P for $A_p^T P + P A_p < 0$ in terms of the Lie algebra. Similarly, we shall give a method to get the common symmetric positive definite matrix P for $A_{ij}^T P + P A_{ij} + P_j < 0$, which makes the problem more complex than Daniel's.

Lemma 5.4 [23] *Let g be a solvable Lie algebra over an algebraically closed field, and let ρ be a representation of a g on a vector space V of finite dimension n. Then, there exists a basis $\{\upsilon_1, \ldots, \upsilon_n\}$ of V such that for each $X \in g$, the matrix of $\rho(X)$ in that basis takes the upper triangular form*

$$\begin{pmatrix} \lambda_1(X) & \cdots & * \\ \vdots & \ddots & \vdots \\ 0 & \cdots & \lambda_n(X) \end{pmatrix},$$

where $\lambda_1(X), \ldots, \lambda_n(X)$ are its eigenvalues.

Considering the problem to find a symmetric positive definite matrix P that satisfies the following inequalities

$$A_{ij}^T P + P A_{ij} + P_j < 0, \, (i, j = 1, \ldots, r), \tag{5.31}$$

where A_{ij} are Hurwitz matrixes and $P_j \geq 0$ is a symmetric matrix.

Theorem 5.14 *If the Lie algebra $\{A_{ij}\}_{La}$ is solvable, there exist a common symmetric positive definite matrix P such that (5.31).*

Proof It is known that Lie algebra $\{A_{ij}\}_{La}$ is solvable, from Lemma 5.4, we can get that there exists a nonsingular complex matrix T such that

$$T^{-1} A_{ij} T = \tilde{A}_{ij}. \tag{5.32}$$

where $\tilde{A}_{ij}$ are upper triangular.

Suppose there exists a positive definite matrix $\tilde{P}$ such that

$$-\tilde{A}_{ij}^T \tilde{P} - \tilde{P} \tilde{A}_{ij} - \tilde{P}_j > 0,$$

where $\tilde{P} = diag\{q_1, \ldots, q_n\}$ is a diagonal matrix and $\tilde{P}_j = (T^T)^{-1} P_j T^{-1}$ is a symmetric matrix.

We obtain

$$\begin{aligned}&-\tilde{A}_{ij}^T \tilde{P} - \tilde{P} \tilde{A}_{ij} - \tilde{P}_j \\ &= \begin{bmatrix} -2q_1(\tilde{A}_{ij})_{11} - (\tilde{P}_j)_{11} & -q_1(\tilde{A}_{ij})_{12} - (\tilde{P}_j)_{12} & \cdots & -q_1(\tilde{A}_{ij})_{1n} - (\tilde{P}_j)_{1n} \\ -q_1(\tilde{A}_{ij})_{12} - (\tilde{P}_j)_{12} & -2q_2(\tilde{A}_{ij})_{22} - (\tilde{P}_j)_{22} & \cdots & -q_1(\tilde{A}_{ij})_{2n} - (\tilde{P}_j)_{2n} \\ \vdots & \vdots & \ddots & \vdots \\ -q_1(\tilde{A}_{ij})_{1n} - (\tilde{P}_j)_{1n} & -q_1(\tilde{A}_{ij})_{2n} - (\tilde{P}_j)_{2n} & \cdots & -2q_n(\tilde{A}_{ij})_{nn} - (\tilde{P}_j)_{nn} \end{bmatrix},\end{aligned} \tag{5.33}$$

5.5 Structures of the Common Hurwitz Matrices

then, we choose

$$q_m > \frac{\max\{(\tilde{P}_j)_{mm}\}}{2\min_{1 \le i,j \le r}\{(-\tilde{A}_{ij})_{mm}\}}.$$

Because $(\tilde{A}_{ij})_{mm} < 0$, we have

$$-2q_m(\tilde{A}_{ij})_{mm} - (\tilde{P}_j)_{mm} > 0, \forall i, j.$$

Now suppose we have chosen $q_1, \ldots q_l$ to make the leading principal minors of the matrix (5.33) up to order l are larger than some positive number α for all i, then we can choose q_{l+1} as follows to make the $(l+1) \times (l+1)$ leading principal minors positive for all i.

$$\begin{cases} q_{l+1} > \dfrac{l!2^{l-1}l\max_{1 \le m, k \le l+1}\left\{\left|q_m(\tilde{A}_{ij})_{mk} + (\tilde{P}_j)_{mk}\right|^{l+1}\right\} + \max\{(\tilde{P}_j)_{l+1,l+1}\}}{2\alpha\min_{1 \le i \le r}\{-(\tilde{A}_{ij})_{l+1,l+1}\}}, \\ q_{l+1} > \dfrac{\max\{(\tilde{P}_j)_{l+1,l+1}\}}{2\min_{1 \le i \le r}\{-(\tilde{A}_{ij})_{l+1,l+1}\}}. \end{cases}$$

We make $(T^{-1})^T \tilde{P} T^{-1} = P$. It is obvious that P is a symmetric positive definite matrix. From (5.33), we have

$$\tilde{A}_{ij}^T \tilde{P} + \tilde{P} \tilde{A}_{ij} + \tilde{P}_j < 0,$$
$$(T)^T A_{ij}^T (T^{-1})^T \tilde{P} + \tilde{P} T^{-1} A_{ij} T + \tilde{P}_j < 0,$$
$$A_{ij}^T (T^{-1})^T \tilde{P} T^{-1} + (T^{-1})^T \tilde{P} T^{-1} A_{ij} + P_j < 0,$$
$$A_{ij}^T P + P A_{ij} + P_j < 0.$$

The proof is completed. □

5.6 Numerical Simulation

In this section, the proposed fuzzy controller is applied on the Model 730 Magnetic Levitation (MagLev) system. The MagLev apparatus may be quickly transformed into a variety of single-input single-output (SISO) and multi-input multi-output (MIMO) configurations. By using repulsive force from the lower coil to levitate a single magnet, an open-loop stable SISO system is created. Attractive levitation via the upper coil affects an open-loop unstable system. Two magnets may be raised by a single coil to produce a SIMO plant. If two coils are used, a MIMO one is produced. These may be locally stable or unstable depending on the selection of the magnet polarities and the nominal magnet positions. The plant has inherently strong nonlinearities due to the natural properties of magnetic fields. Here, we consider SISO system case.

The dynamic equation for MagLev system as follows:

$$m\ddot{y} + c\dot{y} = F_m - mg, \quad (5.34)$$

where m is the mass of the magnet, y is the magnet displacement, c is the friction coefficient, g is the gravity constant, and the magnetic force F_m is defined as $F_m = \frac{u}{a(y+b)^4}$, a, b are positive constant, u is the input as current.

Consider the attrition of the system, the state-space representation (5.34) can be written as

$$\begin{cases} \dot{x}_1 = x_2 + \Delta f_1(x) + \Delta g_1(u), \\ \dot{x}_2 = -\dfrac{c}{m}x_2 + \dfrac{1}{ma(x_1+b)^4}u - g + \Delta f_2(x) + \Delta g_2(u), \\ y = x_1, \end{cases} \quad (5.35)$$

where x_1 denotes the displacement of the magnet, and $\Delta f_1(x)$, $\Delta f_2(x)$, $\Delta g_1(u)$, $\Delta g_2(u)$ are uncertain terms of the MagLev system. By identifying the system parameters, we get $a = 1.04 \times 10^{-4}$ A/N · cm^4, $b = 6.2$ cm, $g = 9.81$ m/s^2, $c = 0.15$ Ns/m, $m = 0.12$ Kg, and the measurable displacement x_1 satisfies $0 \le x_1 \le 0.04$ m. If x_1 is defined as the premise variable, then the (5.35) can be approximated by the following two fuzzy rules:

Rule 1: If $x_1(t)$ is about 0 m
Then
$$\dot{x} = A_1 x + B_1 u + E\Delta A_1 F_a x + E\Delta B_1 F_b u + G,$$
$$y = C_1 x.$$

Rule 2: If $x_1(t)$ is about 0.04 m
Then
$$\dot{x} = A_2 x + B_2 u + E\Delta A_2 F_1 x + E\Delta B_2 F_2 u + G,$$
$$y = C_2 x,$$

where E, F_a, and F_b are known, and they represent the structural information of uncertain parameters ΔA_i and ΔB_i ($i = 1, 2$).

System (5.35) can be inferred as

$$\dot{x} = \sum_{i=1}^{2} \mu_i(x_1)\{A_i x + B_i u + E\Delta A_i F_a x + E\Delta B_i F_b u + G\},$$
$$y = \sum_{i=1}^{2} \mu_i(x_1) C_i x, \quad (5.36)$$

5.6 Numerical Simulation

where $\omega_i(x_1) = \prod_{j=1}^{2} M_j^i(x_{1j})$, $\mu_i(x_1) = \omega_i(x_1) \Big/ \sum_{j=1}^{r} \omega_j(x_1)$ and

$$A_1 = A_2 = \begin{bmatrix} 0 & 1 \\ 0 & -1.25 \end{bmatrix}, B_1 = \begin{bmatrix} 0 \\ 1/ma(0+b) \end{bmatrix} = \begin{bmatrix} 0 \\ 54.23 \end{bmatrix},$$

$$B_2 = \begin{bmatrix} 0 \\ 1/ma(4+b) \end{bmatrix} = \begin{bmatrix} 0 \\ 7.4 \end{bmatrix}, C_1 = C_2 = \begin{bmatrix} 1 & 0 \end{bmatrix}, G = \begin{bmatrix} 0 \\ -9.81 \end{bmatrix}.$$

It is assumed that we have known $E = [0.5\ 0.8]^T$, $F_1 = [0.5\ 0.5]$, and $F_2 = 0.3$. Furthermore, since we consider the lower coil functions in SISO MagLev system, the unknown parameters in the system may be defined as $\Delta A_1 = -0.5$, $\Delta A_2 = 0.3$, $\Delta B_1 = -0.2$, and $\Delta B_2 = -0.8$.

And the membership functions used in this fuzzy model for Rule 1 and Rule 2 are $\omega_1(x_1) = \frac{280e^{-230x_1}}{1+280e^{-230x_1}}$, $\omega_2(x_1) = \frac{1}{1+280e^{-230x_1}}$.

First, we design the fuzzy observer according to Theorem 5.2, and we obtain $L_1 = L_2 = [20\ 1]^T$, $P_1 = \begin{bmatrix} 8 & -1 \\ -1 & 0.625 \end{bmatrix} > 0$, which satisfies (5.10).

Second, the fuzzy controller is designed as $K_1 = [14.8\ 29.6]$, $K_2 = [15.28\ 30.0]$.

We have

$$A_1 - B_1K_1 = \begin{bmatrix} 0 & 1 \\ -802.604 & -1606.458 \end{bmatrix}, A_1 - B_1K_2 = \begin{bmatrix} 0 & 1 \\ -828.634 & -1628.150 \end{bmatrix},$$

$$A_2 - B_2K_1 = \begin{bmatrix} 0 & 1 \\ -109.520 & -220.290 \end{bmatrix}, A_2 - B_2K_2 = \begin{bmatrix} 0 & 1 \\ -113.072 & -223.250 \end{bmatrix}.$$

According to Theorem 5.11, we have

$$A_1 - B_1K_1 = \begin{bmatrix} 2 & 1 \\ 1 & 2 \end{bmatrix}^{-1} \left(\begin{bmatrix} 0 & 0.375 \\ -0.375 & 0 \end{bmatrix} - \frac{1}{2} \begin{bmatrix} 1605.208 & 3209.666 \\ 3209.666 & 6423.832 \end{bmatrix} \right),$$

$$A_1 - B_1K_2 = \begin{bmatrix} 2 & 1 \\ 1 & 2 \end{bmatrix}^{-1} \left(\begin{bmatrix} 0 & 15.559 \\ -15.559 & 0 \end{bmatrix} - \frac{1}{2} \begin{bmatrix} 1657.269 & 3283.419 \\ 3283.419 & 6510.600 \end{bmatrix} \right),$$

$$A_2 - B_2K_1 = \begin{bmatrix} 2 & 1 \\ 1 & 2 \end{bmatrix}^{-1} \left(\begin{bmatrix} 0 & 0.375 \\ -0.375 & 0 \end{bmatrix} - \frac{1}{2} \begin{bmatrix} 219.04 & 437.33 \\ 437.33 & 879.16 \end{bmatrix} \right),$$

$$A_2 - B_2K_2 = \begin{bmatrix} 2 & 1 \\ 1 & 2 \end{bmatrix}^{-1} \left(\begin{bmatrix} 0 & 2.447 \\ -2.447 & 0 \end{bmatrix} - \frac{1}{2} \begin{bmatrix} 226.144 & 447.394 \\ 447.394 & 891.000 \end{bmatrix} \right).$$

So we can choose the common symmetric positive definite matrix as $P_2 = P_2^T = \begin{bmatrix} 2 & 1 \\ 1 & 2 \end{bmatrix} > 0$, $\varepsilon = 0.01$.

Substituting P_2 into (5.17), we have

$$(A_1 - B_1K_1)^T P_2 + P_2(A_1 - B_1K_1) + 2\varepsilon P_2^2 + \varepsilon^{-1} F_1^T E^T E F_1 + \varepsilon^{-1} K_1^T F_2^T E^T E F_2 K_1$$
$$= \begin{bmatrix} -1604.930 & -3209.233 \\ -3209.233 & -6423.028 \end{bmatrix} < 0,$$

$$(A_1 - B_1K_2)^T P_2 + P_2(A_1 - B_1K_2) + 2\varepsilon P_2^2 + \varepsilon^{-1} F_1^T E^T E F_1 + \varepsilon^{-1} K_2^T F_2^T E^T E F_2 K_2$$
$$= \begin{bmatrix} -1659.991 & -3282.9816 \\ -3282.9816 & -6509.796 \end{bmatrix} < 0,$$

$$(A_2 - B_2K_1)^T P_2 + P_2(A_2 - B_2K_1) + 2\varepsilon P_2^2 + \varepsilon^{-1} F_1^T E^T E F_1 + \varepsilon^{-1} K_1^T F_2^T E^T E F_2 K_1$$
$$= \begin{bmatrix} -218.762 & -436.897 \\ -436.897 & -878.356 \end{bmatrix} < 0,$$

$$(A_2 - B_2K_2)^T P_2 + P_2(A_2 - B_2K_2) + 2\varepsilon P_2^2 + \varepsilon^{-1} F_1^T E^T E F_1 + \varepsilon^{-1} K_2^T F_2^T E^T E F_2 K_2$$
$$= \begin{bmatrix} -225.866 & -446.961 \\ -446.961 & -890.196 \end{bmatrix} < 0.$$

It is obvious that the above inequalities satisfy Theorem 5.6.

According to Theorem 5.6, the presented system is stable with above K_1, K_2, P_2.

On the other hand, suppose we know the ΔA_1, ΔA_2, ΔB_1, ΔB_2, and it is easy to get that

$$E\Delta A_1 F_1 - E\Delta B_1 F_2 K_1 = \begin{bmatrix} 0.3190 & 0.7630 \\ 0.5104 & 1.2208 \end{bmatrix}, \quad E\Delta A_1 F_1 - E\Delta B_1 F_2 K_2 = \begin{bmatrix} 0.3334 & 0.7750 \\ 0.5334 & 1.2400 \end{bmatrix},$$

$$E\Delta A_2 F_1 - E\Delta B_2 F_2 K_1 = \begin{bmatrix} 1.8510 & 3.6270 \\ 2.9616 & 5.8032 \end{bmatrix}, \quad E\Delta A_2 F_1 - E\Delta B_2 F_2 K_2 = \begin{bmatrix} 1.9086 & 3.6750 \\ 3.0538 & 5.8800 \end{bmatrix},$$

and the above inequalities can be written as follows:

$$E\Delta A_1 F_1 - E\Delta B_1 F_2 K_1 = \left(\begin{bmatrix} 0 & 0.2345 \\ -0.2345 & 0 \end{bmatrix} - \frac{1}{2} \begin{bmatrix} 0.0833 & -0.3357 \\ -0.3357 & -1.2875 \end{bmatrix} \right) \begin{bmatrix} 2 & 1 \\ 1 & 2 \end{bmatrix},$$

$$E\Delta A_1 F_1 - E\Delta B_1 F_2 K_2 = \left(\begin{bmatrix} 0 & 0.2316 \\ -0.2316 & 0 \end{bmatrix} - \frac{1}{2} \begin{bmatrix} 0.0721 & -0.3478 \\ -0.3478 & -1.2977 \end{bmatrix} \right) \begin{bmatrix} 2 & 1 \\ 1 & 2 \end{bmatrix},$$

$$E\Delta A_2 F_1 - E\Delta B_2 F_2 K_1 = \left(\begin{bmatrix} 0 & 0.8805 \\ -0.8805 & 0 \end{bmatrix} - \frac{1}{2} \begin{bmatrix} -0.0500 & -1.8410 \\ -1.8410 & -5.7632 \end{bmatrix} \right) \begin{bmatrix} 2 & 1 \\ 1 & 2 \end{bmatrix},$$

$$E\Delta A_2 F_1 - E\Delta B_2 F_2 K_2 = \left(\begin{bmatrix} 0 & 0.8690 \\ -0.8690 & 0 \end{bmatrix} - \frac{1}{2} \begin{bmatrix} -0.0948 & -1.8896 \\ -1.8896 & -5.8042 \end{bmatrix} \right) \begin{bmatrix} 2 & 1 \\ 1 & 2 \end{bmatrix}.$$

and

$$\min \lambda(P^{-1} Q_{11} P^{-1} + Q_{11}^1)$$
$$= \min \left(\begin{bmatrix} 2 & 1 \\ 1 & 2 \end{bmatrix}^{-1} \begin{bmatrix} 1605.208 & 3209.666 \\ 3209.666 & 6423.832 \end{bmatrix} \begin{bmatrix} 2 & 1 \\ 1 & 2 \end{bmatrix}^{-1} + \begin{bmatrix} -0.0833 & -0.3357 \\ -0.3357 & -1.2875 \end{bmatrix} \right)$$
$$= 0.7487 > 0,$$

5.6 Numerical Simulation

$$\min \lambda(P^{-1}Q_{12}P^{-1} + Q_{12}^1)$$

$$= min \left(\begin{bmatrix} 2 & 1 \\ 1 & 2 \end{bmatrix}^{-1} \begin{bmatrix} 1657.269 & 3283.419 \\ 3283.419 & 6510.600 \end{bmatrix} \begin{bmatrix} 2 & 1 \\ 1 & 2 \end{bmatrix}^{-1} + \begin{bmatrix} -0.0721 & -0.3478 \\ -0.3478 & -1.2977 \end{bmatrix} \right)$$

$$= 0.6920 > 0,$$

$$\min \lambda(P^{-1}Q_{21}P^{-1} + Q_{21}^1)$$

$$= min \left(\begin{bmatrix} 2 & 1 \\ 1 & 2 \end{bmatrix}^{-1} \begin{bmatrix} 219.04 & 437.33 \\ 437.33 & 879.16 \end{bmatrix} \begin{bmatrix} 2 & 1 \\ 1 & 2 \end{bmatrix}^{-1} + \begin{bmatrix} -0.0500 & -1.8410 \\ -1.8410 & -5.7632 \end{bmatrix} \right)$$

$$= 0.5768 > 0,$$

$$\min \lambda(P^{-1}Q_{22}P^{-1} + Q_{22}^1)$$

$$= min \left(\begin{bmatrix} 2 & 1 \\ 1 & 2 \end{bmatrix}^{-1} \begin{bmatrix} 226.144 & 447.394 \\ 447.394 & 891.00 \end{bmatrix} \begin{bmatrix} 2 & 1 \\ 1 & 2 \end{bmatrix}^{-1} + \begin{bmatrix} -0.0948 & -1.8896 \\ -1.8896 & -5.8042 \end{bmatrix} \right)$$

$$= 0.5601 > 0.$$

From Theorem 5.11, we can know that the system with the disturbance is stable. We use the Simulink, where we choose the initial values $x(0) = [0.02 \ 0]^T$, $\hat{x}(0) = [0 \ 0]^T$, and the input $u(t) = -\sum_{i=1}^{r} \mu_i(x_1) K_i \hat{x}(t) + 0.6$.
The results are shown in Figs. 5.1, 5.2, and 5.3.

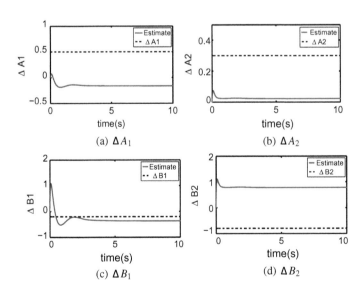

Fig. 5.1 Real parameter and estimation

Fig. 5.2 Real state and estimation

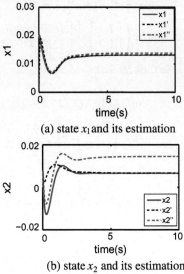

(a) state x_1 and its estimation

(b) state x_2 and its estimation

Fig. 5.3 States Error

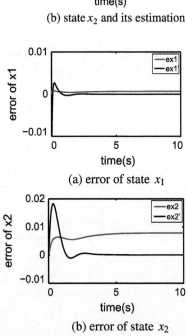

(a) error of state x_1

(b) error of state x_2

5.6 Numerical Simulation

Figure 5.1 shows the estimations of the uncertain parameters ΔA_i and ΔB_i ($i = 1, 2$), where the solid lines are the estimated values, and the dotted lines are the real values. We can get that although the estimations are not equal to the real values, $\Delta \hat{A}_i, \Delta \hat{B}_i$ ($i = 1, 2$) converge to zero by increasing time.

To show the effectiveness of the scheme, the observer with and without adaptive laws were, respectively, applied to the plant model. Figures 5.2 and 5.3 show the comparative results in each case. The dotted black and red lines denote state variables estimated by the fuzzy observer with and without designed adaptive laws, and the solid blue lines denoting the state variables of plant model are shown in Fig. 5.2. Furthermore, Fig. 5.3 shows the error e between plant model state x and observer state $\hat{x}$, where the black lines denote the fuzzy observer with the adaptive laws and the red lines denote the fuzzy observer without adaptive laws. It is obvious that the black lines converge to zero by time increasing in both Fig. 5.3a, b while the red lines do not. This is because the uncertain parts affect the system's stability and do not ignore their effect in analyzing these problems.

It is clear from the figures that the fuzzy observer with the proposed adaptive laws (5.10) can observe the state of the plant model exactly and effectively, and the proposed controller can achieve a good tracking performance.

Furthermore, we use the different initial values for the system, and the results are shown in Figs. 5.4, 5.5, and 5.6,

Figures 5.4 and 5.5 show the real state estimations with different initial values ($0 \leq x_1 \leq 0.04, x_2 = 0$) and Fig. 5.6 shows the error of the states, which will converge to 0 quickly.

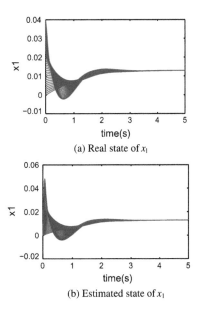

Fig. 5.4 Real state of x_1 and its estimation with different initial values

Fig. 5.5 Real state of x_2 and its estimation with different initial values

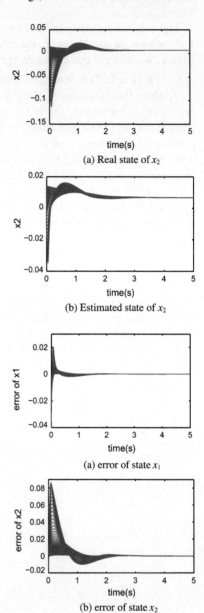

(a) Real state of x_2

(b) Estimated state of x_2

Fig. 5.6 States error with different initial values

(a) error of state x_1

(b) error of state x_2

It is obvious that the above figures show that the proposed observer can estimate the states rapidly, and the designed controller can control the system effectively.

5.7 Conclusion

In this chapter, the fuzzy state observers and fuzzy controllers are developed for a type of uncertain nonlinear systems. The systems are represented by more general fuzzy modeling than T-S fuzzy models. Many important and interesting results are obtained as follows: First, by constructing special Lyapunov function approaches, the adaptive observer laws, i.e., Riccati equations, two differentiators and many solvability conditions about the above equations are presented. Second, based on general Lyapunov function approaches, the proposed controllers are designed to guarantee the stability of the overall closed-loop system, and many solvability conditions on the proposed controllers are carefully analyzed too. More importantly, we give the structures of the common stable matrices for this kind of control problem, including their disturbance matrices structures. Finally, numerical simulations on the SISO magnetic levitation system show the effectiveness of our results.

Acknowledgements This work is Supported by National Key Research and Development Program of China (2017YFF0207400).

References

1. Marino R, Tomei P. Adaptive observers with arbitrary exponential rate of convergence for nonlinear systems. IEEE Trans Autom Control. 1995;40(7):1300–4.
2. Wang OH, Tanaka K, Griffin MF. An approach to fuzzy control of nonlinear systems: stability and design issues. IEEE Trans Fuzzy Syst. 1996;4(1):14–23.
3. Kazuo T, Wang Takayuki I, Hua O. Fuzzy regulators and fuzzy observers: relaxed stabilitu conditions and lmi-based designs. IEEE Trans Fuzzy Syst. 1998;6(2):250–66.
4. Ma XJ, Sun ZQ. Analysis and design of fuzzy controller and fuzzy observer. IEEE Trans Fuzzy Syst. 1998;6(1):41–51.
5. Chen X, Satoshi K, Toshio F. Design of a nonlinear disturbance observer. IEEE Trans Ind Electoron. 2000;47(2):429–37.
6. Precup RE, Doboli S, Preitl S. Stability analysis and development of a class of fuzzy control systems. Eng Appl Artif Intell. 2000;13(3):237–47.
7. Mohamed H. State observer based dynamic fuzzy logic system for a class of SISO nonlinear systems. Int J Autom Comput. 2013;10(2):118–24.
8. Park CW, Cho YW. T-S model based indirect adaptive fuzzy control using online parameter estimation. IEEE Trans Syst Man Cybern. 2004;34(6):2293–302.
9. Lin C, Wang QG, Lee TH, et al. Observer-based control for TCS fuzzy systems with time delay: delay-dependent design method. IEEE Trans Syst Man Cybern Part B Cybern. 2007;37(4):1030–8.
10. Gao Z, Shi X, Ding SX. Fuzzy state/disturbance observer design for TCS fuzzy systems with application to sensor fault estimation. IEEE Trans Syst Man Cybern Part B Cybern. 2008;38(3):875–80.
11. Kamel M, Mohammed C, Mohamed C. Unknown inputs observer for a class of nonlinear uncertain systems: an LMI approach. Int J Autom Comput. 2012;9(3):331–6.
12. Tong S, Zhang X. Observer-based robust fuzzy control of nonlinear systems with parametric uncertainties. Control Theory Appl. 2003;20(2):269–72.
13. Tong S, Li Y. Observer-based fuzzy adaptive control for strict-feedback nonlinear systems. Fuzzy Sets Syst. 2009;160(12):1749–64.

14. Tong S, He X, Zhang H. A combined backstepping and small-gain approach to robust adaptive fuzzy output feedback. IEEE Trans Fuzzy Syst. 2009;17(5):1059–69.
15. Tong SC, Sheng N. Adaptive fuzzy observer backstepping control for a class of uncertain nonlinear systems with unknown time-delay. Int J Autom Comput. 2010;7(2):236–46.
16. Tong SC, Li YM, Feng G, et al. Observer-based adaptive fuzzy backstepping dynamic surface control for a class of MIMO nonlinear systems. IEEE Trans Syst Man Cybern Part B Cybern. 2011;41(4):1124–35.
17. Tong SC, Li Y, Li Y, et al. Observer-based adaptive fuzzy backstepping control for a class of stochastic nonlinear strict-feedback systems. IEEE Trans Syst Man Cybern Part B Cybern. 2011;41(6):1693–704.
18. Liu Y, Zhou N. Observer-based adaptive fuzzy-neural control for a class of uncertain nonlinear systems with unknown dead-zone input. ISA Trans. 2010;49(4):462–9.
19. Mehdi R, Mansoor ZJ, Robert J, Lin T. Unknown nonlinear chaotic gyros synchronization using adaptive fuzzy sliding mode control with unknown dead-zone input. Commun Nonlinear Sci Numer Simul. 2010;15(9):2536–45.
20. Boulkroune A, Saad MM, Farza M. Fuzzy approximation-based indirect adaptive controller for multi-input multi-output non-affine systems with unknown control direction. IET Control Theory Appl. 2012;17:2619–29.
21. Tanaka K, Ohtake H, Seo T, et al. Polynomial fuzzy observer designs: a sum-of-squares approach. IEEE Trans Syst Man Cybern Part B Cybern. 2012;42(5):1330–42.
22. Chen C, Dong D, Lam J, Chu J, Tarn TJ. Control design of uncertain quantum systems with fuzzy estimators. IEEE Trans Fuzzy Syst. 2012;20(5):820–31.
23. Liberzon D, Hespanha JP, Stephen MA. Stability of switched systems: a Lie-algebraic condition. Syst Control Lett. 1999;37(3):117–22.

Chapter 6
Stability of Lurie Time-Varying Systems with Time-Varying Delay Feedbacks

6.1 Introduction

In the real world, working properly with noises is the premise for a complex system, its stability analysis is the most important quality, and Lyapunov stability methods are one of the most important means in analyzing its stability [2]. So how to construct a more suitable Lyapunov function to solve the stability of the practical control system has become a main research target [5].

A practical type of system was introduced in 1957, which later is called as Lurie control system [10]. And, many results about the absolute stability of Lurie system without delay feedback have been obtained [7, 17, 19]. While delay phenomena frequently encountered in many fields of science and engineering, classical means about this problem are to ignore the delay effect in system design, which lead to poor performance [8] and even leads to destabilization. So finding a way to solve the stability problem of Lurie system with delay feedback is attractive. For example,a frequency-domain interpretation of the Lyapunov–Krasovskii functional method was introduced for delay-independent feedback cases [1]. However, delay-independent cases are not an accurate representation on the practice system. Some useful contributions on time-varying delay feedback cases for Lurie systems are derived [3, 4, 6, 13, 14, 16]. Some typical methods are given about the above case, such as a Lyapunov–Razumikhin functional method [6] and Lyapunov–Krasovskii functional method [3, 4, 13, 14, 16].

It is true that most of results are reflected by a linear matrix inequality (LMI) or LMIs [3, 4, 13, 14, 16]. As it is known to all that solving a LMI or LMIs is a difficult procedure, and meanwhile the upper bound of the allowable delay that they obtained is a little conservative [6]. Uncertain phenomena occur in real system, such as the plants, actuators, or controllers in system design,and even simultaneously occur.

The purpose of this chapter is to consider the uncertain phenomena for complex Lurie systems with time-varying delay feedback. First, we considered a more complex situation which approaches the real system such as the uncertain plant internal parameters and the uncertain actuator quantities. Second, by defining solvable matrix inequalities and constructing a Lyapunov–Razumikhn functional, we analyze these solvable matrix inequalities, and give their sufficient and necessary solvable conditions that are easily computed in practical systems. In applications,based on our norm definition,solvable conditions and time-delay bound estimations can be obtained from two cases of solvable norm inequalities. More importantly, we draw a conclusion on the above procedures about the optimal combination about a pair of (τ, Q). Finally, an interested numerical example is presented to illustrate the above results effectively.

6.2 Problem Formulation and Preliminaries

We consider system

$$\begin{cases} \dot{x}(t) = A(t)x(t) + \sum_{i=1}^{p} B_i(t)u_i(t) + Df(\delta(t)), \\ u_i(t) = K_i x(t - \tau_i(t)), \\ \sigma(t) = C^T x(t), \end{cases} \quad (6.1)$$

where $x(t) = \phi(t)$, $t \in [-\tau, 0]$, $x \in \mathbb{R}^n$, $u \in \mathbb{R}^m$, $D \in \mathbb{R}^{n \times m}$, $C \in \mathbb{R}^{n \times m}$, $A(t) \in \mathbb{R}^{n \times n}$, $B_i(t) \in \mathbb{R}^{n \times n}$, $(i = 1, \ldots, p)$, $\sigma = col(\sigma_1, \sigma_2, \ldots, \sigma_m) \in \mathbb{R}^m$, and the nonlinearity function $f(\sigma)$ satisfies $f(\sigma) = col(f_1(\sigma_1), f_2(\sigma_2), \ldots, f_m(\sigma_m)) \in \mathbb{R}^m$, $0 \leqslant \frac{f_i(\sigma_i)}{\sigma_i} \leqslant k_i$, $f(0) = 0$, k_i $(i = 1, \ldots, m)$ are positive constants.

More importantly, uncertain phenomena simultaneously occur in plant $A(t)$ and actuator $B(t)$ in (6.1) as follows:

$$\begin{cases} A(t) = A_0 + \Delta A(t), \\ B_i(t) = B_{i0} + \Delta B_i(t). \end{cases}$$

Without losing generality, let

$$\Delta A(t) = D_1 F_1(t) E_1, \; F_1^T(t) F_1(t) \leqslant I_1,$$
$$\Delta B_i(t) = D_{2i} F_{2i}(t) E_{2i}, \; F_{2i}^T(t) F_{2i}(t) \leqslant I_{2i}, \; (i = 1, 2, \ldots, p),$$

and we need to point out that $D_1, E_1, F_1(t), D_{2i}, F_{2i}(t), E_{2i}$ $(i = 1, 2, \ldots, p)$ are suitable dimension matrices and $I_1, I_{2i} (i = 1, 2, \ldots, p)$ are different unit matrices with different dimensions, and suppose the system (6.1) be minimal.

When $\Delta A(t) = 0$, $\Delta B_i(t) = 0$, $(i = 1, 2, \ldots, p)$, (6.1) is converted into

$$\begin{cases} \dot{x}(t) = A_0 x(t) + \sum_{i=1}^{p} B_{i0} u_i(t) + Df(\delta(t)), \\ u_i(t) = K_i x(t - \tau_i(t)), \\ \sigma(t) = C^T x(t). \end{cases} \quad (6.2)$$

6.2 Problem Formulation and Preliminaries

Equivalently, we can convert the system (6.1) into the following systems

$$\dot{x}(t) = A_1(t)x(t) - \sum_{i=1}^{p} B_i(t) \int_{t-\tau_i(t)}^{t} [A(s)x(s) \\ + \sum_{j=1}^{p} B_j(s)x(s - \tau_j(s)) + Df(\sigma(s))]ds + Df(\sigma(t)). \quad (6.3)$$

where

$$A_1(t) = A(t) + \sum_{i=1}^{p} B_i(t) K_i \\ = A_0 + \sum_{i=1}^{p} B_{i0} K_i + \Delta A(t) + \sum_{i=1}^{p} \Delta B_i(t) K_i,$$

and $A_0 + \sum_{i=1}^{p} B_{i0} K_i$ is Hurwitz stable.

First, we recall the following existing results, which is useful in our subsequent analysis.

Lemma 6.1 *Consider the functional differential equation*

$$\begin{cases} \dot{x}(t) = f(t, x_t), t \geq t_0, \\ x_{t_0}(\theta) = \phi(\theta), \forall \theta \in [-\tau, 0], \end{cases} \quad (6.4)$$

suppose that $u, v, w, p : \mathbb{R}^+ \mapsto \mathbb{R}^+$ *are continuous, nondecreasing functions,* $u(s), v(s), w(s)$ *are positive for* $s > 0$, $u(0) = v(0) = 0$ *and* $p(s) > s$ *for* $s > 0$.

If there is a continuous function $V : \mathbb{R} \times \mathbb{R}^n \mapsto \mathbb{R}$ *such that*
(1) $u(\|x\|) \leq V(t, x) \leq v(\|x\|), t \in \mathbb{R}, x \in \mathbb{R}^n$;
(2) $\dot{V}(t, x(t)) \leq -w(\|x(t)\|)$ *if* $V(t + \theta, x(t + \theta)) < p(V(t, x(t))), \forall \theta \in [-\tau, 0]$;
Then, the trivial solution of the differential equation (6.4) is uniformly asymptotically stable.

We omit its discussion, seeing the reference [7]. □

Especially, if

$$V(x) = x^T P x, \ P^T = P > 0, \\ \alpha = \sqrt{\frac{\lambda_{max}(P)}{\lambda_{min}(P)}}, \\ l > 1, \ V(x(\theta)) < l^2 V(x(t)), \forall \theta \in [t - \tau, t],$$

we outline the result below.

Lemma 6.2 *In Lemma 6.1, if* $x(\theta)^T P x(\theta) < l^2 x(t)^T P x(t), l > 1, \forall \theta \in [t - \tau, t]$, *then we have*

$$\|x(\theta)\| \leq \alpha l \|x(t)\|, \forall \theta \in [t - \tau, t]. \quad (6.5)$$

Proof In fact, based on the above assumptions, we have

$$V(x(\theta)) < l^2 V(x(t)), \\ \lambda_{min}(P) x(\theta)^T x(\theta) \leq x(\theta)^T P x(\theta) = V(x(\theta)), \\ \lambda_{max}(P) x(t)^T x(t) \geq x(t)^T P x(t) = V(x(t)).$$

So,

$$\lambda_{min}(P)x(\theta)^T x(\theta) \le \lambda_{max}(P)x(t)^T x(t).$$

and

$$\|x(\theta)\|^2 \le \frac{\lambda_{max}(P)}{\lambda_{min}(P)}\|x(t)\|^2, \forall \theta \in [t-\tau, t],$$

the desired result is obtained. □

Finally, we outline an important result, which is useful in the upcoming discussion.

Lemma 6.3 *For any positive real number ε, the following inequality*

$$uv \le \varepsilon u^T u + \frac{1}{\varepsilon}v^T v, \forall u, v \in \mathbb{R}^n, \tag{6.6}$$

always holds.

This result appears in most references, hence we omit its discussion. □

Lemma 6.4 *For any positive $Q = Q^T > 0, F^T(t)F(t) \le I$, then*

$$F^T(t)QF(t) - Q < 0 \tag{6.7}$$

always holds.

Proof Since matrix

$$\begin{pmatrix} -Q & F^T Q^{\frac{1}{2}T} \\ FQ^{\frac{1}{2}} & -I \end{pmatrix} \simeq \begin{pmatrix} -Q + F^T Q^{\frac{1}{2}T} Q^{\frac{1}{2}} F & 0 \\ 0 & -I \end{pmatrix}$$
$$= \begin{pmatrix} -Q + F^T Q F & 0 \\ 0 & -I \end{pmatrix}.$$

Simultaneously,

$$\begin{pmatrix} -Q & F^T Q^{\frac{1}{2}T} \\ Q^{\frac{1}{2}} F & -I \end{pmatrix} \simeq \begin{pmatrix} -Q & 0 \\ 0 & -I + Q^{\frac{-1}{2}} Q^{\frac{1}{2}} F F^T Q^{\frac{1}{2}T} Q^{\frac{-1}{2}T} \end{pmatrix}$$
$$= \begin{pmatrix} -Q & 0 \\ 0 & -I + F^T F \end{pmatrix}.$$

From the assumptions $Q = Q^T > 0, F^T(t)F(t) \le I$, we have

$$\begin{pmatrix} -Q & F^T Q^{\frac{1}{2}T} \\ Q^{\frac{1}{2}} F & -I \end{pmatrix} < 0.$$

Further,

$$-Q + F^T Q F < 0.$$

Here $\simeq$ denotes congruent. □

6.3 Some Results on Absolute Stability

Choosing the following Lyapunov function

$$V(x(t)) = x(t)^T P x(t), \quad P = P^T > 0. \tag{6.8}$$

Then

$$\begin{aligned}
&\frac{dV(x(t))}{dt}\Big|_{(6.1)} \\
&= \dot{x}|_{(1)}^T(t)Px(t) + x(t)^T P\dot{x}(t)|_{(6.1)} \\
&= \{A_1(t)x(t) - \sum_{i=1}^p B_i(t)K_i \int_{t-\tau_i(t)}^t [A(s)x(s) \\
&\quad + \sum_{j=1}^p B_j(s)x(s - \tau_j(s)) + Df(\sigma(s))]ds + Df(\sigma(t))\}^T(t)Px(t) \\
&\quad + x(t)^T P\{A_1(t)x(t) \\
&\quad - \sum_{i=1}^p B_i(t)K_i \int_{t-\tau_i(t)}^t [A(s)x(s) + \sum_{j=1}^p B_j(s)K_i x(s - \tau_j(s)) + Df(\sigma(s))]ds + Df(\sigma(t))\} \\
&= x^T(t)[A_1^T(t)P + PA_1(t)]x(t) + 2x^T(t)PDf(\tau(t)) \\
&\quad - 2x^T(t)P\{\sum_{i=1}^p B_i(t)K_i \int_{t-\tau_i(t)}^t [A(s)x(s) + \sum_{j=1}^p B_j(s)K_i x(s - \tau_j(s)) + Df(\sigma(s))]ds\} \\
&= x^T(t)[A_1^T(t)P + PA_1(t)]x(t) + 2x^T(t)PDf(\tau(t)) \\
&\quad - 2x^T(t)P\{\sum_{i=1}^p B_i(t)K_i \int_{t-\tau_i(t)}^t A(s)x(s)ds\} \\
&\quad - 2x^T(t)P\{\sum_{i=1}^p B_i(t)K_i \int_{t-\tau_i(t)}^t \sum_{j=1}^p B_j(s)K_j x(s - \tau_j(s))ds\} \\
&\quad - 2x^T(t)P\{\sum_{i=1}^p B_i(t)K_i \int_{t-\tau_i(t)}^t Df(\sigma(s))ds\} \\
&= x^T(t)[(A_0 + \sum_{i=1}^p B_{i0}K_i)^T P + P(A_0 + \sum_{i=1}^p B_{i0}K_i)]x(t) \\
&\quad + 2x^T(t)P\Delta A(t)x(t) + 2x^T(t)P(\sum_{i=1}^p \Delta B_i(t)K_i)x(t) + 2x^T(t)PDf(\tau(t)) \\
&\quad - 2x^T(t)P\{\sum_{i=1}^p B_{i0}K_i \int_{t-\tau_i(t)}^t A_0 x(s)ds\} \\
&\quad - 2x^T(t)P\{\sum_{i=1}^p B_{i0}K_i \int_{t-\tau_i(t)}^t \Delta A(s)x(s)ds\} \\
&\quad - 2x^T(t)P\{\sum_{i=1}^p \Delta B_i(t)K_i \int_{t-\tau_i(t)}^t A_0 x(s)ds\} \\
&\quad - 2x^T(t)P\{\sum_{i=1}^p \Delta B_i(t)K_i \int_{t-\tau_i(t)}^t \Delta A(s)x(s)ds\} \\
&\quad - 2x^T(t)P\{\sum_{i=1}^p B_{i0}K_i \int_{t-\tau_i(t)}^t \sum_{j=1}^p B_{j0}K_j x(s - \tau_j(s))ds\} \\
&\quad - 2x^T(t)P\{\sum_{i=1}^p B_{i0}K_i \int_{t-\tau_i(t)}^t \sum_{j=1}^p \Delta B_j(s)K_j x(s - \tau_j(s))ds\} \\
&\quad - 2x^T(t)P\{\sum_{i=1}^p \Delta B_i(t)K_i \int_{t-\tau_i(t)}^t \sum_{j=1}^p B_{j0}K_j x(s - \tau_j(s))ds\} \\
&\quad - 2x^T(t)P\{\sum_{i=1}^p \Delta B_i(t)K_i \int_{t-\tau_i(t)}^t \sum_{j=1}^p \Delta B_j(s)K_j x(s - \tau_j(s))ds\} \\
&\quad - 2x^T(t)P\{\sum_{i=1}^p B_{i0}K_i \int_{t-\tau_i(t)}^t Df(\sigma(s))ds\} \\
&\quad - 2x^T(t)P\{\sum_{i=1}^p \Delta B_i(t)K_i \int_{t-\tau_i(t)}^t Df(\sigma(s))ds\}.
\end{aligned} \tag{6.9}$$

Combining Lemmas 6.2, 6.3, and 6.4, for any sufficient small positive real $\varepsilon > 0$, we gradually obtain

$$\begin{aligned}
&2x^T(t)P\Delta A(t)x(t) \\
&\leq \varepsilon x^T(t)P^2 x(t) + \frac{1}{\varepsilon}x^T(t)E_1^T D_1^T D_1 E_1 x(t) \\
&= x^T(t)(\varepsilon P^2 + \frac{1}{\varepsilon}E_1^T D_1^T D_1 E_1)x(t),
\end{aligned} \tag{6.10}$$

$$\begin{aligned}
&2x^T(t)P(\sum_{i=1}^p \Delta B_i(t)K_i)x(t) \\
&\leq p\varepsilon x^T(t)P^2 x(t) + \frac{1}{\varepsilon}\sum_{i=1}^p x^T(t)K_i^T \Delta B_i^T(t)\Delta B_i(t)K_i x(t) \\
&\leq x^T(t)(p\varepsilon P^2 + \frac{1}{\varepsilon}\sum_{i=1}^p K_i^T \Delta B_i^T(t)\Delta B_i(t)K_i)x(t) \\
&\leq x^T(t)(p\varepsilon P^2 + \frac{1}{\varepsilon}\sum_{i=1}^p K_i^T E_{2i}^T D_{2i}^T D_{2i} E_{2i} K_i)x(t),
\end{aligned} \tag{6.11}$$

$$2x^T(t)PDf(\tau(t))$$
$$\leq \varepsilon x^T(t)P^2x(t) + \frac{1}{\varepsilon}(Df(\tau(t)))^T Df(\tau(t)) \quad (6.12)$$
$$\leq x^T(t)(\varepsilon P^2 + \frac{1}{\varepsilon}\phi C D^T D C^T \phi)x(t),$$

where $\phi = diag(k_1, k_2, \ldots, k_m), k_i > 0, i = 1, 2, \ldots, m$.

$$-2x^T(t)P\{\sum_{i=1}^{p} B_{i0}K_i \int_{t-\tau_i(t)}^{t} A_0 x(s)ds$$
$$\leq \sum_{i=1}^{p} \int_{t-\tau_i(t)}^{t} [\varepsilon x^T(t)PB_{i0}K_i K_i^T B_{i0}^T Px(t) + \frac{1}{\varepsilon}x^T(s)A_0^T A_0 x(s)]ds \quad (6.13)$$
$$\leq \tau x^T(t)[\varepsilon \sum_{i=1}^{p} PB_{i0}K_i K_i^T B_{i0}^T P + \frac{p\alpha^2 l^2}{\varepsilon} A_0^T A_0]x(t),$$

$$-2x^T(t)P\{\sum_{i=1}^{p} B_{i0}K_i \int_{t-\tau_i(t)}^{t} \Delta A(s)x(s)ds\}$$
$$\leq \sum_{i=1}^{p} \int_{t-\tau_i(t)}^{t} [\varepsilon x^T(t)PB_{i0}K_i K_i^T B_{i0}^T Px(t) + \frac{1}{\varepsilon}x^T(s)\Delta A^T(s)\Delta A(s)x(s)]ds$$
$$\leq \tau x^T(t)[\varepsilon \sum_{i=1}^{p} PB_{i0}K_i K_i^T B_{i0}^T P + \frac{p\alpha^2 l^2}{\varepsilon} E_1^T D_1^T D_1 E_1]x(t), \quad (6.14)$$

$$-2x^T(t)P\{\sum_{i=1}^{p} \Delta B_i(t)K_i \int_{t-\tau_i(t)}^{t} A_0(s)x(s)ds\}$$
$$\leq \sum_{i=1}^{p} \int_{t-\tau_i(t)}^{t} [\varepsilon x^T(t)P\Delta B_i(t)K_i K_i^T \Delta B_i^T(t)Px(t) + \frac{1}{\varepsilon}x^T(s)A_0^T A_0 x(s)]ds$$
$$\leq \tau x^T(t)[\varepsilon \sum_{i=1}^{p} PD_{2i}E_{2i}K_i K_i^T E_{2i}^T D_{2i}^T P + \frac{p\alpha^2 l^2}{\varepsilon} A_0^T A_0]x(t), \quad (6.15)$$

$$-2x^T(t)P\{\sum_{i=1}^{p} \Delta B_i(t)K_i \int_{t-\tau_i(t)}^{t} \Delta A(s)x(s)ds\}$$
$$\leq \sum_{i=1}^{p} \int_{t-\tau_i(t)}^{t} [\varepsilon x^T(t)P\Delta B_i(t)K_i K_i^T B_i^T(t)Px(t) + \frac{1}{\varepsilon}x^T(s)\Delta A^T(s)\Delta A(s)x(s)]ds \quad (6.16)$$
$$\leq \tau x^T(t)[\sum_{i=1}^{p} \varepsilon PD_{2i}E_{2i}K_i K_i^T E_{2i}^T D_{2i}^T P + \frac{p\alpha^2 l^2}{\varepsilon} E_1^T D_1^T D_1 E_1]x(t),$$

$$-2x^T(t)P\{\sum_{i=1}^{p} B_{i0}K_i \int_{t-\tau_i(t)}^{t} \sum_{j=1}^{p} B_{j0}K_j x(s-\tau_j(s))ds\}$$
$$\leq \sum_{i=1,j=1}^{p} \int_{t-\tau_i(t)}^{t} 2x^T(t)PB_{i0}K_i B_{j0}K_j x(s-\tau_j(s))ds$$
$$\leq \sum_{i=1,j=1}^{p} \int_{t-\tau_i(t)}^{t} [\varepsilon x^T(t)PB_{i0}K_i K_i^T B_{i0}^T Px(t) + \frac{1}{\varepsilon}x^T(s-\tau_j(s))K_j^T B_{j0}^T B_{j0}K_j x(s-\tau_j(s))]ds$$
$$\leq \sum_{i=1,j=1}^{p} \int_{t-\tau_i(t)}^{t} [\varepsilon x^T(t)PB_{i0}K_i K_i^T B_{i0}^T Px(t) + \frac{1}{\varepsilon}x^T(s-\tau_j(s))K_j^T B_{j0}^T B_{j0}K_j x(s-\tau_j(s))]ds$$
$$\leq \sum_{i=1,j=1}^{p} \int_{t-\tau_i(t)}^{t} [\varepsilon x^T(t)PB_{i0}K_i K_i^T B_{i0}^T Px(t) + \frac{\alpha^2 l^2}{\varepsilon}x^T(t)K_j^T B_{j0}^T B_{j0}K_j x(t)]ds$$
$$\leq \sum_{i=1,j=1}^{p} \int_{t-\tau_i(t)}^{t} [\varepsilon x^T(t)PB_{i0}K_i K_i^T B_{i0}^T Px(t) + \frac{\alpha^2 l^2}{\varepsilon}x^T(t)K_j^T B_{j0}^T B_{j0}K_j x(t)]ds$$
$$\leq \tau p x^T(t)[\sum_{i=1}^{p} \varepsilon PB_{i0}K_i K_i^T B_{i0}^T P + \sum_{i=1}^{p} \frac{\alpha^2 l^2}{\varepsilon} K_j^T B_{j0}^T B_{j0}K_j]x(t),$$
$$(6.17)$$

$$-2x^T(t)P\{\sum_{i=1}^{p} B_{i0}K_i \int_{t-\tau_i(t)}^{t} \sum_{j=1}^{p} \Delta B_j(s)K_j x(s-\tau_j(s))ds\}$$
$$\leq \sum_{i=1,j=1}^{p} \int_{t-\tau_i(t)}^{t} 2x^T(t)PB_{i0}K_i \Delta B_j(s)K_j x(s-\tau_j(s))ds$$
$$\leq \sum_{i=1,j=1}^{p} \int_{t-\tau_i(t)}^{t} [\varepsilon x^T(t)PB_{i0}K_i K_i^T B_{i0}^T Px(t)$$
$$+ \frac{1}{\varepsilon}x^T(s-\tau_j(s))K_j^T \Delta B_j^T(s)\Delta B_j(s)K_j x(s-\tau_j(s))]$$
$$\leq \tau p x^T(t)[\sum_{i=1}^{p} \varepsilon PB_{i0}K_i K_i^T B_{i0}^T P + \sum_{i=1}^{p} \frac{\alpha^2 l^2}{\varepsilon} K_i^T E_{2i}^T D_{2i}^T D_{2i} E_{2i}K_i]x(t),$$
$$(6.18)$$

6.3 Some Results on Absolute Stability

$$\begin{aligned}
&-2x^T(t)P\{\sum_{i=1}^{p}\Delta B_i(t)K_i\int_{t-\tau_i(t)}^{t}\sum_{j=1}^{p}B_{j0}K_jx(s-\tau_j(s))ds\}\\
&\leq \sum_{i=1,j=1}^{p}\int_{t-\tau_i(t)}^{t}2x^T(t)P\Delta B_i(t)K_iB_{j0}K_jx(s-\tau_j(s))ds\\
&\leq \sum_{i=1,j=1}^{p}\int_{t-\tau_i(t)}^{t}[\varepsilon x^T(t)P\Delta B_i(t)K_iK_i^T\Delta B_i(t)^TPx(t)\\
&+\frac{1}{\varepsilon}x^T(s-\tau_j(s))K_j^TB_{j0}^TB_{j0}K_jx(s-\tau_j(s))]ds\\
&\leq p\tau x^T(t)[\varepsilon PD_{2i}E_{2i}K_iK_i^TE_{2i}^TD_{2i}^TP\\
&+\frac{\alpha^2l^2}{\varepsilon}K_i^TB_{i0}^TB_{i0}K_i]x(t),
\end{aligned} \quad (6.19)$$

$$\begin{aligned}
&-2x^T(t)P\{\sum_{i=1}^{p}\Delta B_i(t)K_i\int_{t-\tau_i(t)}^{t}\sum_{j=1}^{p}\Delta B_j(s)K_jx(s-\tau_j(s))ds\}\\
&\leq \sum_{i=1,j=1}^{p}\int_{t-\tau_i(t)}^{t}[\varepsilon x^T(t)P\Delta B_i(t)K_iK_i^T\Delta B_i(t)^TPx(t)\\
&+\frac{1}{\varepsilon}x^T(s-\tau_j(s))K_j^T\Delta B_j(s)^T\Delta B_j(s)K_jx(s-\tau_j(s))]ds\\
&\leq \tau p x^T(t)[\varepsilon PD_{2i}E_{2i}K_iK_i^TE_{2i}^TD_{2i}^TP+\frac{\alpha^2l^2}{\varepsilon}K_i^TE_{2i}^TD_{2i}^TD_{2i}E_{2i}K_i]x(t),
\end{aligned} \quad (6.20)$$

$$\begin{aligned}
&-2x^T(t)P\{\sum_{i=1}^{p}B_{i0}K_i\int_{t-\tau_i(t)}^{t}Df(\sigma(s)ds\}\\
&\leq \sum_{i=1}^{p}\int_{t-\tau_i(t)}^{t}[\varepsilon x^T(t)PB_{i0}K_iK_i^TB_{i0}^TPx(t)+\frac{1}{\varepsilon}f^T(\sigma(s))D^TDf(\sigma(s))]ds\\
&\leq \tau x^T(t)[\sum_{i=1}^{p}\varepsilon PB_{i0}K_iK_i^TB_{i0}^TP+\frac{p\alpha^2l^2}{\varepsilon}\phi CD^TDC^T\phi]x(t),
\end{aligned} \quad (6.21)$$

$$\begin{aligned}
&-2x^T(t)P\{\sum_{i=1}^{p}\Delta B_i(t)K_i\int_{t-\tau_i(t)}^{t}Df(\sigma(s)ds\}\\
&\leq \sum_{i=1}^{p}\int_{t-\tau_i(t)}^{t}[\varepsilon x^T(t)PD_{2i}E_{2i}K_iK_i^TE_{2i}^TD_{2i}^TPx(t)+\frac{1}{\varepsilon}f^T(\sigma(s))D^TDf(\sigma(s))]ds\\
&\leq \tau x^T(t)[\sum_{i=1}^{p}PD_{2i}E_{2i}K_iK_i^TE_{2i}^TD_{2i}^TP+\frac{p\alpha^2l^2}{\varepsilon}\phi CD^TDC^T\phi]x(t),
\end{aligned} \quad (6.22)$$

Substituting (6.10)–(6.22) for (6.9), we have

$$\begin{aligned}
&\frac{dV(x(t))}{dt}\Big|_{(6.1)}\\
&= x^T(t)[(A_0+\sum_{i=1}^{p}B_{i0}K_i)^TP+P(A_0+\sum_{i=1}^{p}B_{i0}K_i)]x(t)\\
&+x^T(t)(\varepsilon P^2+\frac{1}{\varepsilon}E_1^TD_1^TD_1E_1)x(t)\\
&+x^T(t)(p\varepsilon P^2+\frac{1}{\varepsilon}\sum_{i=1}^{p}K_i^TE_{2i}^TD_{2i}^TD_{2i}E_{2i}K_i)x(t)\\
&+x^T(t)(\varepsilon P^2+\frac{1}{\varepsilon}\phi CD^TDC^T\phi)x(t)\\
&+\tau x^T(t)[\varepsilon \sum_{i=1}^{p}PB_{i0}K_iK_i^TB_{i0}^TP+\frac{p\alpha^2l^2}{\varepsilon}A_0^TA_0]x(t)\\
&+\tau x^T(t)[\varepsilon \sum_{i=1}^{p}PB_{i0}K_iK_i^TB_{i0}^TP+\frac{p\alpha^2l^2}{\varepsilon}E_1^TD_1^TD_1E_1]x(t)\\
&+\tau x^T(t)[\varepsilon \sum_{i=1}^{p}PD_{2i}E_{2i}K_iK_i^TE_{2i}^TD_{2i}^TP+\frac{p\alpha^2l^2}{\varepsilon}A_0^TA_0]x(t)\\
&+\tau x^T(t)[\sum_{i=1}^{p}\varepsilon PD_{2i}E_{2i}K_iK_i^TE_{2i}^TD_{2i}^TP+\frac{p\alpha^2l^2}{\varepsilon}E_1^TD_1^TD_1E_1]x(t)\\
&+\tau p x^T(t)[\sum_{i=1}^{p}\varepsilon PB_{i0}K_iK_i^TB_{i0}^TP+\sum_{i=1}^{p}\frac{\alpha^2l^2}{\varepsilon}K_j^TB_{j0}^TB_{j0}K_j]x(t)\\
&+\tau p x^T(t)[\sum_{i=1}^{p}\varepsilon PB_{i0}K_iK_i^TB_{i0}^TP+\sum_{i=1}^{p}\frac{\alpha^2l^2}{\varepsilon}K_i^TE_{2i}^TD_{2i}^TD_{2i}E_{2i}K_i]x(t)\\
&+p\tau x^T(t)[\varepsilon PD_{2i}E_{2i}K_iK_i^TE_{2i}^TD_{2i}^TP+\frac{\alpha^2l^2}{\varepsilon}K_i^TB_{i0}^TB_{i0}K_i]x(t)\\
&+\tau p x^T(t)[\varepsilon PD_{2i}E_{2i}K_iK_i^TE_{2i}^TD_{2i}^TP+\frac{\alpha^2l^2}{\varepsilon}K_i^TE_{2i}^TD_{2i}^TD_{2i}E_{2i}K_i]x(t)\\
&+\tau x^T(t)[\sum_{i=1}^{p}\varepsilon PB_{i0}K_iK_i^TB_{i0}^TP+\frac{p\alpha^2l^2}{\varepsilon}\phi CD^TDC^T\phi]x(t)\\
&+\tau x^T(t)[\sum_{i=1}^{p}PD_{2i}E_{2i}K_iK_i^TE_{2i}^TD_{2i}^TP+\frac{p\alpha^2l^2}{\varepsilon}\phi CD^TDC^T\phi]x(t).
\end{aligned} \quad (6.23)$$

Further, we obtain

$$\frac{dV(x(t))}{dt}|_{(6.1)}$$
$$= x^T(t)[(A_0 + \sum_{i=1}^{p} B_{i0}K_i)^T P + P(A_0 + \sum_{i=1}^{p} B_{i0}K_i) +$$
$$\varepsilon(p+2)P^2 + \frac{1}{\varepsilon}[E_1^T D_1^T D_1 E_1 + \sum_{i=1}^{p} K_i^T E_{2i}^T D_{2i}^T D_{2i} E_{2i} K_i + \phi C D^T D C^T \phi]]x(t)$$
$$+ \tau x^T(t)[\varepsilon(2p+3) \sum_{i=1}^{p} (PB_{i0}K_i K_i^T B_{i0}^T P + PD_{2i}E_{2i}K_i K_i^T E_{2i}^T D_{2i}^T P)$$
$$+ \frac{2p\alpha^2 l^2}{\varepsilon}(A_0^T A_0 + E_1^T D_1^T D_1 E_1$$
$$+ \sum_{i=1}^{p} K_i^T B_{i0}^T B_{i0} K_i + \sum_{i=1}^{p} K_i^T E_{2i}^T D_{2i}^T D_{2i} E_{2i} K_i + \phi C D^T D C^T \phi)]x(t)$$
$$= x^T(t)\{-Q + \varepsilon(p+2)P^2 + \frac{1}{\varepsilon}[E_1^T D_1^T D_1 E_1 + \sum_{i=1}^{p} K_i^T E_{2i}^T D_{2i}^T D_{2i} E_{2i} K_i + \phi C D^T D C^T \phi]$$
$$+ \tau[\varepsilon(2p+3) \sum_{i=1}^{p} (PB_{i0}K_i K_i^T B_{i0}^T P + PD_{2i}E_{2i}K_i K_i^T E_{2i}^T D_{2i}^T P)$$
$$+ \frac{2p\alpha^2 l^2}{\varepsilon}(A_0^T A_0 + E_1^T D_1^T D_1 E_1$$

$$+ \sum_{i=1}^{p} K_i^T B_{i0}^T B_{i0} K_i + \sum_{i=1}^{p} K_i^T E_{2i}^T D_{2i}^T D_{2i} E_{2i} K_i + \phi C D^T D C^T \phi)]\}x(t). \quad (6.24)$$

Defining

$$W = (A_0 + \sum_{i=1}^{p} B_{i0}K_i)^T P + P(A_0 + \sum_{i=1}^{p} B_{i0}K_i)$$
$$+ \varepsilon(p+2)P^2 + \frac{1}{\varepsilon}[E_1^T D_1^T D_1 E_1 + \sum_{i=1}^{p} K_i^T E_{2i}^T D_{2i}^T D_{2i} E_{2i} K_i + \phi C D^T D C^T \phi]$$
$$+ \tau[\varepsilon(2p+3) \sum_{i=1}^{p} (PB_{i0}K_i K_i^T B_{i0}^T P + PD_{2i}E_{2i}K_i K_i^T E_{2i}^T D_{2i}^T P)$$
$$+ \frac{2p\alpha^2 l^2}{\varepsilon}(A_0^T A_0 + E_1^T D_1^T D_1 E_1$$
$$+ \sum_{i=1}^{p} K_i^T B_{i0}^T B_{i0} K_i + \sum_{i=1}^{p} K_i^T E_{2i}^T D_{2i}^T D_{2i} E_{2i} K_i + \phi C D^T D C^T \phi)],$$
$$W_1 = (A_0 + \sum_{i=1}^{p} B_{i0}K_i)^T P + P(A_0 + \sum_{i=1}^{p} B_{i0}K_i) + \varepsilon(p+2)P^2$$
$$+ \frac{1}{\varepsilon}[E_1^T D_1^T D_1 E_1$$
$$+ \sum_{i=1}^{p} K_i^T E_{2i}^T D_{2i}^T D_{2i} E_{2i} K_i + \phi C D^T D C^T \phi],$$
$$W_2 = \varepsilon(2p+3) \sum_{i=1}^{p} (PB_{i0}K_i K_i^T B_{i0}^T P + PD_{2i}E_{2i}K_i K_i^T E_{2i}^T D_{2i}^T P)$$
$$+ \frac{2p\alpha^2 l^2}{\varepsilon}(A_0^T A_0 + E_1^T D_1^T D_1 E_1$$
$$+ \sum_{i=1}^{p} K_i^T B_{i0}^T B_{i0} K_i + \sum_{i=1}^{p} K_i^T E_{2i}^T D_{2i}^T D_{2i} E_{2i} K_i + \phi C D^T D C^T \phi).$$
(6.25)

It is clear that
$$W_2^T = W_2 \geq 0, \ W_1^T = W_1. \quad (6.26)$$

If $W = W_1 + \tau W_2 < 0$, based on Lemma 6.1, it is clear that the trivial solution of (6.1) is absolute stable. □

From the above discussion, we know how to decide the matrix inequality $W_1 + \tau W_2 < 0$ is an important precondition.

Above all, we introduce the following lemma.

Lemma 6.5 Let $= A_0 + \sum_{i=1}^{p} B_{i0}K_i$ be Hurwitz stable, then for a given matrix $Q = Q^T > 0$, there always exists a $P = P^T > 0$ and $L = L^T > 0$ such that

$$\widetilde{A}_0^T P + P\widetilde{A}_0 = -Q = -LL < 0,$$

where $\widetilde{A}_0 = A_0 + \sum_{i=1}^{p} B_{i0}K_i, L = S^T \Lambda S, L^T = L, \Lambda = diag[\lambda_1^{\frac{1}{2}}, \lambda_2^{\frac{1}{2}}, \ldots, \lambda_n^{\frac{1}{2}}]$, $\lambda_i, (i = 1, 2, \ldots, n)$ is the eignvalue of $Q, S^T S = I$.

Based on Lyapunov theorem, the desired result is obtained. □

6.3 Some Results on Absolute Stability

Further, the following results to $W_1 + \tau W_2 < 0$ are outlined too.

Theorem 6.1 $W_1 < 0$ is a sufficient and necessary condition toward $W = W_1 + \tau W_2 < 0$, where $\forall \tau > 0$ and $W_2^T = W_2 \geq 0$.

Proof If $W_1 < 0$, we can obtain $-W_1 > 0$. Based on Lemma 6.5, there exist a nonsingular matrix Q $Q = -W_1 > 0$, such that

$$-W_1 + \tau W_2 = -Q_1 Q_1 + \tau Q_2 Q_2$$
$$= Q_1(-I + \tau(Q_1^{-1} Q_2)(Q_2 Q_1^{-1})) Q_1.$$

It is clear that there exists at least an enough small positive real number τ such that

$$-I + \tau (Q_2 Q_1^{-1})^T (Q_2 Q_1^{-1}) < 0.$$

On the other hand, if
$$W = W_1 + \tau W_2 < 0, \tau > 0,$$
$$W_2^T = W_2 \geq 0,$$
$$W_3^T = W_3 < 0,$$

then we obtain
$$-\tau W_2^T = -\tau W_2 \leq 0,$$
$$W_3^T = W_3 < 0.$$

Further, we have
$$W_1 = W_3 - \tau W_2 < 0.$$

$\square$

Combining Lemmas 6.1, 6.5 with Theorem 6.1, we obtain the following result.

Theorem 6.2 *If there exists some real number $\varepsilon > 0$ such that $W_1 < 0$, then there must exist some real $\tau > 0$ such as $W_1 + \tau W_2 < 0$, and simultaneously the trivial solution of (6.1) is absolutely stable.*

Note 6.1 Similarly to (6.10)–(6.22), we may obtain other different inequalities, for example,

$$\begin{aligned} & 2x^T(t) P \Delta A(t) x(t) \\ & \leq x^T(t)(\varepsilon P D_1^T D_1 P + \tfrac{1}{\varepsilon} E_1^T E_1) x(t), \end{aligned} \qquad (6.27)$$

$$\begin{aligned} & 2x^T(t) P (\sum_{i=1}^{p} \Delta B_i(t) K_i) x(t) \\ & \leq x^T(t)(\varepsilon \sum_{i=1}^{p} P D_{2i}^T D_{2i} P + \tfrac{1}{\varepsilon} \sum_{i=1}^{p} K_i^T E_{2i}^T E_{2i} K_i) x(t), \end{aligned} \qquad (6.28)$$

$$\begin{aligned} & 2x^T(t) P D f(\tau(t)) \\ & \leq x^T(t)(\varepsilon P D^T D P + \tfrac{1}{\varepsilon} \phi C D^T D C^T \phi) x(t), \end{aligned} \qquad (6.29)$$

where $\phi = diag(k_1, k_2, \ldots, k_m), k_i > 0, i = 1, 2, \ldots, m$.

$$-2x^T(t)P\{\sum_{i=1}^p B_{i0}K_i \int_{t-\tau_i(t)}^t A_0 x(s)ds\}$$
$$\leq \tau x^T(t)[\varepsilon \sum_{i=1}^p PB_{i0}B_{i0}^T P + \frac{p\alpha^2 l^2}{\varepsilon} A_0^T K_i^T K_i A_0]x(t), \quad (6.30)$$

$$-2x^T(t)P\{\sum_{i=1}^p B_{i0}K_i \int_{t-\tau_i(t)}^t \Delta A(s)x(s)ds\}$$
$$\leq \tau x^T(t)[\varepsilon \sum_{i=1}^p PB_{i0}K_i D_1 D_1^T K_i^T B_{i0}^T P + \frac{p\alpha^2 l^2}{\varepsilon} E_1^T E_1]x(t), \quad (6.31)$$

$$-2x^T(t)P\{\sum_{i=1}^p \Delta B_i(t)K_i \int_{t-\tau_i(t)}^t A_0(s)x(s)ds\}$$
$$\leq \tau x^T(t)[\varepsilon \sum_{i=1}^p PD_{2i}D_{2i}^T P + \frac{p\alpha^2 l^2}{\varepsilon} A_0^T E_{2i}^T K_i^T K_i E_{2i} A_0]x(t), \quad (6.32)$$

$$-2x^T(t)P\{\sum_{i=1}^p \Delta B_i(t)K_i \int_{t-\tau_i(t)}^t \Delta A(s)x(s)ds\}$$
$$\leq \tau x^T(t)[\sum_{i=1}^p \varepsilon PD_{2i}D_{2i}^T P + \frac{p\alpha^2 l^2}{\varepsilon} E_1^T D_1^T K_i^T E_{2i}^T E_{2i} K_i D_1 E_1]x(t), \quad (6.33)$$

$$-2x^T(t)P\{\sum_{i=1}^p B_{i0}K_i \int_{t-\tau_i(t)}^t \sum_{j=1}^p B_{j0}K_j x(s-\tau_j(s))ds\}$$
$$\leq \tau x^T(t)[\sum_{i=1,j=1}^p \varepsilon PB_{i0}K_i B_{j0}B_{j0}^T K_i^T B_{i0}^T P + p\sum_{i=1}^p \frac{\alpha^2 l^2}{\varepsilon} K_j^T K_j]x(t), \quad (6.34)$$

$$-2x^T(t)P\{\sum_{i=1}^p B_{i0}K_i \int_{t-\tau_i(t)}^t \sum_{j=1}^p \Delta B_j(s)K_j x(s-\tau_j(s))ds\}$$
$$\leq \tau x^T(t)[\sum_{i=1,j=p}^p \varepsilon PB_{i0}K_i D_{2j}D_{2j}^T K_i^T B_{i0}^T P + p\sum_{i=1}^p \frac{\alpha^2 l^2}{\varepsilon} K_i^T E_{2i}^T E_{2i} K_i]x(t), \quad (6.35)$$

$$-2x^T(t)P\{\sum_{i=1}^p \Delta B_i(t)K_i \int_{t-\tau_i(t)}^t \sum_{j=1}^p B_{j0}K_j x(s-\tau_j(s))ds\}$$
$$\leq \tau x^T(t)[p\varepsilon \sum_{i=1}^p PD_{2i}D_{2i}^T P + \frac{\alpha^2 l^2}{\varepsilon} \sum_{i=1,j=1}^p K_i^T B_{i0}^T E_{2j} K_j K_j^T E_{2j}^T B_{i0} K_i]x(t), \quad (6.36)$$

$$-2x^T(t)P\{\sum_{i=1}^p \Delta B_i(t)K_i \int_{t-\tau_i(t)}^t \sum_{j=1}^p \Delta B_j(s)K_j x(s-\tau_j(s))ds\}$$
$$\leq \tau x^T(t)[\varepsilon \sum_{i=1,j=1}^p PD_{2i}E_{2i}K_i D_{2j}D_{2j}^T K_i^T E_{2i}^T D_{2i}^T P + p\frac{\alpha^2 l^2}{\varepsilon} \sum_{i=1}^p K_i^T E_{2i}^T E_{2i} K_i]x(t), \quad (6.37)$$

$$-2x^T(t)P\{\sum_{i=1}^p B_{i0}K_i \int_{t-\tau_i(t)}^t Df(\sigma(s)ds\}$$
$$\leq \tau x^T(t)[\sum_{i=1}^p \varepsilon PB_{i0}B_{i0}^T P + \frac{\alpha^2 l^2}{\varepsilon} \sum_{i=1}^p \phi CD^T K_i^T K_i DC^T \phi]x(t), \quad (6.38)$$

$$-2x^T(t)P\{\sum_{i=1}^p \Delta B_i(t)K_i \int_{t-\tau_i(t)}^t Df(\sigma(s)ds\}$$
$$\leq \tau x^T(t)[\sum_{i=1}^p PD_{2i}D_{2i}^T P + \frac{\alpha^2 l^2}{\varepsilon} \sum_{i=1}^p \phi CD^T K_i^T E_{2i}^T E_{2i} K_i DC^T \phi]x(t). \quad (6.39)$$

In this chapter, we only discuss the cases on the combination (6.10)–(6.22), (6.24), the other results on many combinations among these possible inequalities (6.10)–(6.22), (6.27)–(6.39) are similar. So, we omit these discussions.

6.4 Some Discussions on Solvable Conditions

From the above discussions, it is clear that

6.4 Some Discussions on Solvable Conditions

$$W_1 = (A_0 + \sum_{i=1}^{p} B_{i0}K_i)^T P + P(A_0 + \sum_{i=1}^{p} B_{i0}K_i) + \varepsilon(p+2)P^2$$

$$+ \frac{1}{\varepsilon}[E_1^T D_1^T D_1 E_1 + \sum_{i=1}^{p} K_i^T E_{2i}^T D_{2i}^T D_{2i} E_{2i} K_i + \phi C D^T D C^T \phi] < 0.$$

Then, that how to realize $W_1 < 0$ is an interested attempt.

For this attempt, we first outline the below important notions.

Definition 6.1 For the system (6.1), we call some positive real number ε, such that $W_1 = \widetilde{A_0}^T P + P \widetilde{A_0} + \varepsilon(p+2)P^2 + \frac{1}{\varepsilon}[E_1^T D_1^T D_1 E_1 + \sum_{i=1}^{p} K_i^T E_{2i}^T D_{2i}^T D_{2i} E_{2i} K_i + \phi C D^T D C^T \phi] < 0$ as solvable conditions, and call W_1 as a solvable inequality.

Above all, we give several solvable conditions on $W_1 < 0$ as follows.
For the convenience of discussion, let

$$\begin{aligned} P_1 &= (p+2)P^2, \\ P_2 &= E_1^T D_1^T D_1 E_1 + \sum_{i=1}^{p} K_i^T E_{2i}^T D_{2i}^T D_{2i} E_{2i} K_i + \phi C D^T D C^T \phi. \end{aligned} \quad (6.40)$$

Theorem 6.3 *Solvable conditions on $W_1 < 0$ are sufficient and necessary to*

$$\begin{aligned} &\tfrac{1}{4} Q P_1^{-2} Q - P_2 > 0, \\ &(\varepsilon I - \tfrac{1}{2} P_1^{-1} Q P_1^{-1})^2 < (\tfrac{1}{2} P_1^{-1} Q P_1^{-1})^2 - P_1^{-1} P_2 P_1^{-1}. \end{aligned} \quad (6.41)$$

Proof From the above assumptions on W_1 and (6.40), we know the following facts that

$$\begin{aligned} P_1^T &= P_1 > 0, \\ P_2^T &= P_2 \geq 0, \\ -Q + \varepsilon P_1^2 &+ \tfrac{1}{\varepsilon} P_2 < 0. \end{aligned}$$

Combining the fact $\varepsilon > 0$ and the above inequality $W_1 < 0$, we have

$$\varepsilon^2 P_1^2 - \varepsilon Q + P_2 < 0.$$

Further, we have

$$P_1(\varepsilon^2 I - \varepsilon P_1^{-1} Q P_1^{-1} + \tfrac{1}{4} P_1^{-1} Q P_1^{-1} P_1^{-1} Q P_1^{-1}) P_1 - \tfrac{1}{4} Q P_1^{-2} Q + P_2 < 0.$$

Further

$$P_1(\varepsilon I - \tfrac{1}{2} P_1^{-1} Q P_1^{-1})^2 P_1 < \tfrac{1}{4} Q P_1^{-2} Q - P_2.$$

So, we have

$$\begin{aligned} &\tfrac{1}{4} Q P_1^{-2} Q - P_2 > 0, \\ &(\varepsilon I - \tfrac{1}{2} P_1^{-1} Q P_1^{-1})^2 < (\tfrac{1}{2} P_1^{-1} Q P_1^{-1})^2 - P_1^{-1} P_2 P_1^{-1}. \end{aligned}$$

Conversely, if (6.41) holds, then we easily obtain the fact $W_1 < 0$. □

Based on Theorem 6.3, Theorem 6.1 can be convert into the following result.

Theorem 6.4 *For (6.1), defining P_1, P_2 as (6.40),*
(1) if (6.41) holds, then there must exist some real $\tau > 0$ such as $W_1 + \tau W_2 < 0$, and simultaneously the trivial solution of (6.1) is absolute stable.

(2) if
$$\tfrac{1}{4} Q P_1^{-2} Q - P_2 > 0,$$
$$(\varepsilon I - \tfrac{1}{2} P_1^{-1} Q P_1^{-1})^2 \geq (\tfrac{1}{2} P_1^{-1} Q P_1^{-1})^2 - P_1^{-1} P_2 P_1^{-1},$$

then there does not exist any real $\tau > 0$ such as $W_1 + \tau W_2 < 0$, and the trivial solution of (6.1) is unstable.
(3) if $\tfrac{1}{4} Q P_1^{-2} Q - P_2 \leq 0$, then (6.41) does not hold and the trivial solution of (6.1) is unstable.

For (6.2), we can easily obtain a similar result to (6.41) as follows:

$$\tfrac{1}{4} Q P_1^{-2} Q - \phi C D^T D C^T \phi > 0,$$
$$(\varepsilon I - \tfrac{1}{2} P^{-1} Q P^{-1})^2 < (\tfrac{1}{2} P^{-1} Q P^{-1})^2 - P^{-1} \phi C D^T D C^T \phi P^{-1}. \quad (6.42)$$

Theorem 6.5 *For (6.2),*
(1) if (6.42) holds, then there must exist some real $\tau > 0$ such as

$$-Q + \varepsilon P^2 + \tfrac{1}{\varepsilon} \phi C D^T D C^T \phi$$
$$+ \tau [\varepsilon(p+2) \sum_{i=1}^{p} P B_{i0} K_i K_i^T B_{i0}^T P + \tfrac{p \alpha^2 l^2}{\varepsilon} (A_0^T A_0 + \sum_{i=1}^{p} K_i^T B_{i0}^T B_{i0} K_i + \phi C D^T D C^T \phi)]$$
$$< 0,$$

and simultaneously the trivial solution of (6.2) is absolute stable.
(2) if
$$\tfrac{1}{4} Q P_1^{-2} Q - \phi C D^T D C^T \phi > 0,$$
$$(\varepsilon I - \tfrac{1}{2} P^{-1} Q P^{-1})^2 \geq (\tfrac{1}{2} P^{-1} Q P^{-1})^2 - P^{-1} \phi C D^T D C^T \phi P^{-1},$$

then there does not exist any real $\tau > 0$ such that the trivial solution of (6.2) is unstable.
(3) if $\tfrac{1}{4} Q P_1^{-2} Q - P_2 \leq 0$, then the trivial solution of (6.2) is unstable.

Definition 6.2 For any matrix $A \in \mathbb{R}^{n \times m}$, and for any vector $\gamma \in \mathbb{R}^n$, we define some norms as follows:

$$\|A\| = \sqrt{\lambda_{\max}(A^T A)}, \quad \|\gamma\| = \sqrt{(\gamma^T \gamma)}.$$

Based on Definitions 6.1, 6.2, we shall discuss (6.23)'s solvable inequalities and their corresponding solvable conditions as interested norm applications.

First, we consider one case on W as follows.

6.4 Some Discussions on Solvable Conditions

Case 6.1

$$\lambda_{min}(Q - \varepsilon(p+2)P^2) - \frac{1}{\varepsilon}[\|D_1E_1\|^2 + \sum_{i=1}^{p}\|D_{2i}E_{2i}K_i\|^2 + \|DC^T\phi\|^2]$$
$$-\tau[\varepsilon(2p+3)\sum_{i=1}^{p}\|PB_{i0}K_i\|^2 + \varepsilon(2p+3)\|PD_{2i}E_{2i}K_i\|^2 + \frac{2p\alpha^2 l^2}{\varepsilon}[\|A_0\|^2 + \|D_1E_1\|^2$$
$$+\sum_{i=1}^{p}\|B_{i0}K_i\|^2 + \sum_{i=1}^{p}\|D_{2i}E_{2i}K_i\|^2 + \|DC^T\phi\|^2]] > 0.$$
(6.43)

Using Lemma 6.5, we obtain

$$\begin{aligned}
\lambda_{min}(Q - \varepsilon(p+2)P^2) \\
&= \lambda_{min}(L(I - \varepsilon(p+2)L^{-1}P^2L^{-1})L) \\
&\geq \lambda_{min}(L)\lambda_{min}(I - \varepsilon(p+2)L^{-1}P^2L^{-1})\lambda_{min}(L) \\
&\geq \lambda_{min}(Q)\lambda_{min}(I - \varepsilon(p+2)L^{-1}P^2L^{-1}) \\
&\geq \lambda_{min}(Q)\lambda_{min}(I - \varepsilon(p+2)(L^{-1}P^2L^{-1})) \\
&= \lambda_{min}(Q)[1 - \varepsilon(p+2)\|L^{-1}P\|^2].
\end{aligned}$$
(6.44)

From Case 6.1, we consider its corresponding solvable inequality as follows:

$$\lambda_{min}(Q)[1 - \varepsilon(p+2)\|L^{-1}P\|^2] - \frac{1}{\varepsilon}[\|D_1E_1\|^2 + \sum_{i=1}^{p}\|D_{2i}E_{2i}K_i\|^2 + \|DC^T\phi\|^2] > 0.$$
(6.45)

For the above inequality, we give the following solvable results.

Theorem 6.6 *If*

$$\Delta_1 = \lambda_{min}(Q)^2 - 4(p+2)\|L^{-1}P\|^2\lambda_{min}(Q)$$
$$\cdot(\|D_1E_1\|^2 + \sum_{i=1}^{p}\|D_{2i}E_{2i}K_i\|^2 + \|DC^T\phi\|^2) > 0,$$
$$\varepsilon_{11}^0 = max(0, \frac{\lambda_{min}(Q)-\Delta_1^{\frac{1}{2}}}{2(p+2)\|L^{-1}P\|^2\lambda_{min}(Q)}), \varepsilon_{12}^0 = \frac{\lambda_{min}(Q)+\Delta_1^{\frac{1}{2}}}{2(p+2)\|L^{-1}P\|^2\lambda_{min}(Q)},$$
(6.46)

1) For any $\varepsilon \in D_1(Q) = \{\varepsilon|\varepsilon \in (\varepsilon_{11}^0, \varepsilon_{12}^0), Q^T = Q > 0\}$, then

$$\tau_1 < \frac{\lambda_{min}(Q-\varepsilon(p+2)P^2)-\frac{1}{\varepsilon}[\|D_1E_1\|^2+\sum_{i=1}^{p}\|D_{2i}E_{2i}K_i\|^2+\|DC^T\phi\|^2]}{\varepsilon(2p+3)\sum_{i=1}^{p}(\|PB_{i0}K_i\|^2 + \|PD_{2i}E_{2i}K_i\|^2)}.$$
$$+\frac{2p\alpha^2 l^2}{\varepsilon}(\|A_0\|^2 + \|D_1E_1\|^2 + \sum_{i=1}^{p}\|B_{i0}K_i\|^2 + \sum_{i=1}^{p}\|D_{2i}E_{2i}K_i\|^2 + \|DC^T\phi\|^2)$$
(6.47)

Further, the origin of (6.1) is absolutely stable.
(2) For any $0 \leq \varepsilon \leq \varepsilon_{11}^0$ or $\varepsilon \geq \varepsilon_{12}^0, \Delta_1 > 0$, (6.45) does not hold, and the origin of (6.1) is unstable;
(3) If $\Delta_1 \leq 0$, that is to say, there does not exist any ε to satisfy (6.45), and the origin of (6.1) is unstable.

From the above discussion, we easily obtain the proof, and we omit its discussion. □

When $\Delta A(t) = 0, \Delta B_i(t) = 0$ in (6.1), which means (6.2), we consider the following inequality

$$\lambda_{min}(Q - \varepsilon P^2) - \frac{1}{\varepsilon}\|DC^T\phi\|^2$$
$$-\tau[\varepsilon(p+2)\sum_{i=1}^{p}\|PB_{i0}K_i\|^2 + \frac{p\alpha^2 l^2}{\varepsilon}(\|A_0\|^2 + \sum_{i=1}^{p}\|B_{i0}K_i\|^2 + \|DC^T\phi\|^2) > 0,$$
(6.48)

and its corresponding solvable inequality

$$\lambda_{min}(Q - \varepsilon P^2) - \tfrac{1}{\varepsilon}\|DC^T\phi\|^2 > 0. \tag{6.49}$$

Theorem 6.7 *If*

$$\begin{aligned}\Delta_{10} &= \lambda_{min}(Q)^2 - 4\|L^{-1}P\|^2\|DC^T\phi\|^2 > 0,\\ \varepsilon_{11}^1 &= max(0, \tfrac{\lambda_{min}(Q)-\Delta_{10}^{\frac{1}{2}}}{2\|L^{-1}P\|^2\lambda_{min}(Q)}),\ \varepsilon_{12}^2 = \tfrac{\lambda_{min}(Q)+\Delta_{10}^{\frac{1}{2}}}{2\|L^{-1}P\|^2\lambda_{min}(Q)},\end{aligned} \tag{6.50}$$

(1) For any $\varepsilon \in D_{10}(Q) = \{\varepsilon|(\varepsilon_{11}^1, \varepsilon_{12}^2),\ Q^T = Q > 0\}$, then

$$\tau_{10} < \frac{\lambda_{min}(Q-\varepsilon P^2)-\tfrac{1}{\varepsilon}\|DC^T\phi\|^2}{\varepsilon(p+2)\sum_{i=1}^{p}\|PB_{i0}K_i\|^2 + \tfrac{p\alpha^2 l^2}{\varepsilon}(\|A_0\|^2 + \sum_{i=1}^{p}\|B_{i0}K_i\|^2 + \|DC^T\phi\|^2)}. \tag{6.51}$$

Further, the origin of (6.2) is absolutely stable.
(2) For any $0 \le \varepsilon \le \varepsilon_{11}^1$ or $\varepsilon \ge \varepsilon_{12}^2$, $\Delta_1 > 0$, (6.49) does not hold, and the origin of (6.2) is unstable;
(3) If $\Delta_{10} \le 0$, that is to say, there does not exist any ε to satisfy (6.49), and the origin of (6.2) is unstable.

Similarly, we consider another case on W as follows.

Case 6.2

$$\begin{aligned}&\lambda_{min}(Q) - \varepsilon(p+2)P^2 - \tfrac{1}{\varepsilon}[\|D_1E_1\|^2 + \sum_{i=1}^{p}\|D_{2i}E_{2i}K_i\|^2 + \|DC^T\phi\|^2]\\ &-\tau[\varepsilon(2p+3)(\sum_{i=1}^{p}\|PB_{i0}K_i\|^2 + \|PD_{2i}E_{2i}K_i\|^2) + \tfrac{2p\alpha^2 l^2}{\varepsilon}(\|A_0\|^2 + \|D_1E_1\|^2\\ &+ \sum_{i=1}^{p}\|B_{i0}K_i\|^2 + \sum_{i=1}^{p}\|D_{2i}E_{2i}K_i\|^2 + \|DC^T\phi\|^2)] > 0.\end{aligned} \tag{6.52}$$

From Case 6.2, we consider its corresponding solvable inequality as follows

$$\lambda_{min}(Q) - \varepsilon(p+2)P^2 - \tfrac{1}{\varepsilon}[\|D_1E_1\|^2 + \sum_{i=1}^{p}\|D_{2i}E_{2i}K_i\|^2 + \|DC^T\phi\|^2] > 0. \tag{6.53}$$

Theorem 6.8 *If*

$$\begin{aligned}\Delta_2 &= \lambda_{min}^2(Q) - 4(p+2)\|P\|^2(\|D_1E_1\|^2 + \sum_{i=1}^{p}\|D_{2i}E_{2i}K_i\|^2 + \|DC^T\phi\|^2) > 0,\\ \varepsilon_{21}^0 &= max(0, \tfrac{\lambda_{min}(Q)-\Delta_2^{\frac{1}{2}}}{2(p+2)\|P\|^2}),\ \varepsilon_{22}^0 = \tfrac{\lambda_{min}(Q)+\Delta_2^{\frac{1}{2}}}{2(p+2)\|P\|^2},\end{aligned} \tag{6.54}$$

(1) For any $\varepsilon \in D_2(Q) = \{\varepsilon|(\varepsilon_{21}^0, \varepsilon_{22}^0),\ Q^T = Q > 0\}$, then

$$\tau_2 < \frac{\lambda_{min}(Q)-\varepsilon(p+2)P^2-\tfrac{1}{\varepsilon}[\|D_1E_1\|^2 + \sum_{i=1}^{p}\|D_{2i}E_{2i}K_i\|^2 + \|DC^T\phi\|^2]}{\varepsilon(2p+3)\sum_{i=1}^{p}(\|PB_{i0}K_i\|^2 + \|PD_{2i}E_{2i}K_i\|^2) + \tfrac{2p\alpha^2 l^2}{\varepsilon}(\|A_0\|^2 + \|D_1E_1\|^2 + \sum_{i=1}^{p}\|B_{i0}K_i\|^2 + \sum_{i=1}^{p}\|D_{2i}E_{2i}K_i\|^2 + \|DC^T\phi\|^2)}. \tag{6.55}$$

6.4 Some Discussions on Solvable Conditions

and further, the origin of (6.1) is absolutely stable.

(2) For any $0 < \varepsilon \le \varepsilon_{21}^0$ or $\varepsilon \ge \varepsilon_{22}^0$, $\Delta_1 > 0$, (6.53) does not hold, and the origin of (6.1) is unstable;

(3) If $\Delta_2 \le 0$, that is to say, there does not exist any ε to satisfy (6.53), and the origin of (6.1) is unstable.

When $\Delta A(t) = 0$, $\Delta B_i(t) = 0$ in (6.1), which means (6.2), we obtain the following inequality

$$\lambda_{min}(Q) - \varepsilon P^2 - \tfrac{1}{\varepsilon}\|DC^T\phi\|^2$$
$$-\tau[\varepsilon(p+2)\sum_{i=1}^{p}\|PB_{i0}K_i\|^2 + \tfrac{p\alpha^2 l^2}{\varepsilon}(\|A_0\|^2 \sum_{i=1}^{p}\|B_{i0}K_i\|^2 + \|DC^T\phi\|^2)] > 0, \quad (6.56)$$

and its corresponding solvable inequality

$$\lambda_{min}(Q) - \varepsilon P^2 - \tfrac{1}{\varepsilon}\|DC^T\phi\|^2 > 0. \quad (6.57)$$

Theorem 6.9 *If*

$$\Delta_{20} = \lambda_{min}^2(Q) - 4\|P\|^2\|DC^T\phi\|^2 > 0,$$
$$\varepsilon_{21}^1 = max(0, \tfrac{\lambda_{min}(Q)-\Delta_2^{\frac{1}{2}}}{2\|P\|^2}), \varepsilon_{22}^2 = \tfrac{\lambda_{min}(Q)+\Delta_2^{\frac{1}{2}}}{2\|P\|^2} \ge 0, \quad (6.58)$$

(1) For any $\varepsilon \in D_{20}(Q) = \{\varepsilon|(\varepsilon_{21}^1, \varepsilon_{22}^2), Q^T = Q > 0\}$, then

$$\tau_{20} < \frac{\lambda_{min}(Q)-\varepsilon P^2 - \tfrac{1}{\varepsilon}\|DC^T\phi\|^2}{\varepsilon(p+2)\sum_{i=1}^{p}\|PB_{i0}K_i\|^2 + \tfrac{p\alpha^2 l^2}{\varepsilon}(\|A_0\|^2 \sum_{i=1}^{p}\|B_{i0}K_i\|^2 + \|DC^T\phi\|^2)}. \quad (6.59)$$

Further, the origin of (6.2) is absolutely stable.

(2) For any $0 < \varepsilon \le \varepsilon_{21}^1$ or $\varepsilon \ge \varepsilon_{22}^2$, $\Delta_1 > 0$, (6.58) does not hold, and the origin of (6.2) is unstable;

(3) If $\Delta_2 \le 0$, that is to say, there does not exist any ε to satisfy (6.58), and the origin of (6.2) is unstable.

From a Lyapunov equation $A^T P + PA = -Q$, $Q^T = Q > 0$, we learn the fact that Q is not unique in general.

Note 6.2 From the above discussions, we learn the standard procedure on solvable conditions as follows:

First, we need to compute all possible inequalities and to find its corresponding solvable inequalities;

Second, we need to decide their all solvable conditions $D(Q) = \cup_i D(Q_i)$;

Third, using all possible inequalities, we obtain all possible delay $F = \tau|\tau \in [\tau_a, \tau_b]$, where $\tau_a = min(0, \tau(D_i))$, $\tau_b = max\tau(D_i)$, $\tau_b \ge \tau_a$;

Finally, by choosing the next different Q_i and repeating the above process, we may obtain more and more accurate τ.

Fig. 6.1 $\varepsilon(Q_1) - \tau$ function

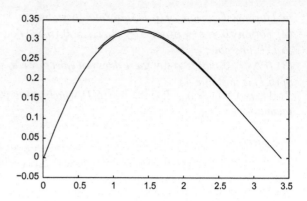

In this chapter, if we adapt other vector norm's definitions, such as $\|\alpha\|_1, \|\alpha\|_2, \|\alpha\|_\infty$ ($\forall \alpha \in \mathbb{R}^n$) as applications to deal with the above solvable problem, the obtained results are similar to the existing ones in our chapter, and we omit these discussions.

6.5 Numerical Examples

In order to illustrate the above results, we only consider the following Lurie system with time-varying delay

$$\begin{cases} \dot{x}(t) = A(t)x(t) + B(t)u(t) + Df(\delta(t)), \\ u(t) = x(t - \tau(t)), \\ \sigma(t) = Cx(t). \end{cases} \quad (6.60)$$

with constant matrices $A = \begin{bmatrix} -2 & 0 \\ -1 & -2 \end{bmatrix}$, $B = \begin{bmatrix} -0.2 & -0.5 \\ 0.5 & -0.2 \end{bmatrix}$, $C = \begin{bmatrix} 0.6 \\ 0.8 \end{bmatrix}$, $D = \begin{bmatrix} -0.2 \\ -0.3 \end{bmatrix}$, $\sigma(t) = 0.6x_1(t) + 0.8x_2(t)$, $f \in F_{[0,0.5]}$.

From the above results, we know that τ is the function about matrix Q.

We investigate $\varepsilon(Q) - \tau$ by choosing the different Q as following according to Theorems 6.7 and 6.9 (the above curve is τ_1 from (6.51) and the below curve is τ_2 from (6.59))

(1) Let $Q_1 = \begin{bmatrix} -4.4 & -1 \\ -1 & -4.4 \end{bmatrix}$, then it follows that

$$P_1 = I, \tau_{1,max} = 0.3268, \tau_{2,max} = 0.3228,$$

and Fig. 6.1.

(2) Let $Q_2 = \begin{bmatrix} 1 & 0 \\ 0 & 1 \end{bmatrix}$, then it follows that

6.5 Numerical Examples

Fig. 6.2 $\varepsilon(Q_2) - \tau$ function

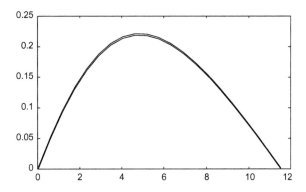

Fig. 6.3 $\varepsilon(Q_3) - \tau$ function

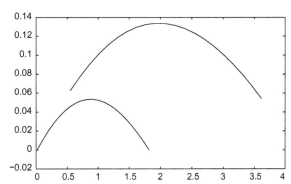

$$P_2 = \begin{bmatrix} 0.2397 & -0.0545 \\ -0.0545 & 0.2397 \end{bmatrix}, \tau_{1,max} = 0.2212, \tau_{2,max} = 0.2186,$$

and Fig. 6.2.

(3) Let $Q_3 = \begin{bmatrix} 2.2 & 1.0 \\ 1.0 & 4.4 \end{bmatrix}$, then it follows that

$$P_3 = \begin{bmatrix} 0.4864 & -0.0599 \\ -0.0599 & 0.9864 \end{bmatrix}, \tau_{1,max} = 0.1336, \tau_{2,max} = 0.0534,$$

and Fig. 6.3.

(4) Let $Q_4 = \begin{bmatrix} 2.2 & 1.0 \\ 1.0 & 6.4 \end{bmatrix}$, then it follows that

$$P_4 = \begin{bmatrix} 0.4988 & -0.054 \\ -0.054 & 1.4533 \end{bmatrix}, \tau_{1,max} = 0.0819, \tau_{2,max} = 0.0210,$$

and Fig. 6.4.

(5) Let $Q_5 = \begin{bmatrix} 1.2 & 1.0 \\ 1.0 & 8.4 \end{bmatrix}$, then it follows that

Fig. 6.4 $\varepsilon(Q_4) - \tau$ function

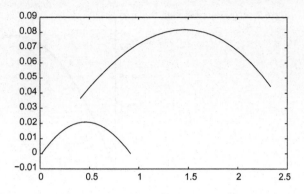

Fig. 6.5 $\varepsilon(Q_5) - \tau$ function

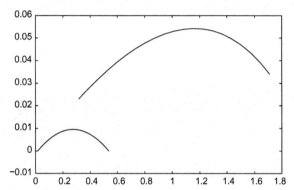

$$P_5 = \begin{bmatrix} 0.2777 & -0.0218 \\ -0.0218 & 1.9140 \end{bmatrix}, \tau_{1,max} = 0.0176, \tau_{2,max} = 9.3925 \times 10^4,$$

and Fig. 6.5.

(6) Let $Q_6 = \begin{bmatrix} 0.8000 & 1.000 \\ 1.000 & 1.9116 \end{bmatrix}$, then it follows that

$$P_6 = \begin{bmatrix} 0.1483 & -0.0109 \\ -0.0109 & 1.9116 \end{bmatrix}, \tau_{1,max} = 0.0061, \tau_{2,max} = 9.0757 \times 10^4,$$

and Fig. 6.6.

6.6 Conclusion

In order to get the reliable results, this chapter provides many absolute stable criteria of Lurie nonlinear system with time-varying plant and time-varying actuator under time-varying delay feedback. Many interesting results are outlined as follow: First,

Fig. 6.6 $\varepsilon(Q_6) - \tau$ function

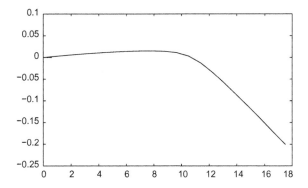

by defining solvable matrix inequalities and constructing a Lyapunov–Razumikhn functional, we analyze these solvable matrix inequalities, and give their sufficient and necessary solvable conditions that are easily computed in practical systems. In applications, based on our norm definition, solvable conditions and time-delay bound estimations can be obtained from two cases of solvable norm inequalities. More importantly, we draw a conclusion on the above procedures about the optimal combination about a pair of (τ, Q). Finally, an interested numerical example is presented to illustrate the above results effectively.

Acknowledgements This work is Supported by National Key Research and Development Program of China (2017YFF0207400).

References

1. Bliman PA. Lyapunov-Krasovskii functionals and frequency domain: delay-independent absolute stability criteria for delay systems. Int J Robust Nonlinear Control. 2001;11(8):771–88.
2. Bliman PA. Stability criteria for delay systems with sector-bounded nonlinearities. In: American control conference, proceedings of the IEEE, vol. 1; 2001. p. 402–7.
3. Cao J, Zhong S, Hu Y. Delay-dependent condition for absolute stability of Lurie control systems with multiple time delays and nonlinearities. J Math Anal Appl. 2006;338(1):497–504.
4. Gao JF, Pan HP, Ji XF. A new delay-dependent absolute stability criterion for Lurie systems with time-varying delay. Acta Automatica Sinica. 2010;36(6):845–50.
5. Gu K, Chen J, Kharitonov VL. Stability of time-delay systems. Berlin: Springer; 2003.
6. Guo SL, Chu TG, Huang L. Absolute stability of the origin of Lurie-type nonlinear systems with MIMO bounded time-delays. J Eng Math. 2002;19(3):1–6.
7. Hale JK. Introduction to functional differential equations, vol. 99. Berlin: Springer; 1993.
8. Hale JK, Lunel SMV. Effects of small delays on stability and control. In: Operator theory and analysis. Birkhauser, Basel; 2001. p. 275–301.
9. Han QL. Absolute stability of time-delay systems with sector-bounded nonlinearity. Automatica. 2005;41(12):2171–6.
10. Lur'e AI. Some non-linear problems in the theory of automatic control. HM Stationery Office; 1957.
11. Malek-Zavarei M, Jamshidi M. Time-delay systems: analysis, optimization and applications. Amsterdam: Elsevier Science Inc; 1987.

12. Mukhija P, Kar IN, Bhatt RKP. Delay-distribution-dependent robust stability analysis of uncertain Lurie systems with time-varying delay. Acta Automatica Sinica. 2012;38(7):1100–6.
13. Qiu F, Zhang Q. Absolute stability analysis of Lurie control system with multiple delays: an integral-equality approach. Nonlinear Anal Real World Appl. 2011;12(3):1475–84.
14. Tian J, Zhong S, Xiong L. Delay-dependent absolute stability of Lurie control systems with multiple time-delays. Appl Math Comput. 2007;188(1):379–84.
15. Wang JZ, Duan ZS, Yang Y, Huang L. Analysis and control of nonlinear systems with stationary sets: time-domain and frequency-domain methods. Singapore: World Scientific; 2009.
16. Xue M, Fei S, Li T, Pan J. Delay-dependent absolute stability criterion for Lurie system with probabilistic interval time-varying delay. J Control Theory Appl. 2012;10(4):477–82.
17. Yakubovich VA, Leonov GA, Gelig AK. Stability of stationary sets in control systems with discontinuous nonlinearities. Singapore: World Scientific; 2004.
18. Yin C, Zhong SM, Chen WF. On delay-dependent robust stability of a class of uncertain mixed neutral and Lure dynamical systems with interval time-varying delays. J Franklin Inst. 2010;347(9):1623–42.
19. Yu P. Absolute stability of nonlinear control systems, vol. 25. Berlin: Springer; 2008.
20. Yu L. On the absolute stability of a class of time-delay systems. Acta Automatica Sinica. 2003;5(29):780–4.
21. Yu L, Han QL, Yu S, Gao J. Delay-dependent conditions for robust absolute stability of uncertain time-delay systems. In: 42nd IEEE conference on decision and control, 2003. Proceedings, vol. 6; 2003. p. 6033–6037.
22. Zhu S, Zhang C, Cheng Z, Feng JE. Delay-dependent robust stability criteria for two classes of uncertain singular time-delay systems. IEEE Trans Autom Control. 2007;52(5):880–5.

Chapter 7
Stability Criteria on a Type of Differential Inclusions with Nonlinear Integral Delays

7.1 Introduction

Generally, it is difficult to describe the dynamic behaviors of real systems precisely using mathematical equations that consist only of known constant parameters since systems often have an inherent uncertainty. For governing many dynamical systems, a stabilization method of mathematical system models that include uncertain parameters must be developed. For this reason, a considerable attention has been paid for robust stabilization of uncertain systems [1–24]. In most instances, physical, chemical, biological, and economical phenomena depend naturally not only on the present state but also on past occurrences. The importance of time delays in the stability analysis is well recognized in a wide range of applications [7, 10]. Therefore, this chapter specifically examines the stability problem of uncertain differential inclusions with delays. For nonlinear systems with delays, the Lyapunov stability approach with the Krasovskii-based or Razumikhin-based method is a commonly used tool. Generally, the above problem has been reduced to solving linear matrix inequalities (LMI) [7, 10], which necessitates complicated numerical calculations on the algorithm, even though a convenient software package is available. Moreover, LMI conditions fall into the first category; for this reason, they are often used to determine the permissible ranges of uncertain parameters for the stabilization. When the ranges of uncertain parameter values exceed a certain value, LMI solver becomes infeasible. In such cases, guidelines for redesigning other style Lyapunov functions are usually lacking. In this chapter, we give a special-type Lyapunov functional to solve the delay differential inclusion problem. Time delays are important components of many dynamical systems that describe coupling or interconnection between dynamics, propagation or transport phenomena, and heredity and competition in population dynamics. It presents a wide and self-contained panorama of analytical methods and computational algorithms using a unified eigenvalue-based approach illustrated by examples and applications in electrical and mechanical engineering, biology, and complex network analysis. It bridges the fields of control (analysis and feedback design, robustness, and uncertainty) and numerical analysis (explicit algorithms and methods). Solutions of the (robust) stability analysis and stabilization problem of linear time-delay systems are present, which are the result of this cross-fertilization of control theory, numerical linear algebra, numerical bifurcation analysis, and optimization.

The Piola–Kirchhoff stress tensors in mechanical engineering are used to express the stress relative to the reference configuration. For infinitesimal deformations or rotations, the Cauchy and Piola–Kirchoff tensors are identical. We investigate that the domain of definition of the stress response function is denoted by the integral delays. In generality, the above function may been described into the following dynamical systems [7]

$$\dot{x}(t) = Ax(t) + \int_0^T B(\tau)x(t-\tau)d\tau. \tag{7.1}$$

On the other hand, discontinuities are present in a wide variety of systems; e.g., friction in mechanical systems, diodes in electrical circuits, or switching control systems. Models of such systems are commonly formulated as differential equations with discontinuous right-hand sides. Such systems can be formulated as differential inclusions with a set-valued right-hand side [1, 4, 6–10, 14–17, 19–22]. That is to say

$$\dot{x}(t) = A_s x(t), A_s \in \{A_1 \ A_2 \ \cdots \ A_Q\}. \tag{7.2}$$

We consider the following system that denotes longitudinal oscillation cases on the switched plant with beams or some mechanical systems with friction and unilateral contact, electrical networks with diodes, and switches or discontinuities in embedded control systems originating from switches in the software [7]

$$\dot{x}(t) = A_s x(t) + \int_0^T B(\tau)x(t-\tau)d\tau, A_s \in \{A_1 \ A_2 \ \cdots \ A_Q\}. \tag{7.3}$$

In general cases, equilibria, periodic and quasiperiodic solutions of differential inclusions represent important classes of limit solutions for such systems (7.2) or (7.3). The equilibria may range from one or multiple isolated equilibrium points, either on or outside the discontinuity surfaces to equilibrium sets located on the discontinuity surfaces. Likewise, the (quasi-)periodic solutions can either lie entirely in the smooth domain of the state-state space or hit (or partially slide along) the discontinuity surface. The stability properties of these limit solutions largely influence the global dynamics of the system. Therefore, the development of a stability theory for such limit solutions is a key problem. Now, we only need to discuss the asymptotically stable of the differential inclusions (7.2) or (7.3) by Lyapunov functional method, which is one of important methods [1–22] such as

$$V(t,x) = x^T P x + \int_0^T x(t-\tau)^T Sx(t-\tau)d\tau, \tag{7.4}$$

7.1 Introduction

where $P^T = P > 0, S^T = S > 0$. Further the differential inclusion (7.2) or (7.3) can be converted into LMI problems using the Lyapunov functional method (7.4). But when the ranges of uncertain parameter values exceed a certain value, LMI solver becomes infeasible. In such cases, guidelines for redesigning other style Lyapunov functions are usually lacking.

In this chapter, we shall discuss that there exists the following type polyhedron Lyapunov functional

$$V(t, x) = v_m(l, x) + \int_0^T x(t-\tau)^T Sx(t-\tau)d\tau, \tag{7.5}$$

where $v_m(l, x) = \max\limits_{1 \le i \le m} \langle l, x \rangle^2$ (seeing [19]) and $S(-\tau)$ is variable parameter-positive matrix. And it is clear that the polyhedron Lyapunov functional (7.5) may turn the differential inclusion problems (7.2) or (7.3) into the scalar computation problems which are easily computed and does not need to discuss the solution of LMI. Moreover, the above conclusions are similarly generalized to the linear matrix inequalities cases.

According to definitions of papers [1, 8], we give out the following definitions on the asymptotically stable of the differential inclusions.

Definition 7.1 The asymptotically stable of the zero solution of the differential inclusion (7.2) is that

(1) For every $\varepsilon > 0$, there exists $\delta(\varepsilon) > 0$ such that when $t \ge t_0$ and $\|x(t_0)\| < \delta(\varepsilon)$, $\|x(t)\| < \varepsilon$ holds;

(2) If there exists $\theta > 0$ such that $\|x_0\| < \theta(x_0 = x(t_0))$, then $\lim\limits_{t \to \infty} x(t) = 0$ holds.

Theorem 7.1 ([18]) *The asymptotically stable of the zero solution of the differential inclusion (7.2) is sufficient and necessary for the asymptotically stable of the following differential inclusion*

$$\dot{x} \in F(x), F(x) = \{y : y = Ax, A \in co(\Delta), co(\Delta) = co(A_1, A_2, \ldots, A_q)\}, \tag{7.6}$$

where $A_1, A_2, \ldots, A_q$ denote the points of the polyhedron convex cone $co(\Delta)$.
Let its derivative with respect to time be

$$\mu = \max_{y \in F(x)} \frac{dv(x)}{dy},$$

where $\frac{dv(x)}{dy} = \lim\limits_{t \to 0^+} \frac{v(x+hy)}{h}$.

Theorem 7.2 ([14]) *We consider the asymptotically stable of the zero solution of the following differential inclusion (7.2) is sufficient and necessary that there exists a strictly convex singular polyhedron Lyapunov functional that satisfies*

$$\begin{cases} v(x) = x^T P(x)x, \\ P(x) = (l_{ij}(x))_{i,j=1}^n, \\ P^T(x) = P(x) = P(\tau x), x \neq 0, \|\tau\| = 1, \\ v(0) = 0, \end{cases} \quad (7.7)$$

whose derivative with respect to time is

$$\mu = \max_{y \in F(x)} \frac{dv(x)}{dy} \leq -\gamma \|x\|^2, \gamma > 0. \quad (7.8)$$

7.2 Algebraic Criteria of Asymptotic Stability

In this section, we shall consider asymptotically stable of the zero solution of the weighted integrated nonlinear differential inclusions as follows:

$$\begin{cases} \dot{x} \in F_s(t, x), s = 1, 2, \ldots, q, \\ F_s(t, x) = \{y : y = A_s x(t) + \int_0^T B(\tau) x(t - \tau) d\tau\}, \\ A_s \in \{A_1 \ A_2 \ \cdots \ A_q\}, \end{cases} \quad (7.9)$$

where $A_s, B(\tau) \in \mathbb{R}^{n \times n}, s = 1, 2, \ldots, q$.

Theorem 7.3 ([18]) *We consider the asymptotically stable of the zero solution of the following differential inclusion*

$$\dot{x} \in F(x), F(x) = \{y : y = Ax, A \in co(\Delta) = co(A_1, A_2, \ldots, A_q)\},$$

which is sufficient and necessary that there exists a number $m \geq n$, an $n \times m$ matrix L, and an $m \times m$ matrix

$$\tau_s = (\gamma_{ij}^{(s)})_{i,j=1}^m, s = 1, \ldots, q,$$

whose every column strictly satisfies

$$\gamma_{ii}^{(s)} + \sum_{j \neq i} \left|\gamma_{ij}^{(s)}\right| < 0, i = 1, \ldots, m, s = 1, \ldots, q,$$

and

$$A_s^T L = L \tau_s^T, s = 1, 2, \ldots, q. \quad (7.10)$$

7.2 Algebraic Criteria of Asymptotic Stability

Now, we choose the following polyhedron Lyapunov functional

$$v_m(l, x) = \max_{1 \le i \le m} \langle l, x \rangle^2, \text{ or } V = x^T L L^T x(t),$$

where l_i, $(i = 1, \ldots m)$ are all n dimensions constant vectors that satisfy (7.10) and L is defined as the above.

We can line out the following theorem using the above polyhedron Lyapunov functionals.

Theorem 7.4 *The asymptotically stable of the zero solution of the differential inclusion (7.2) is sufficient and necessary that there exists a strictly convex polyhedron Lyapunov functional $v_m(l, x)$ (or $V = x^T L L^T x(t)$) which satisfies*

$$\mu_m(l, x) = \max_{y \in F(x)} \frac{dv_m(l,x)}{dy} < 0, \text{ (or } \dot{V}(x) = \frac{dV(L,x)}{dt} < 0). \tag{7.11}$$

The proof, which is readily obtained via Theorems 7.1 and 7.2, and noting Definition 7.1 we omitted their discussion here. □

Theorem 7.5 *Given differential inclusion (7.9), let*
(1) *There exists an $n \times m$ matrix L (whose order is n) and a $m \times m$ matrix $\tau_s = (\gamma_{ij}^{(s)})_{i,j=1}^m$, $s = 1, \ldots, q$ which satisfies (7.10);*
(2) *There exists a differential matrix*

$$Q^T(t) = Q(t) > 0,$$
$$Q(T) = 0,$$
$$\frac{dQ(t)}{dt} = -C^T(t)C(t).$$

Let a Lyapunov functional $V : C([-\tau, 0], \mathbb{R}^n) \to \mathbb{R}^+$ satisfy

$$V(\varphi) = v_m(l, \varphi) + \int_{-T}^0 \varphi^T(\theta) Q(-\theta) \varphi(\theta) d\theta,$$
$$\varepsilon_0 = \varepsilon + m \frac{\lambda_{\max}(Q(0))}{\lambda_{\min}(LL^T)} + \max_{1 \le i \le m} (l_i^T \int_0^T B^T(\tau) C^{-1}(\tau) C^{-T}(\tau) B(\tau) d\tau l_i) < 0,$$

$$\tag{7.12}$$

where $\varepsilon = \max_{s,i}(\tau(\gamma_{ii}^{(s)}) + \sum_{j \ne i} \left| \tau(\gamma_{ij}^{(s)}) \right|) < 0$, then the zero solution of (7.9) is uniformly asymptotically stable.

Proof Let a Lyapunov functional

$$V(\varphi) = v_m(l, \varphi) + \int_{-T}^0 \varphi^T(\theta) Q(-\theta) \varphi(\theta) d\theta, \tag{7.13}$$

whose derivative with respect to time along (7.9) is

$$\frac{dV(x(t))}{dt}\bigg|_{((7.9))}$$
$$= \max_{y \in F(x)} \frac{\partial v_m(l,x)}{\partial y} + x^T(t)Q(0)x(t) - x^T(t-T)Qx(t-T)$$
$$+ \int_{t-T}^{t} x^T(\tau) \frac{\partial Q(t-\tau)}{\partial t} x(t-\tau) d\tau$$
$$\leq \max_{1 \leq s \leq q, 1 \leq i \leq m} (\gamma_{ii}^{(s)} + \sum_{j \neq i} |\gamma_{ij}^{(s)}|) v_m(l,x) + \max_{1 \leq s \leq q, 1 \leq i \leq m} \frac{\lambda_{\max}(Q(0))}{\lambda_{\min}(LL^T)} \sum_{I=1}^{M} \langle l_i, x \rangle^2 \quad (7.14)$$
$$+ 2 \max_{1 \leq s \leq q, 1 \leq i \leq m} \langle l_i, x \rangle \langle l_i, \int_0^T B(\tau) x(t-\tau) d\tau \rangle$$
$$- \int_0^T x^T(\tau) C(t-\tau) C^T(t-\tau) x(\tau) d\tau.$$

Let
$$\xi(t,\tau) = x^T(t-\tau)C(\tau) - x^T(t) \max_{1 \leq s \leq q, 1 \leq i \leq m} (l_i l_i^T B(\tau) C^{-1}(\tau)).$$

Then (7.14) can be converted into

$$\frac{dV(x(t))}{dt}\bigg|_{(7.9)}$$
$$\leq \max_{1 \leq s \leq q, 1 \leq i \leq m} (\gamma_{ii}^{(s)} + \sum_{j \neq i} |\gamma_{ij}^{(s)}|) v_m(l,x) + \max_{1 \leq s \leq q, 1 \leq i \leq m} \frac{\lambda_{\max}(Q(0))}{\lambda_{\min}(LL^T)} \sum_{I=1}^{M} \langle l_i, x \rangle^2$$
$$+ \max_{1 \leq s \leq q, 1 \leq i \leq m} x^T(t) l_i l_i^T \int_0^T B(\tau) C^{-1}(\tau) C^{-T}(\tau) B(\tau) d\tau l_i l_i^T x(t) \quad (7.15)$$
$$- \int_0^T \xi^T(t-\tau) \xi(t-\tau) d\tau$$
$$=: \varepsilon_0 v_m(l,x) - \int_0^T \xi^T(t-\tau) \xi(t-\tau) d\tau.$$

So
$$\frac{dV(x(t))}{dt}\bigg|_{(7.9)} \leq \varepsilon_0 v_m(l,x).$$

Associating the definitions and basic theorems of functional equations, we know that the solution of (7.9) is uniformly asymptotically stable. □

Theorem 7.6 *Given (7.9), let*
(1) The assumption (1) of Theorem 7.5 holds;
(2) There exists a common polyhedron Lyapunov function $v_m(l,x)$, a positive definite symmetric matrix R and nonsingular function matrix $C(.) : [0,T] \to \mathbb{R}^{n \times n}$ that satisfy

$$\left\| \int_0^T C^T(t)C(t) dt \right\| < \infty,$$
$$\varepsilon_0 = \varepsilon + m \frac{\lambda_{\max}(Q(0))}{\lambda_{\min}(LL^T)} + \max_{1 \leq i \leq m} (l_i^T \int_0^T B(\tau) C^{-1}(\tau) C^{-T}(\tau) B(\tau) d\tau l_i) < 0.$$

then the zero solution of (7.9) is uniformly asymptotically stable.

7.2 Algebraic Criteria of Asymptotic Stability

Proof From the procedure of Theorem 7.5 and the definition of $Q(0)$, we can obtain

$$\frac{dV(x(t))}{dt}\bigg|_{(7.9)} \leq (\varepsilon + m \frac{\lambda_{\max}(Q(0))}{\lambda_{\min}(LL^T)} + \max_{1 \leq i \leq m}(l_i^T \int_0^T B(\tau)C^{-1}(\tau)C^{-T}(\tau)B(\tau)\,d\tau l_i))v_m(l, x)$$
$$- \int_0^T \xi^T(t-\tau)\xi(t-\tau)d\tau,$$

Associating with Theorem 7.5, the desired result can be obtained. □

Definition 7.2 Let B be a positive definite symmetric matrix, the zero solution of (7.9) is called B-robust asymptotically stable when all delays belong to the region $[0, T]$, and the weighted positive definite matrix function $B(.) : [0, T] \to \mathbb{R}^n$ satisfies

$$\min_{\beta(.)} \int_0^T \frac{B(\tau)B^T(\tau)}{\beta(\tau)}d\tau \int_0^T \beta(\tau)d\tau \leq B, \quad (7.16)$$

the zero solution of (7.9) is always asymptotically stable, where $\beta(\tau)$ is a positive definite function.

Given system (7.9), assume that (A)

$$\int_0^T \beta(\tau)d\tau = 1, \beta(\tau) \in \mathbb{R}^+, \tau \in \mathbb{R},$$
$$C(t) = \sqrt{\beta(t)}I > 0.$$

We can obtain the following results.

Theorem 7.7 *Under the conditions of Theorem 7.5 or 7.6, let*

$$\varepsilon + m \frac{\lambda_{\max}(\int_0^T C(\tau)C^T(\tau)d\tau)}{\lambda_{\min}(LL^T)} + \max_{1 \leq i \leq m}(l_i^T B l_i) < 0, \quad (7.17)$$

then the zero solution of (7.9) is B-robust asymptotically stable.

Proof Let $C(t) = \sqrt{\beta(t)}I > 0$, associated the Assumption (A) and Definition 7.2, we easily get

$$l_i^T B l_i \geq l_i^T \int_0^T \frac{B(\tau)B^T(\tau)}{\beta(\tau)}d\tau l_i$$
$$\geq l_i^T \int_0^T B(\tau)C^T(\tau)C(\tau)B(\tau)d\tau l_i. \quad (7.18)$$

It obviously obtained the desired result. □

Similarly, given A_s is a $n \times n$ matrix in (7.9) and L is defined as (7.10), there exists a positive definite symmetric matrix $P^T = P > 0$ (for example, $P = LL^T$) such that $A_s^T P + PA_s < 0, A_s \in co(A)$, we easily outline the following results.

Theorem 7.8 *Under the conditions of Theorem 7.6, let*

$$Q^T(t) = Q(t) > 0,$$
$$Q(T) = 0,$$
$$\frac{dQ(t)}{dt} = -C^T(t)C(t),$$

and a Lyapunov functional $V : C([-\tau, 0], \mathbb{R}^n) \to \mathbb{R}^+$ *as follows:*

$$V(\varphi) = \varphi^T(0)P\varphi(0) + \int_{-\tau}^{0} \varphi^T(\theta)Q(-\theta)\varphi(\theta)d\theta,$$

whose derivative with respect to time along (7.9) gives

$$\frac{dV(x(t))}{dt}\Big|_{(7.9)} \leq -x^T(t)R_s x(t),$$

where

$$-R_s = A_s^T P + P A_s + C^T P C + Q(0) + P \int_0^T B(\tau)C^T(\tau)C(\tau)B(\tau)d\tau P,$$

$s = 1, \ldots, q$, *then the zero solution of (7.9) is B-robust asymptotically stable.*

Theorem 7.9 *For Theorem 7.7, if the sufficient conditions of Theorem 7.8 hold and if* $\left\| \int_0^T C^T(t)C(t)dt \right\| < \infty$ *and*

$$0 = A_s^T P + P A_s + C^T P C + Q(0) + P \int_0^T B(\tau)C^T(\tau)C(\tau)B(\tau)d\tau P + R_s,$$

where $s = 1, \ldots, q$, *then the zero solution of (7.9) is B-robust asymptotically stable.*

Theorem 7.10 *For Theorem 7.9, if*

$$A_s^T P + P A_s + I + P B P < 0,$$

where $s = 1, \ldots, q$, *then the zero solution of (7.9) is B-robust asymptotically stable.*

7.3 Numerical Example Analysis

Example 7.1 We consider the following system

$$\begin{cases} \dot{x} \in F_s(t, x), s = 1, 2, 3, 4, \\ F_s(t, x) = \left\{ y : y = A_s x(t) + \int_0^T B(\tau)x(t-\tau)d\tau \right\}, \\ A_s \in \{A_1, A_2, A_3, A_4\}, \end{cases} \quad (7.19)$$

7.3 Numerical Example Analysis

where

$$A_1 = \begin{bmatrix} -3 & 0 \\ 0 & -3 \end{bmatrix}, \tau_1 = \begin{bmatrix} -3 & 0 & 0 \\ 0 & -3 & 0 \\ 0 & 0 & -3 \end{bmatrix},$$

$$A_2 = \begin{bmatrix} 14 & 40 \\ -15 & -36 \end{bmatrix}, \tau_2 = \begin{bmatrix} -6 & 0 & -2 \\ 0 & -6 & 0 \\ 0 & 0 & -16 \end{bmatrix},$$

$$A_3 = \begin{bmatrix} 24 & 56 \\ -21 & -46 \end{bmatrix}, \tau_3 = \begin{bmatrix} -8 & 0 & -2 \\ 0 & -4 & 0 \\ 0 & 0 & -18 \end{bmatrix},$$

$$A_4 = \begin{bmatrix} 33 & 72 \\ -27 & -57 \end{bmatrix}, \tau_4 = \begin{bmatrix} -6 & 0 & -3 \\ 0 & -3 & 0 \\ 0 & 0 & -21 \end{bmatrix},$$

$$L = \begin{bmatrix} l_1 & l_2 & l_3 \end{bmatrix} = \begin{bmatrix} 1 & 3 & 5 \\ 2 & 4 & 10 \end{bmatrix}, B = \begin{bmatrix} 0.05e^{-0.5t} & 0 \\ 0 & 0 \end{bmatrix},$$

$$T = 5, 0 \leq t \leq T,$$

which satisfy $A_s^T L = L\tau_s^T$, $s = 1, 2, 3, 4$.

We suppose

$$x(t) = x(0), -T \leq t \leq 0,$$
$$Q(t) = \begin{bmatrix} 0.1(T-t) & 0 \\ 0 & 0.1(T-t) \end{bmatrix}, 0 \leq t \leq T$$

which satisfies

$$Q(t) = Q^T(t) > 0, (0 \leq t < T),$$
$$Q(T) = 0,$$
$$\frac{dQ(t)}{dt} = -C^T(t)C(t) = -0.1I.$$

We obtain

$$\varepsilon = \max_{4,i}(\tau(\gamma_{ii}^{(s)}) + \sum_{j \neq i} |\tau(\gamma_{ij}^{(s)})|) = -3,$$

$$m \frac{\lambda_{\max}(Q(0))}{\lambda_{\min}(LL^T)} = \frac{0.3T}{0.6739},$$

$$\max_{1 \leq i \leq m} (l_i^T \int_0^T B^T(\tau) C^{-1}(\tau) C^{-T}(\tau) B(\tau) d\tau l_i)$$

$$= \max_{1 \leq i \leq m} \left(l_i^T \begin{bmatrix} -\frac{1}{40}e^{-T} + \frac{1}{40} & 0 \\ 0 & 0 \end{bmatrix} l_i \right)$$

$$= \max(-\frac{1}{40}e^{-T} + \frac{1}{40}, -\frac{9}{40}e^{-T} + \frac{9}{40}, -\frac{5}{8}e^{-T} + \frac{5}{8})$$

$$= -\frac{5}{8}e^{-T} + \frac{5}{8}.$$

So

$$\varepsilon_0 = \varepsilon + m\frac{\lambda_{\max}(Q(0))}{\lambda_{\min}(LL^T)} + \max_{1\leq i\leq m}(l_i^T \int_0^T B^T(\tau)C^{-1}(\tau)C^{-T}(\tau)B(\tau)d\tau l_i)$$
$$= -3 + \frac{0.3T}{0.6739} - \frac{5}{8}e^{-T} + \frac{5}{8}$$
$$= -0.153$$
$$< 0.$$

According to the Theorem 7.5, if we choose the Lyapunov function as $V(\varphi) = v_m(l, \varphi) + \int_{-T}^{0} \varphi^T(\theta)Q(-\theta)\varphi(\theta)d\theta$ then the zero solution of system (7.19) is uniformly asymptotically stable.

We illustrate the above results by given different $A_i (i = 1, 2, 3, 4)$ cases using MATLAB simulation.

(i) Let

$$A = \begin{cases} A_1, & 0 < t \leq 1 \text{ and } 5 < t \leq 6, \\ A_2, & 1 < t \leq 2 \text{ and } 6 < t \leq 7, \\ A_3, & 2 < t \leq 3, \\ A_4, & else. \end{cases}$$

Figures 7.1, 7.2 and 7.3 show phase portrait $x_2(x_1)$ and trajectories $x_1(t), x_2(t)$ passing initial the points(red dots), which satisfy $x_1^2 + x_2^2 = 0.64$.

(ii) Let

$$A = \begin{cases} A_3, & 1 < t \leq 2 \text{ and } 5 < t \leq 6, \\ A_4, & 2 < t \leq 4, \\ A_1, & 6 < t \leq 7, \\ A_2, & else. \end{cases}$$

Figures 7.4, 7.5, and 7.6 show phase portrait $x_2(x_1)$ and trajectories $x_1(t), x_2(t)$ passing initial the points (red dots), which satisfy $x_1^2 + x_2^2 = 0.64$.

Fig. 7.1 Trajectories of x_1

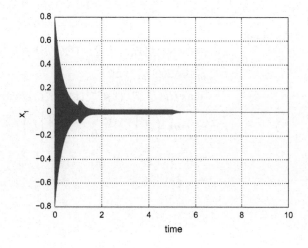

7.3 Numerical Example Analysis

Fig. 7.2 Trajectories of x_2

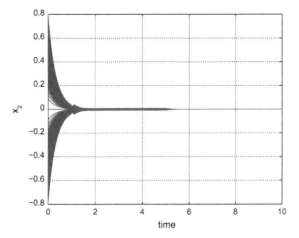

Fig. 7.3 Phase trajectories

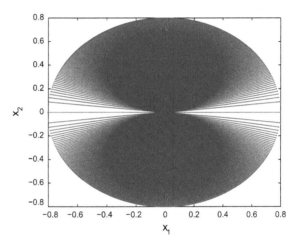

Fig. 7.4 Trajectories of x_1

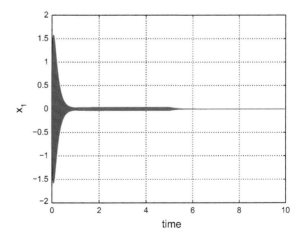

Fig. 7.5 Trajectories of x_2

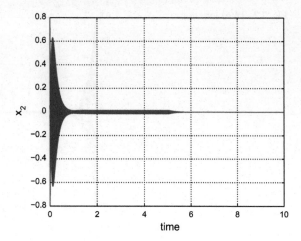

Fig. 7.6 Phase trajectories

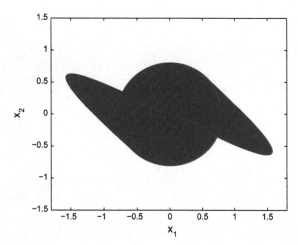

(iii) Let

$$A = \begin{cases} A_4, & 0 < t \leq 1 \text{ and } 3 < t \leq 4, \\ A_3, & 1 < t \leq 2 \text{ and } 6 < t \leq 7, \\ A_2, & 2 < t \leq 3 \text{ and } 5 < t \leq 6, \\ A_1, & else. \end{cases}$$

Figures 7.7, 7.8, and 7.9 show phase portrait $x_2(x_1)$ and trajectories $x_1(t)$, $x_2(t)$ passing initial the points(red dots), which satisfy $x_1{}^2 + x_2{}^2 = 0.64$.

7.3 Numerical Example Analysis

Fig. 7.7 Trajectories of x_1

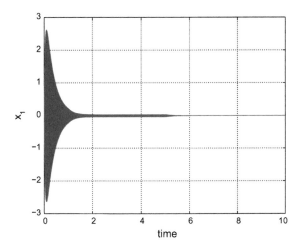

Fig. 7.8 Trajectories of x_2

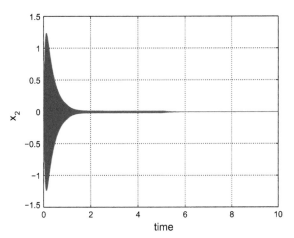

(iv) Let

$$A = \begin{cases} A_2, & 0 < t \le 1 \text{ and } 3 < t \le 4, \\ A_4, & 2 < t \le 3 \text{ and } 5 < t \le 6, \\ A_1, & 4 < t \le 5 \text{ and } 6 < t \le 7, \\ A_3, & else. \end{cases}$$

Figures 7.10, 7.11, and 7.12 show phase portrait $x_2(x_1)$ and trajectories $x_1(t)$, $x_2(t)$ passing initial the points(red dots), which satisfy $x_1^2 + x_2^2 = 0.64$.

Fig. 7.9 Phase trajectories

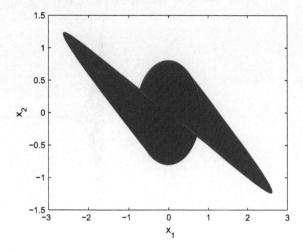

Fig. 7.10 Trajectories of x_1

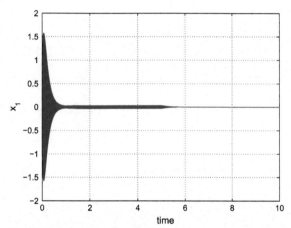

Fig. 7.11 Trajectories of x_2

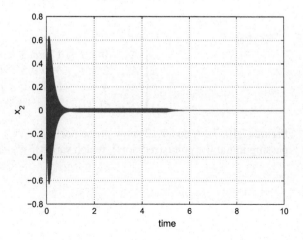

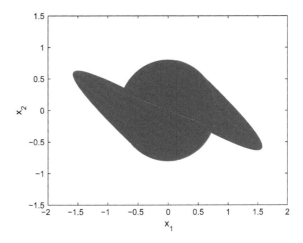

Fig. 7.12 Phase trajectories

7.4 Conclusion

It is very difficult to directly analyze the stability of differential inclusions. If all points of the convex polyhedron are found, then we can analyze the stability of convex points to obtain the desired results of the differential inclusions. The new method of constructing Lyapunov functional for a nonlinear delays differential inclusion is presented. The sufficient conditions on the asymptotic stability of the origin about the above differential inclusion is analyzed. Finally, an interesting example is at length presented to illustrate the main result effectively.

Acknowledgements This work is Supported by National Key Research and Development Program of China (2017YFF0207400).

References

1. Aubin JP, Cellina A. Differential inclusion. Berlin: Springer; 1984.
2. Bacciotti A, Rosier L. Regularity of Lyapunov functions for stable systems. Syst Control Lett. 2000;41(2):265–70.
3. Battilotti S. Robust stabilization of nonlinear systems with point wise norm-bounded uncertainties: a control Lyapunov function approach. IEEE Trans Autom Control. 1999;44(1):3–17.
4. Branicky MS. Multiple Lyapunov functions and other analysis tools for switched and hybrid systems. IEEE Trans Autom Control. 1998;43(4):475–82.
5. Boyd SP, Yang Q. Structured and simultanous Lyapunov functions for systems stability problem. Int J Control. 1995;49:2215–40.
6. Dayawansa WP, Martin CF. A converse Lyapunov theorem for a class of dynamical systems which undergo switching. IEEE Trans Autom Control. 1999;44(4):751–60.
7. Dugard L, Verrist EI. Stability and control of time-delay systems. Berlin: Springer; 1998. p. 303–17.
8. Filippov AF. Differential equations with discontinous right-hand side. Moscow: Nauda; 1985.

9. Guo S, Irene M, Si L, Han L. Several integral inequalities and their applications in nonlinear differential systems. Appl Math Comput. 2013;219:4266–77.
10. Guo S, Yan S, Wen S. Absolute Stability of Nonlinear Systems with MIMO Bounded Time-Delays. *American Society of Mechanical Engineers(ASME)*, Design Engineering Division, 2003,**116**(2).
11. Guo S, Irene M, Han L, Xin WF, Feng XJ. Commuting matrices, equilibrium points for control systems with single saturated input. Applied Mathematics and Computation. preprint.
12. Joaquain MC. Stability of the convex hull of commuting matrixes. In: Proceedings of the 32nd conference on decision and control. San Antonio, Texas; 1993.
13. Huang L. Stability theory. Beijing: Peking University Press; 1992. p. 235–83.
14. Horn RA, Johnson CR. Topics in matrix analysis. Cambridge: Cambridge University Press; 1991.
15. Kumpati SN, Jeyendran B. A common lyapunov function for stable LTI systems with commuting A-matrixes. IEEE Trans Autom Control. 1994;39(12):2469–71.
16. Lee SH, Kim TH, Lim JT. A new stability analysis of switched systems. Automatica. 2000;36:917–22.
17. Liberzon D, Hespanta JP, Morse AS. Stability of switched systems: a Lie-algebraic condition. Syst Control Lett North-Holland. 1999;37:117–22.
18. Mareada KS, Balakrishan J. A common Lyapunov function for stable LTI systems with commuting A-martices. IEEE Trans Autom Control. 1994;39(12):2469–71.
19. Molchanov AP, Pyatnitskiy YS. Criteria of asymptotic stability of differential and difference inclusions encountered in control theory. Syst Control Lett. 1989;13:59–64.
20. La Salle J, Lefschetz S. Stability by Lyapunov's direct method. New York: Academic Press; 1961.
21. Polanski K. On absolute stability analysis by polyhydric Lypunov functions. Automatica. 2000;36:573–8.
22. Schmitendorf WE, Barmish BR. Null controllability of linear systems with constrained controls. SIAM J Control Optim. 1980;18:327–45.
23. Shevitz D, Paden B. Lyapunov stability theory of non-smooth systems. In: Proceeding of the 32nd conference on decision and control. San Antonio, Texas; 1993, pp. 416–421.
24. Tatsushi O, Yasuyuki F. Two conditions concerning common quadratic lyapunov functions for linear systems. IEEE Trans Autom Control. 1997;42(5):719–21.

Part III
Integral Inequalities and Their Applications in Time-Varying Nonlinear Systems

Part III

Chapter 8
Several Integral Inequalities

8.1 Introduction

In the study of ordinary differential equations and integral equations, one often deals with certain integral inequalities. The Gronwall–Bellman inequality and its various linear and nonlinear generalizations are crucial in the discussion of the existence, uniqueness, continuation, boundedness, oscillation and stability, and other qualitative properties of solutions of differential and integral equations. The literature on such inequalities and their applications is vast; see [1–12] and the references given therein. This field on integral inequalities has developed significantly over the past decade, and has yielded many new results and powerful applications in numerical integration, probability theory and stochastics, statistics, information theory, integral operator theory, approximation theory, probability theory, stochastic, numerical analysis, applied mathematics, and even control theory and corresponding engineering problems [10–12]. The main aim of this chapter is to present a kind of integral inequalities, and the results for the stability analysis of nonlinear differential systems for both simple and multiple integrals are given.

In the chapter, several new integral inequalities are built in the first section, and it is easily found that the proposed integral inequalities are generalized to the existing integral inequalities, whose variable different moduli are more than 1. Compared with the existing integral inequalities in the references [1–11], our all proposed integral inequalities may deal with more complicated differential equations.

8.2 Two Integral Inequalities

The following theorem is one of the main results which is obtained in this chapter.

Theorem 8.1 *Assume that the following conditions for* $t \geq t_0$ *hold.*
(1) the constant $u_0 > 0$.
(2)

$$u(t) \leq u_0 + \int_{t_0}^t g_1(s)ds + \int_{t_0}^t g_2(s)u(s)^{\alpha_1}ds \\ + \sum_{j=1}^l \int_{t_0}^t b_j(s) \int_{t_0}^s c_j(\theta)u(\theta)d\theta ds \\ + \sum_{k=1}^m \int_{t_0}^t e_k(s) \int_{t_0}^s f_k(\theta)u(\theta)^{\alpha_1}d\theta ds, \quad (8.1)$$

where $g_i(t)(i=1,2), b_j(t), c_j(t)(j=1,2,\ldots,l), e_k(t), f_k(t)(k=1,2,\ldots,m)$ *and* $u(t)$ *are both nonnegative continuous functions and constant* $\alpha_1 > 0$.
(3) Let

$$p(t) = g_1(t) + \sum_{j=1}^l b_j(t) \int_{t_0}^t c_j(s)ds, \quad q(t) = g_2(t) + \sum_{k=1}^m e_k(t) \int_{t_0}^t f_k(s)ds$$

and if

$$1 + (1-\alpha_1)u_0^{\alpha_1-1} \int_{t_0}^t q(s) \exp((\alpha_1-1) \int_{t_0}^t p(\theta)d\theta)ds > 0$$

holds, then we have

$$u(t) \leq u_0 \exp(\int_{t_0}^t p(s)ds)\{1 \\ +(1-\alpha_1)u_0^{(\alpha_1-1)} \int_{t_0}^t q(\theta) \exp((\alpha_1-1) \int_{t_0}^t p(\theta))d\theta ds\}^{\frac{1}{1-\alpha_1}}, t \geq t_0. \quad (8.2)$$

Proof Let

$$M(t) = u_0 + \int_{t_0}^t g_1(s)u(s)ds + \int_{t_0}^t g_1(s)u(s)^{\alpha_1}ds \\ + \sum_{j=1}^l \int_{t_0}^t b_j(s) \int_{t_0}^s c_j(\theta)u(\theta)d\theta ds + \sum_{k=1}^m \int_{t_0}^t e_k(s) \int_{t_0}^s f_k(\theta)u(\theta)^{\alpha_1}d\theta ds, \quad (8.3)$$

then from the above assumption and Theorem 8.1(2), we clearly know that $u(t) \leq M(t), M(t_0) = u_0$, $M(t)$ is nondecreasing and

$$M'(t) = g_1(t)u(t) + g_2(t)u^{\alpha_1}(t) + \sum_{j=1}^l b_j(t) \int_{t_0}^t c_j(s)u(s)ds \\ + \sum_{k=1}^m e_k(s) \int_{t_0}^t f_k(s)u(s)^{\alpha_1}ds. \quad (8.4)$$

Since $M(t)$ is nondecreasing, (8.4) may be written as

8.2 Two Integral Inequalities

$$M'(t) = g_1(t)M(t) + g_2(t)M^{\alpha_1}(t) + \sum_{j=1}^{l} b_j(t) \int_{t_0}^{t} c_j(s)M(s)ds \\ + \sum_{k=1}^{m} e_k(s) \int_{t_0}^{t} f_k(s)M(s)^{\alpha_1}ds. \tag{8.5}$$

We have, from observing Theorem 8.1(3),

$$M'(t) \leq p(t)M(t) + q(t)M^{\alpha_1}(t). \tag{8.6}$$

Integrating (8.6) yields

$$M(t) \leq u_0 + \int_{t_0}^{t} p(s)M(s)ds + \int_{t_0}^{t} q(s)M^{\alpha_1}(s)ds. \tag{8.7}$$

Corollary 8.2 *Assume that for Theorem 8.1, if $b_j(t) = 0(c_j(t) = 0)(j = 1, \ldots, l)$, $e_k(t) = 0$, $(f_k(t) = 0)$, $(k = 1, \ldots, m)$, we obtain Lemma of [5].*

For Theorem 8.1, if $g_2(t) = 0$, $e_k(t) = 0$, $(f_k(t) = 0)(k = 1, \ldots, m)$, we obtain the following results:

Theorem 8.3 *Assume that the following conditions for $t \geq t_0$ hold*
(1) the constant $u_0 > 0$;
(2)

$$u(t) \leq u_0 + \int_{t_0}^{t} g_1(s)ds + \sum_{j=1}^{l} \int_{t_0}^{t} b_j(s) \int_{t_0}^{s} c_j(\theta)u(\theta)d\theta ds \tag{8.8}$$

where $g_1(t), b_j(t), c_j(t)(j = 1, 2, \ldots, l)$ and $u(t)$ are both nonnegative continuous functions;
(3) Let

$$p(t) = g_1(t) + \sum_{j=1}^{l} b_j(t) \int_{t_0}^{t} c_j(s)ds,$$

then we have

$$u(t) \leq u_0 \exp\left(\int_{t_0}^{t} p(s)ds\right), t \geq t_0.$$

Corollary 8.4 *Especially for Theorem 8.3, if $l = 2$, we easily obtain Lemma 3.3 of [3].*

Theorem 8.5 *Assume that the following conditions for $t \geq t_0$ hold*
(1) the constant $u_0 > 0$;
(2)

$$u(t) \leq u_0 + \int_{t_0}^{t} g_1(s)u(s)ds + \sum_{i=2}^{N} \int_{t_0}^{t} g_i(s)u(s)^{\alpha_i}ds + \int_{t_0}^{t} b_1(s) \int_{t_0}^{s} c_1(\theta)u(\theta)d\theta ds \\ + \sum_{j=1}^{L} \int_{t_0}^{t} b_j(s) \int_{t_0}^{s} c_j(\theta)u(\theta)^{\beta_j}d\theta ds, \tag{8.9}$$

where $g_i(t)$ $(i=1,\ldots N), b_j(t), c_j(t), (j=1,2,\ldots,l)$, and $u(t)$ are all nonnegative continuous functions and the constants $1 \leq \alpha_1 \leq \alpha_2 \leq \ldots \leq \alpha_N, 1 \leq \beta_1 \leq \beta_2 \leq \ldots \leq \beta_L$;

(3) Let

$$A(t) = \sum_{i=1}^{N} g_i(t) + \sum_{j=1}^{L} b_j(t) \int_{t_0}^{t} c_j(s)ds + \sum_{j=2}^{L} b_j(t) \int_{t_0}^{t} c_j(s)ds,$$
$$B(t) = \sum_{i=2}^{N} g_i(t) + \sum_{j=2}^{L} b_j(t) \int_{t_0}^{t} c_j(s)ds, \tag{8.10}$$
$$\bar{\alpha} = max\{\alpha_N, \beta_L\}, \underline{\beta} = \{\alpha_2, \beta_2\}.$$

Suppose that

$$u(t) \leq u_0 + \int_{t_0}^{t} g_1(s)u(s)ds + \sum_{i=2}^{N} \int_{t_0}^{t} g_i(s)u(s)^{\alpha_i}ds + \int_{t_0}^{t} b_1(s) \int_{t_0}^{s} c_1(\theta)u(\theta)d\theta ds$$
$$+ \sum_{j=1}^{L} \int_{t_0}^{t} b_j(s) \int_{t_0}^{s} c_j(\theta)u(\theta)^{\beta_j} d\theta ds, \tag{8.11}$$

where $g_i(t)$ $(i=1,\ldots N), b_j(t), c_j(t)(j=1,2,\ldots,l)$ and $u(t)$ are all nonnegative continuous functions and the constants $1 \leq \alpha_1 \leq \alpha_2 \leq \ldots \leq \alpha_N, 1 \leq \beta_1 \leq \beta_2 \leq \ldots \leq \beta_L$;
and let

$$\int_{t_0}^{t} A(s)ds < \infty, 1 + (1-\bar{\alpha})u_0^{\bar{\alpha}-1} \int_{t_0}^{t} B(s)\exp((\bar{\alpha}-1)\int_{t_0}^{s} A(\theta)d\theta)ds > 0, \tag{8.12}$$

(4) Let

$$C(t) = g_1(t) + b_1(t)\int_{t_0}^{t} c_1(s)ds,$$
$$D(t) = \sum_{i=2}^{N} g_i(t)M^{\alpha_i - \underline{\beta}} + \sum_{j=2}^{L} b_j(t)M^{\beta_j - \underline{\beta}} \int_{t_0}^{t} c_j(s)ds, \tag{8.13}$$
$$1 + (1-\underline{\beta})u_0^{\underline{\beta}-1} \int_{t_0}^{t} D(s)\exp((\underline{\beta}-1)\int_{t_0}^{s} C(\theta)d\theta)ds > 0,$$

where M is some positive constant, then

$$u(t) \leq u_0 \exp\left(\int_{t_0}^{t} C(s)ds\right)\{1 + (1-\underline{\beta})u_0^{\underline{\beta}-1} \int_{t_0}^{t} D(s) \exp\left((\underline{\beta}-1)\int_{t_0}^{s} C(\theta)d\theta\right)ds\} > 0. \tag{8.14}$$

The proof is similar to that of Theorem 8.1, and we omit these details.

Note 8.1 For Theorem 8.1, we can obtain some different integral inequalities when $g_i(t)$ $(i=1,\ldots N), b_j(t), c_j(t), (j=1,2,\ldots,l)$ in (8.1) take some special type functions.

We omit these discussions.

8.3 Generalization of Two Integral Inequalities

We shall first consider new generations about those integral inequalities in (8.1),(8.9).

$$u(t) \leq u_0 + \int_{t_0}^{t} g_i(s)u(s)ds + \int_{t_0}^{t} \sum_{i=2}^{N} g_i(s)u(s)^{\alpha_i} ds \\ + \sum_{j=2}^{N} \int_{t_0}^{t} b_i(s) \int_{t_0}^{t} c_j(\theta)u(\theta)ds + \sum_{k=1}^{M} \int_{t_0}^{t} e_k(s) \int_{t_0}^{t} f_k(\theta)u(\theta)^{\beta_k} d\theta ds, \quad (8.15)$$

where $\alpha_i (i = 2, , N)$, $\beta_k (k = 1, \ldots, M)$ are constants greater than 1, and $\alpha_2 \leq \alpha_3 \leq \ldots \leq \alpha_N$, $\beta_1 \leq \beta_2 \leq \ldots \leq \beta_M$, $g_i(t)(i = 1, \ldots, N)$, $b_j(t), c_j(t)(j = 1, \ldots, L)$, $e_k(t), f_k(t)(k = 1, \ldots, M)$, $u(t)$ are nonnegative continuous functions in $\mathbb{I} = [t_0, +\infty)$, $t_0 \geq 0$.

Clearly, the integral inequality (8.15) further generalizes those integral inequalities appearing in Theorems 8.1 and 8.5. For example,

1. $u(t) \leq u_0 + \int_{t_0}^{t} g_1(s)u(s)ds + \int_{t_0}^{t} g_2(s)u(s)^{\alpha_1} ds + \sum_{j=1}^{l} \int_{t_0}^{t} b_j(s) \int_{t_0}^{s} c_j(\theta)u(\theta)d\theta ds$
$+ \sum_{k=1}^{m} \int_{t_0}^{t} e_k(s) \int_{t_0}^{s} f_k(\theta)u(\theta)^{\alpha_1} d\theta ds, \alpha_1 > 0;$

2. $u(t) \leq u_0 + \int_{t_0}^{t} g_1(s)u(s)ds + \int_{t_0}^{t} \sum_{i=2}^{N} g_i(s)u(s)^{\alpha_i} ds + \int_{t_0}^{t} b_1(s) \int_{t_0}^{s} c_1(\theta)u(\theta)d\theta ds$
$+ \int_{t_0}^{t} \sum_{j=2}^{L} b_j(s) \int_{t_0}^{s} c_j(\theta)u(\theta)^{\beta_j} d\theta ds, 1 \leq \alpha_2 \leq \alpha_3 \leq \cdots \leq \alpha_N, 1 \leq \beta_2 \leq \beta_3 \leq \cdots \leq \beta_L.$

We first present the following new results.

Theorem 8.6 *For the nonnegative continuous function $M(t)$ in $\mathbb{I}$, if $M'(t) \geq 0$, constants $\alpha, \beta, \alpha \geq \beta \geq 1$ hold, then we have*

$$M(t)^{\beta} \leq \beta M(t) + \frac{\beta}{\alpha} M(t)^{\alpha}. \quad (8.16)$$

Proof Constructing the following function

$$F(t) = \frac{\beta}{\alpha} M(t)^{\alpha} + \beta M(t) - M(t)^{\beta}. \quad (8.17)$$

Clearly, $F(t)$ is continuously differentiable

$$F'(t) = \beta(M(t)^{\alpha-1} + 1 - \beta M(t) - M(t)^{\beta-1})M'(t). \quad (8.18)$$

Combination of $\alpha \geq \beta \geq 1$, for any $M(t) \geq 1$ or $M(t) \leq 1$, we know that

$$M(t)^{\alpha-1} + 1 - M(t)^{\beta-1} \geq 0.$$

We have, from (8.18),

$$F'(t) = \beta(M(t)^{\alpha-1} + 1 - M(t)^{\beta-1})M'(t) \geq 0. \tag{8.19}$$

Let us say $M(t_0) = 0$, then

$$F(t_0) = \beta M(t_0) + \frac{\beta}{\alpha}M(t_0)^\alpha - M(t_0)^\beta \geq 0. \tag{8.20}$$

We conclude that

$$F(t) \geq 0, t \in \mathbb{I}.$$

The proof is completed. □

Combination of Theorem 8.5 and the integral inequality (8.15) yields the following result.

Theorem 8.7 *Assume that the following conditions for $t \geq t_0$ hold.*
(I) u_0 is a nonnegative constant;
(II) The integral inequality (8.15) holds where $g_i(t)$ ($i = 1, \cdots, N$), $b_j(t)$, $c_j(t)$ ($j = 1, \cdots, L$), $e_k(t)$, $f_k(t)$ ($k = 1, \cdots, M$), $u(t)$, α_i ($i = 2, \cdots, N$), β_k ($k = 1, \cdots, M$) are defined as the above;
(III) let

$$P(t) = g_1(t) + \sum_{i=2}^{N} \alpha_i g_i(t)$$
$$+ \sum_{j=1}^{L} b_j(t) \int_{t_0}^{t} c_j(s)ds + \sum_{k=1}^{M} \int_{t_0}^{t} \beta_k e_k(t) \int_{t_0}^{t} f_k(s)ds,$$
$$Q(t) = \frac{1}{\alpha}[\sum_{i=2}^{N} \alpha_i g_i(t) + \sum_{k=1}^{M} \int_{t_0}^{t} \beta_k e_k(t) \int_{t_0}^{t} f_k(s)ds]$$
$$\alpha = \max(\alpha_N, \beta_M),$$

Suppose that

$$1 + (1-\alpha)u_0^{(\alpha-1)} \int_{t_0}^{t} Q(s) \exp((\alpha-1) \int_{t_0}^{s} P(\theta))ds > 0, \tag{8.21}$$

then

$$u(t) \leq u_0 \exp(\int_{t_0}^{t} p(s)ds)$$
$$\cdot \{1 + (1-\alpha)u_0^{(\alpha-1)} \int_{t_0}^{t} Q(s) \exp((\alpha-1) \int_{t_0}^{s} P(\theta))ds\}^{\frac{1}{1-\alpha}}, t \geq t_0. \tag{8.22}$$

Proof Let

8.3 Generalization of Two Integral Inequalities

$$M(t) = u_0 + \int_{t_0}^t g_1(s)u(s)ds + \int_{t_0}^t \sum_{i=2}^N g_i(s)u(s)^{a_i}ds$$
$$+ \sum_{j=1}^L \int_{t_0}^t b_i(s) \int_{t_0}^s c_j(\theta)u(\theta)ds + \sum_{k=1}^M \int_{t_0}^t e_k(s) \int_{t_0}^s f_k(\theta)u(\theta)^{\beta_k}d\theta ds. \tag{8.23}$$

It is obvious that
$$u(t) \leq M(t).$$

Differentiating (8.23)'s sides, we have

$$M'(t) = g_1(t)u(t) + \sum_{i=2}^N g_i(t)u(t)^{a_i} + \sum_{j=1}^L b_j(t) \int_{t_0}^s c_j(s)u(s)ds$$
$$+ \sum_{k=1}^M e_k(t) \int_{t_0}^s f_k(s)u(s)^{\beta_k}ds. \tag{8.24}$$

Combining Theorem 8.5(II) and $M(t)$ is nondecreasing function in t, (8.24) may be written as

$$M'(t) \leq g_1(t)M(t) + \sum_{i=2}^N g_i(t)M(t)^{a_i}$$
$$+ \sum_{j=1}^L b_j(t) \int_{t_0}^t c_j(s)ds M(t) + \sum_{k=1}^M e_k(t) \int_{t_0}^t f_k(s)M(t)^{\beta_k}ds. \tag{8.25}$$

Using Theorem 8.5, (8.25) can be written as

$$M'(t) = g_1(t)M(t) + \sum_{i=2}^N g_i(t)(\alpha_i M(t) + \tfrac{\alpha}{\alpha_i}M(t)^\alpha)$$
$$+ \sum_j = 1^L b_j(t) \int_{t_0}^t c_j(s)ds M(t) + \sum_{k=1}^M e_k(t) \int_{t_0}^t f_k(s)ds(\beta_k M(t) \tfrac{\beta_k}{\alpha} M(t)^\alpha)$$
$$= [g_1(t) + \sum_{i=2}^N \alpha_i g_i(t) + \sum_{j=2}^L b_j(t) \int_{t_0}^t c_j(s)ds$$
$$+ \sum_{k=2}^M e_k(t) \int_{t_0}^t f_k(s)ds]M(t) + [\sum_{j=1}^L \tfrac{\alpha_i}{a} g_i(t) + \sum_{k=1}^M \tfrac{\beta_i}{\alpha} e_k(t) \int_{t_0}^t f_k(s)ds]M(t)^\alpha. \tag{8.26}$$

Further, combining Theorem 8.5(III), we have

$$M'(t) \leq P(t)M(t) + Q(t)M(t)^\alpha. \tag{8.27}$$

Integrating (8.27) yields

$$M'(t) \leq u_0 + \int_{t_0}^t P(t)M(t)ds + \int_{t_0}^t Q(s)M(s)^\alpha ds.$$

In view of the inequality of [4], the desired result is obtained. □

Note 8.2 Theorem 8.7 is the generalization of those integral inequalities on Theorem 8.3.

Note 8.3 It is obvious that Theorem 8.7 is the generalization of Theorem 8.5.

Now, we consider more general and complicated integral inequalities as follows:

$$u(t) \leq u_0 + \int_{t_0}^{t} b_1^{(1)}(s)u(s)ds + \int_{t_0}^{t} b^{(H)}{}_1 \int_{t_0}^{t(H-1)} b_H^{(H)}(s)u(s)dsdt+$$
$$\int_{t_0}^{t} b_1^{(I)}(t_1) \int_{t_0}^{t} b_2^{(I)}(t_2) \int_{t_0}^{t(I-1)} b_I^{(I)}(s)u(s)dsdt_{(I-1)}dt_1 + \sum_{i=1}^{N_1} \int_{t_0}^{t} a_{1i}^{(1)} u(s)^{a_{1i}} ds+$$
$$\sum_{j=1}^{N_M} \int_{t_0}^{t} a_{1j}^{(M)}(t_1) \int_{t_0}^{t} a_{2j}^{(M)}(t_2) \int_{t_0}^{t_{M-1}} a_{Mj}^{(M)}(s)u(s)^{a_{Mj}} dsdt_1+ \quad (8.28)$$
$$\sum_{K=1}^{N_L} \int_{t_0}^{t} a_{1K}^{(L)}(t_1) \int_{t_0}^{t} a_{2K}^{(L)}(t_2) \int_{t_0}^{t_{L-1}} a_{LK}^{(L)}(s)u(s)a^{L,K} dsdt_1,$$

where $b_1^{(1)}(t); \cdots ; b_1^{(H)}(t); \cdots ; b_H^{(H)}; b_1^{(I)}(t); \cdots ; b_I^{(I)}(t); a_{1i}^1(t)(i=1,\cdots,N_1;$
$a_{1j}^{(M)}(t); \cdots ; a_{Mj}^{(M)}(t)$ $(j=1,\cdots,N_M); a_{1K}^{(L)}(t), \cdots , a_{LK}^{(L)}(t), (k=1,\cdots,N_L),$
$u(t)$ are all nonnegative continuous functions in $\mathbb{I}$, and $1 < \alpha_{11} \leq \alpha_{12} \leq \cdots \leq \alpha_{1N_1};$
$1 < \alpha_{M1} \leq \alpha_{M2} \leq \cdots \leq \alpha_{MN_M} \cdots ; 1 < \alpha_{L1} \leq \cdots \leq \alpha_{LN_L}$ are constants.

Theorem 8.8 *Assume the following conditions hold for $t \geq t_0$.*
I) u_0 is a nonnegative constant;
II) The integral inequality (8.28) holds, where all functions are defined as (8.28);
III) Let

$$P(t) \triangleq \int_{t_0}^{t} b_1^{(1)}(s)ds + \cdots + \int_{t_0}^{t} b_1^{(I)}(t_1) \int_{t_0}^{t} b_2^{(I)}(t_2) \int_{t_0}^{t(I-1)} b_I^{(I)}(s)ds \cdots dt_1$$
$$+ \sum_{i=1}^{N_1} \alpha_{1i} \int_{t_0}^{t} a_{1i}^{(1)}(s)ds + \cdots$$
$$+ \sum_{j=1}^{N_M} \alpha_{Mj} \int_{t_0}^{t} a_{1j}^{(M)}(t_1) \int_{t_0}^{t_1} a_{2j}^{(M)}(t_2) \cdots \int_{t_0}^{t_{M-1}} a_{M,j}^{(M)}(s)ds \ldots dt_1$$
$$+ \cdots + \sum_{K=1}^{N_L} \alpha_{L,k} \int_{t_0}^{t} a_{1K}^{(L)}(t_1) \cdots \int_{t_0}^{t_{L-1}} a_{L,K}^{(L)}(s)ds \cdots dt_1,$$

$$Q(t) \triangleq \frac{1}{\alpha} [\sum_{i=1}^{N_1} \alpha_{1i} \int_{t_0}^{t} \alpha_{1i}^1(s)ds + \cdots$$
$$+ \sum_{j=1}^{N_M} \alpha_{Mj} \int_{t_0}^{t} a_{1j}^{(M)}(t_1) \int_{t_0}^{t_1} a_{2j}^{(M)}(t_2) \cdots \int_{t_0}^{t_{M-1}} a_{M,j}^{(M)}(s)ds \cdots dt_1$$
$$+ \cdots + \sum_{K=1}^{N_L} \alpha_{L,k} \int_{t_0}^{t} a_{1K}^{(L)}(t_1) \cdots \int_{t_0}^{t_{L-1}} a_{L,K}^{(L)}(s)ds \cdots dt_1],$$
$$\alpha \triangleq \max(\alpha_{1,N_1}, \alpha_{L,N_L}).$$

Suppose that

$$1 + (1-\alpha)u_0^{(\alpha-1)} \int_{t_0}^{t} Q(s) \exp((\alpha-1) \int_{t_0}^{s} P(\theta)) ds > 0,$$

then

$$u(t) \leq u_0 \exp(\int_{t_0}^{t} p(s)ds)\{1+(1-\alpha)u_0^{(\alpha-1)} \int_{t_0}^{t} Q(s)e^{((\alpha-1)\int_{t_0}^{s} P(\theta))ds}\}^{\frac{1}{1-\alpha}}, t \geq t_0. \tag{8.29}$$

The proof is similar to that of Theorem 8.5, and we omit these details. □

8.4 Conclusion

Several new integral inequalities are built, and it is easily found that the proposed integral inequalities are generalized to the existing integral inequalities whose variable different modulus are more than 1. Compared with the existing integral inequalities in the references [1–11], our proposed integral inequalities may deal with more complicated differential equations.

Acknowledgements This work is Supported by National Key Research and Development Program of China (2017YFF0207400).

References

1. Guo S, Irene M, Si L, Han L. Several integral inequalities and their applications in nonlinear differential systems. Appl Math Comput. 2013;219:4266–77.
2. Hong J. A result on stability of time-varying delay differential equation. Acta Math Sin. 1983;26(3):257–61.
3. Si L. Boundness, stability of the solution of time-varying delay neutral differential equation. Acta Math Sin. 1974;17(3):197–204.
4. Si L. Stability of delay neutral differential equation. Huhhot: Inner Mongolia Educational Press; 1994. p. 106–41.
5. Alekseev VM. An estimate for the perturbation of the solution of ordinary differential equation. Vestnik Moskovskogo Universiteta. Seriya I. Matematika, Mekhanika. 1961;2:28–36.
6. Li Y. Boundness, stability and error estimate of the solution of nonlinear different equation. Acta Math Sin. 1962;12(1):28–36 In Chinese.
7. Brauer F. Perturbations of nonlinear systems of differential equations. J Math Anal Appl. 1966;14:198–206.
8. Elaydi S, Rao M, Rama M. Lipschitz stability for nonlinear Volterra integro differential systems. Appl Math Comput. 1988;27(3):191–9.
9. Giovanni A, Sergio V. Lipschitz stability for the inverse conductivity problem. Adv Appl Math. 2005;35(2):207–41.
10. Hale JK. Ordinary differential equations. Interscience, New York: Wiley; 1969.
11. Jiang F, Meng F. Explicit bounds on some new nonlinear integral inequalities with delay. J Comput Appl Math. 2007;205(1):479–86.
12. Soliman AA. Lipschitz stability with perturbing Liapunov functionals. Appl Math Lett. 2004;17(8):939–44.
13. Soliman AA. On Lipschitz stability for comparison systems of differential equations via limiting equation. Appl Math Comput. 2005;163(3):1061–7.

14. Huang L. Stability theory. Beijing: Peking University Press; 1992. p. 235–83.
15. Horn RA, Johnson CR. Topics in matrix analysis. Cambridge University press; 1991.
16. Kumpati SN, Jeyendran B. A common lyapunov function for stable LTI systems with commuting A-matrixes. IEEE Trans Autom Control. 1994;39(12):2469–71.
17. Lee SH, Kim TH, Lim JT. A new stability analysis of switched systems. Automatica. 2000;36:917–22.
18. Liberzon D, Hespanta JP, Morse AS. Stability of switched systems: a Lie-algebraic condition. Syst Control Lett N-Holl. 1999;37:117–22.
19. Mareada KS, Balakrishan J. A common Lyapunov function for stable LTI systems with commuting A-martices. IEEE Trans Autom Control. 1994;39(12):2469–71.
20. Molchanov AP, Pyatnitskiy YS. Criteria of asymptotic stability of differential and difference inclusions encountered in control theory. Syst Control Lett N-Holl. 1989;13:59–64.
21. La Salle J, Lefschetz S. Stability by Lyapunov's direct method. New York, N.Y.: Academic Press; 1961.
22. Polanski K. On absolute stability analysis by polyhydric Lypunov functions. Automatica. 2000;36:573–8.
23. Schmitendorf WE, Barmish BR. Null controllability of linear systems with constrained controls. SIAM J Control Optim. 1980;18:327–45.
24. Shevitz D, Paden B. Lyapunov stability theory of non-smooth systems. In: Proceeding of the 32nd conference on decision and control, San Antonio, Texas; 1993. p. 416–421.
25. Tatsushi O, Yasuyuki F. Two conditions concerning common quadratic lyapunov functions for linear systems. IEEE Trans Autom Control. 1997;42(5):719–21.

Chapter 9
Lipschitz Stability Analysis on a Type of Nonlinear Perturbed System

9.1 Introduction

It is well known that nonlinear differential equation theory has been an active area of research for many years [1–17]. There are many methods or tools to deal with different nonlinear systems, including integral inequalities methods. The subject has attracted great interest and been investigated thoroughly by many authors [1–7, 9–11, 13, 14]. For example, Lipschitz stability of nonlinear differential equations was first put forward in [3, 4], and these concepts and the corresponding methods are developed by other authors [6, 7, 9]. Some existing integral inequalities and the corresponding applications are presented (see paper [10, 11, 13]).

The main results in this chapter are arranged as follows. In the second section, some basic definitions of Lipschitz stability are presented, and some general results about perturbed systems are outlined. In the third section, several sufficient conditions such as uniform Lipschitz stability, uniform Lipschitz asymptotic stability, and exponential stability behind T of nonlinear differential systems are established based on the proposed integral inequality, the concepts in the reference [7], and nonlinear variation of constants formula of the reference [1]. The above conditions may similarly be generalized to linear perturbed differential systems. Finally, the simple example of uniform Lipschitz asymptotic stability of the origin of nonlinear differential systems is illustrated by the above-obtained results.

9.2 Basic Notions and General Results

Given the nonlinear differential system

$$\tfrac{dx}{dt} = f(t,x), \tag{9.1}$$

where $f \in C[\mathbb{I} \times \mathbb{R}^n, \mathbb{R}^n]$ and $\mathbb{I} = [t_0, \infty), t_0 \geq 0$.

9 Lipschitz Stability Analysis on a Type of Nonlinear Perturbed System

Suppose that the function f be smooth enough to guarantee existence, uniqueness, and continuous dependence of solution $x(t) = x(t, t_0, x_0)$ of (9.1).

The associated variation system

$$\frac{dz}{dt} = f_x(t, x(t, t_0, x_0))z, \qquad (9.2)$$

where $f_x = (\frac{\partial f}{\partial x})_{n \times n}$ exists and is continuous on $\mathbb{I} \times \mathbb{R}^n$. $f(t, 0) = 0, x(t) := x(t, t_0, x_0)$ is the solution of (9.1) with $x(t_0, t_0, x_0) = x_0$.

The fundamental matrix solution $\phi(t, t_0, x_0)$ of (9.2) is given in [2] by

$$\phi(t, t_0, x_0) = \frac{\partial x(t, t_0, x_0)}{\partial x_0}, \qquad (9.3)$$

where $\phi(t_0, t_0, x_0) = I$.

In addition to the above hypothesis, we shall always assume that

(A) There exists a continuous function $\alpha(t)$ on $\mathbb{I}$ such that the largest eigenvalue $\lambda(f_x(t, x))$ of $\frac{1}{2}[f_x(t, x) + f_x^T(t, x)]$ satisfies

$$\lambda(f_x(t, x)) \leq \alpha(t), \qquad (9.4)$$

where $t \geq t_0, x \in \mathbb{D}$, $\mathbb{D}$ is some opener, $f_x^T(t, x)$ is the transpose of $f_x(t, x)$.

Let

$$\delta = \lim_{t \to \infty} \sup_{\forall t_0 \in I} \frac{1}{t - t_0} \int_{t_0}^{t} \alpha(s)ds < 0 \qquad (9.5)$$

holds.

According to [2], it is well known that δ is independent of t_0 and that there exists a sufficient larger T such that

$$\int_{t_0}^{t} \alpha(s)ds \leq \frac{\delta}{2}(t - t_0), t > T >> t_0 > 0. \qquad (9.6)$$

Lemma 9.1 Assume that (A) holds and $x_0, y_0 \in \overline{\mathbb{D}} \subset \mathbb{D}$, $x(t, t_0, x_0), x(t, t_0, y_0) \in \mathbb{D}$, then

$$\|x(t, t_0, x_0) - x(t, t_0, y_0)\| \leq \|y_0 - x_0\| \exp\left(\int_{t_0}^{t} \alpha(s)ds\right), t \geq t_0. \qquad (9.7)$$

Consider the system (9.1)'s perturbed system

$$\frac{dy}{dt} = f(t, y) + g(t, y), \qquad (9.8)$$

where $f, g \in C[\mathbb{I} \times \mathbb{R}^n, \mathbb{R}^n], \mathbb{I} = [t_0, \infty], t_0 \geq 0$. Suppose that $y(t) = y(t, t_0, y_0)$ be solution to (9.8) through (t_0, y_0) and $x(t) = x(t, t_0, y_0)$ is the one to (9.1) through (t_0, y_0). Combination of (9.1) and (9.1), the following lemma from [2] is obtained.

Lemma 9.2 If $y_0 \in \mathbb{D}$, then for all $t \geq t_0$ such that $x(t, t_0, y_0) \in \mathbb{D}, y(t, t_0, y_0) \in \mathbb{D}$, we have

9.2 Basic Notions and General Results

$$y(t, t_0, y_0) = x(t, t_0, y_0) + \int_{t_0}^{t} \Phi(t, s, y(s, t_0, y_0))g(s, y(s, t_0, y_0))ds, \quad (9.9)$$

where $\phi(t, t_0, x_0)$ is the fundamental matrix $\dot{z} = f_x(t, x)z$.

Combination of Lemmas 9.1 and 9.2 yields the following results.

Lemma 9.3 *If (A) holds and if $t > t_0$, x_0, $y_0 \in \overline{\mathbb{D}} \subset \mathbb{D}$, $y(t, t_0, x_0)$, $x(t, t_0, y_0) \in \mathbb{D}$, we have*

$$\begin{aligned}\|y(t, t_0, y_0) - x(t, t_0, x_0)\| &\leq \|y_0 - x_0\| \exp(\int_{t_0}^{t} \alpha(s)ds) \\ &+ \int_{t_0}^{t} \exp\left(\int_{s}^{t} \alpha(\theta)d\theta\right) \|g(s, t_0, y_0)\| ds, \ t \geq t_0.\end{aligned} \quad (9.10)$$

In fact, based on Lemma 9.2, we have

$$\|y(t, t_0, y_0) - x(t, t_0, y_0)\| \leq \int_{t_0}^{t} \exp(\int_{x}^{t} \alpha(\theta)d\theta) \|g(s, t_0, y_0)\|ds, t \geq t_0. \quad (9.11)$$

On the other hand, we have, along with Lemma 9.1,

$$\|x(t, t_0, y_0) - x(t, t_0, x_0)\| \leq \|y_0 - x_0\| \exp\left(\int_{t_0}^{t} \alpha(s)ds\right), t \geq t_0.$$

According to the following inequality

$$\|y(t, t_0, y_0) - x(t, t_0, x_0)\| \leq \|y(t, t_0, y_0) - x(t, t_0, y_0)\| + \|y(t, t_0, y_0) - x(t, t_0, x_0)\|,$$

the desired result is obtained.

Before giving further more details, we discuss several nonlinear differential equations.

Example 9.1 Consider the following scalar differential equation

$$\dot{x} = (6 \sin t - 2t)x, t \in \mathbb{E} = [0, \infty), \quad (9.12)$$

with solution

$$x(t) = x_0 exp[6 \sin t - 6t \cos t - t^2 - 6 \sin t_0 + 6 t_0 \cos t_0 + t^2].$$

Lemma 9.3 gives

$$\|x(t)\| \leq \|x_0\| exp[12 + (t + t_0)(6 - t + t_0)] = M(t, t_0)\|x_0\|, t \in \mathbb{I}.$$

It is well known that $\lim_{t \to \infty} M(t, t_0) = 0$ holds uniformly for all $t_0 \in \mathbb{E}$ (See Yoshizawa [17]).

From Yoshizawa [17], it is known that the zero solution of (9.12) is neither uniformly Lyapunov stable nor asymptotically stable, and at the same time not uniform Lipschitz stable.

Example 9.2 Consider the following scalar differential equation

$$\dot{x}(t) = -\frac{x}{1+t}, t \in \mathbb{I} = [0, \infty). \tag{9.13}$$

From Yoshizawa [17], (9.13)'s solution through (t_0, x_0) is

$$x(t) = -\frac{1+t_0}{1+t}x_0.$$

It is known that the zero of the above system (9.13) is uniformly Lyapunov stable.

We define $M(t, t_0) = \frac{1+t_0}{1+t}$, it is clearly $\lim_{t \to \infty} M(t, t_0) = 0$, but it does not hold uniformly for t_0.

Example 9.3 Consider the following scalar differential equation

$$\begin{aligned} \dot{x}(t) &= y, \\ \dot{y} &= -\frac{g}{l}sinx, \end{aligned} \tag{9.14}$$

choosing

$$V(x, y) = \frac{1}{2}y^2 + \frac{g}{l}(1 - cosx),$$

it is obvious that $V(x, y) \in C[\mathbb{R} \times \mathbb{R}, \mathbb{R}^+], V(0, 0) = 0$.

It is well known that the zero of (9.14) is uniformly stable, but the solution of the above system (9.14) is an ellipsoid integral function, which is not denoted through the elementary function. So the zero solution is not uniformly Lipschitz stable.

From the above discussion, there are strong differences between Lyapunov stability, Lyapunov asymptotic stability, and uniform Lipschitz stability. Besides stability, the asymptotic behavior of a zero solution indicates that solutions of the differential equations will ultimately tend to the zero solution, while they may leave the neighborhood of the zero solution at first. Consequently, conditions under which attractivity is guaranteed will be formulated. So we have to generalize those notions on uniform Lipschitz stability (See Dannan and Elaydi [3]) and to study the asymptotic behavior of solutions of original systems (9.1).

Before giving further details, we give some of the main definitions that we need in the sequel.

Definition 9.1 The zero solution of (9.1) is said to be uniformly Lipschitz stable if there exists $M > 0$ and $\delta > 0$ such that $\|x(t, t_0, x_0)\| \leq M\|x_0\|$, whenever $\|x_0\| < \delta$ and $t \geq t_0 \geq 0$.

(See Dannan and Elaydi [3]).

Definition 9.2 The zero solution of (9.1) is said to be uniform Lipschitz quasi-asymptotic stable if there exists nonnegative continuous functions $M(t, t_0)$ on $\mathbb{I}$ such that $\lim_{t \to \infty} M(t, t_0) = 0$ uniformly holds to t_0 and $\|x(t, t_0, x_0)\| \leq M(t, t_0)\|x_0\|$, whenever $\|x_0\| < \delta$, $M(t, t_0)$ is independent on t_0.

9.2 Basic Notions and General Results

Definition 9.3 The zero solution of (9.1) is said to be uniformly Lipschitz stable if the zero solution of the (9.1) is uniformly Lipschitz stable, and at the same time, the zero solution is uniform Lipschitz quasi-asymptotic stable.

Definition 9.4 The zero solution of (9.1) is said to be uniformly Lipschitz stable in variation if there exists $M > 0$ and $\delta > 0$ such that $\|Y(t, t_0, x_0)\| \leq M$, whenever $\|x_0\| < \delta$ and $Y(t, t_0, x_0) = (\frac{\partial f}{\partial x})_{n \times n}$, $Y(t_0, t_0, x_0) = I$, $t \geq t_0 > 0$.

Consider the nonlinear differential system

$$\dot{x} = A(t)x, \tag{9.15}$$

where $A(t) \in C[\mathbb{E}]$ and $\mathbb{E} = [t_0, \infty)$, $t_0 \geq 0$.

We easily obtain the following results.

Theorem 9.4 *If the zero solution of the system (9.15) is uniformly Lipschitz stable (uniformly Lipschitz asymptotically stable), then the zero solution of the system (9.15) is also uniformly stable (uniformly asymptotically stable).*

The proof procedure is easily obtained from Definition 9.1–9.4. We omit the details.

Theorem 9.5 *Let $S_\delta = \{x \in \mathbb{R}^n : \|x\| < \delta\}$ and $\mathbb{E} = [t_0, \infty)$, $t_0 \geq 0$. Suppose there exist two functions $V(t, x)$ and $g(t, x)$ satisfying the following conditions:*

(i) $g(t, x) \in C[\mathbb{E} \times S_\delta, \mathbb{R}]$ and $g(t, 0) = 0$.
(ii) $V(t, x) \in C[\mathbb{E} \times \mathbb{R}^+, \mathbb{R}]$, $V(t, 0) = 0$, and $V(t, x)$ is locally Lipschitz in x and satisfies

$$V(t, x) \geq b(\|x\|), \tag{9.16}$$

where $b(r) \in C[\mathbb{E} \times \mathbb{R}^+, \mathbb{R}]$, $b(0) = 0$ and $b(r)$ is strictly monotonically increasing in r such that

$$b^{-1}(\alpha r) \leq rq(\alpha), \tag{9.17}$$

for some function q with $q(\alpha) \geq 1$ if $\alpha \geq 1$.
(iii) *For $(t, x) \in \mathbb{E} \times S_\delta$,*

$$\dot{V}(t, x)|_{(9.15)} \leq g(t, V(t, x)). \tag{9.18}$$

If the zero solution of

$$\dot{u} = g(t, u)), u(t_0) = u_0 \geq 0 \tag{9.19}$$

is uniformly Lipschitz stable (uniformly Lipschitz asymptotically stable), then so is the zero solution of (9.15).

We omit the proof which is similar to that of Theorem 2.1 in [4].

Note 9.1 Theorem 9.5 generalizes Theorem 2.1 of [3].

9.3 Lipschitz Stability on Nonlinear Perturbed Systems

In this section, we discuss some sufficient conditions for Lipschitz stability of nonlinear perturbed differential equations.

Consider (9.1)'s perturbed system

$$\tfrac{dy}{dt} = f(t, y) + g(t, Ty) + h(t, y, Ly), \tag{9.20}$$

where T and L are continuous operators with map $\mathbb{R}^n$ into $\mathbb{R}^n$, $f \in C[\mathbb{I} \times \mathbb{R}^n, \mathbb{R}^n]$, $h \in C[\mathbb{I} \times \mathbb{R}^n \times \mathbb{R}^n, \mathbb{R}^n]$, $\mathbb{I} = [t_0, \infty), t_0 \geq 0, f(t, 0) = 0, f_x(t, x)$ exists and is continuous on $\mathbb{I} \times \mathbb{R}^n$.

Suppose that $y(t) = y(t, t_0, y_0)$ be solution to (9.20) through (t_0, y_0) and $x(t) = x(t, t_0, y_0)$ be solution to (9.1) through (t_0, y_0). The matrix $\phi(t, t_0, x_0)$ is the fundamental matrix of $\dot{z} = f_x(t, x)z$.

Theorem 9.6 *Assume that the trivial solution of (9.1) is uniform Lipschitz stable in variation and*

(1)

$$\|g(t, y)\| \leq \sum_{j=1}^{L} b_j(t) \int_{t_0}^{t} c_j(s) \|y\| ds,$$

$$\|h(t, y, Ly)\| \leq g_1(t) \|y(t)\| + g_2(t) \|y(t)\|^{\alpha_1} + \sum_{k=1}^{M} e_k(t) \int_{t_0}^{t} f_k(s) \|y(s)\|^{\alpha_1} ds,$$

where $b_j(s), c_j(s) \in C^+[\mathbb{I}], j = 1, \ldots, L, g_i(t) \in C^+[\mathbb{I}], i = 1, 2, e_k(t), f_k(t) \in C^+[\mathbb{I}], k = 1, \ldots, M$ and the constant $\alpha_1 \geq 1$.

(2) Let

$$P(t) = g_1(t) + \sum_{j=1}^{L} b_j(t) \int_{t_0}^{t} c_j(s) ds,$$

$$Q(t) = g_2(t) + \sum_{k=1}^{M} e_k(t) \int_{t_0}^{t} f_k(s) ds,$$

and if, for each $t_0 \in \mathbb{I}$,

$$\int_{t_0}^{\infty} P(s) ds < \infty, \quad \int_{t_0}^{\infty} Q(s) \exp((\alpha_1 - 1) \int_{t_0}^{s} P(\theta) d\theta) ds < \infty$$

hold, then the trivial solution of (9.20) is uniform Lipschitz stable.

Proof Let $y(t, t_0, y_0)$ be the solution of (9.1) with (t_0, y_0), then by the nonlinear variation of constants formula on Aleskeev [1], we have

$$y(t, t_0, y_0) = x(t, t_0, y_0) + \int_{t_0}^{t} \phi(t, s, y(t, t_0, y_0)) g(s, Ty) ds$$
$$+ \int_{t_0}^{t} \phi(t, s, y(t, t_0, y_0)) h(s, y(s), Ly) ds.$$

9.3 Lipschitz Stability on Nonlinear Perturbed Systems

In view of the assumption of Theorem 9.6(1), we have

$$\|y(t, t_0, y_0)\| \leq \|x(t, t_0, y_0)\| + \int_{t_0}^{t} \|\phi(t, s, y(t, t_0, y(t, t_0, y_0)))\| \|g(s, Ty)\| ds$$
$$+ \int_{t_0}^{t} \|\phi(t, s, y(t, t_0, y_0))\| \|h(s, y(s), Ly)\| ds$$
$$\leq M \|y_0\| + M \int_{t_0}^{t} g_1(s) \|y(s)\| + M \int_{t_0}^{t} g_2(s) \|y(s)\|^{\alpha_1} ds$$
$$+ M \int_{t_0}^{t} \sum_{j=1}^{l} b_j(s) \int_{t_0}^{s} c_j(\theta) \|y(\theta)\| d\theta ds + M \int_{t_0}^{t} \sum_{k=1}^{M} e_k(s) \int_{t_0}^{s} \|f_k(s)\| \|y(\theta)\|^{\alpha_1} d\theta ds,$$

where $\|\phi(t, t_0, y_0)\| \leq M$, M is some Lipschitz stable constant.

We obtain the desired result along with Theorem 9.6(2) and Theorem 8.1. □

If $g_2(t) = 0, e_k(t) = 0, f_k(t) = 0, (k = 1, \ldots, m), L = 2$, we obtain the following results.

Theorem 9.7 *Assume that the trivial solution of (9.1) is uniform Lipschitz stable in variation and (1)*

$$\|g(t, y)\| \leq \sum_{j=1}^{L} b_j(t) \int_{t_0}^{t} c_j(s) \|y\| ds,$$

$$\|h(t, y, Ly)\| \leq g_1(t) \|y(t)\| + \sum_{j=M+1}^{L} b_j(t) \int_{t_0}^{t} c_j(s) \|y(s)\| ds,$$

where $b_j(s), c_j(s) \in C^+[\mathbb{I}], j = 1, \ldots, L, g_1(t) \in C^+[\mathbb{I}]$; (2) Let

$$P_1(t) = g_1(t) + \sum_{j=1}^{L} b_j(t) \int_{t_0}^{t} c_j(s) ds,$$

$$P_2(t) = g_2(t) + \sum_{j=M+1}^{L} b_j(t) \int_{t_0}^{t} c_j(s) ds,$$

and if, for each $t_0 \in \mathbb{I}$,

$$\int_{t_0}^{\infty} (P_1(s) + P_2(s)) ds < \infty,$$

holds, then the trivial solution of (9.20) is uniform Lipschitz stable.

Note 9.2 If $M = 1, L = 2$ in Theorem 9.7, we easily obtained the Lemma 3.5 of [3].

Theorem 9.8 *Assume that*

(1) $\lambda(f_x(t, x)) \leq \alpha(t) < 0$;
(2) *The assumption (1) in Theorem 9.6 holds;*
(3) *Let*

$$P_1(t) = g_1(t) + \sum_{j=1}^{L} b_j(t) \int_{t_0}^{t} \exp(-\int_{x}^{t} \alpha(\theta) d\theta) c_j(s) ds,$$

$$Q_1(t) = g_2(t) + \sum_{k=1}^{M} e_k(t) \int_{t_0}^{t} \exp(-\int_{a}^{t} \alpha(\theta) d\theta) f_k(s) ds,$$

and if, for each $t_0 \in \mathbb{I}$,

$$\int_{t_0}^{\infty} P_1(s)ds < \infty,$$
$$\int_{t_0}^{\infty} Q_1(s)\exp((\alpha_1 - 1)\int_{t_0}^{s} P_1(\theta)d\theta)ds < \infty.$$

Then for each $t_0 \in \mathbb{I}$, if

$$\lim_{t \to \infty} \exp(\int_{t_0}^{t} \alpha(s)ds) = 0,$$

the trivial solution of (9.20) is uniform Lipschitz asymptotically stable.

Proof Let $y(t, t_0, y_0)$ is the solution of (9.20) through (t_0, y_0). By the nonlinear variation of constants formula of [1], we have

$$\begin{aligned} y(t, t_0, y_0) &= x(t, t_0, y_0) + \int_{t_0}^{t} \phi(t, s, y(t, t_0, y_0))g(s, Ty)ds \\ &+ \int_{t_0}^{t} \phi(t, s, y(t, t_0, y_0))h(s, y(s), Ly)ds, \end{aligned} \qquad (9.21)$$

where $y(t)$ and $x(t)$ are defined the same as the above assumptions.

In view of (9.21), we have

$$\|y(t, t_0, y_0)\| \leq \|x(t, t_0, y_0)\| + \int_{t_0}^{t} \|\phi(t, s, y(t, t_0, y(t, t_0, y_0)))\| \|g(s, Ty)\| ds$$
$$+ \int_{t_0}^{t} \|\phi(t, s, y(t, t_0, y_0))\| \|h(s, y(s), Ly)\| ds. \qquad (9.22)$$

According to Theorem 9.8(1) and the relation between (9.1) and (9.20), we obtain

$$\|\phi(t, s, y(s, t_0, y_0))\| \leq \exp(\int_{t_0}^{t} \alpha(s)ds), t \geq t_0, \qquad (9.23)$$

and

$$\begin{aligned} \|x(t, t_0, y_0)\| &\leq \int_0^1 \|\phi(t, s, \lambda y_0)d\lambda\| \|y_0\| \\ &\leq \max_{0 \leq \lambda \leq 1} \|\phi(t, s, \lambda y_0)\| \|y_0\| \\ &= \|\phi(t, s, \varepsilon y_0)\| \|y_0\|, \exists \varepsilon \in [0, 1] \\ &\leq \exp(\int_{t_0}^{t} \alpha(s)ds \|y_0\|, t \geq t_0. \end{aligned} \qquad (9.24)$$

By combining (9.4) and (9.24), (9.22) can be written as

$$\begin{aligned} \|y(t, t_0, y_0)\| &\leq \exp(\int_{t_0}^{t} \alpha(s)ds) \|y_0\| + \int_{t_0}^{t} \exp(\int_{s}^{t} \alpha(\theta)d\theta)g_1(s) \|y(t, t_0, y_0)\| ds \\ &+ \int_{t_0}^{t} \exp(\int_{s}^{t} \alpha(\theta)d\theta)g_2(s) \|y(t, t_0, y_0)\|^{\alpha_1} ds \\ &+ \sum_{j=1}^{L} \int_{t_0}^{t} \exp(\int_{s}^{t} \alpha(\theta)d\theta)b_j(s) \int_{t_0}^{s} c_j(\theta) \|y(t, t_0, y_0)\| d\theta ds \\ &+ \sum_{k=1}^{M} \int_{t_0}^{t} \exp(\int_{s}^{t} \alpha(\theta)d\theta)b_k(s) \int_{t_0}^{s} f_k(\theta) \|y(t, t_0, y_0)\|^{\alpha_1} d\theta ds. \end{aligned} \qquad (9.25)$$

Now, in view of the following

9.3 Lipschitz Stability on Nonlinear Perturbed Systems

$$\exp((1-\alpha_1)\int_{t_0}^t \alpha(s)ds) \geq 1, t \geq t_0,$$

where $\alpha_1 \geq 1$.

So, (9.25) can be written as

$$\|y(t,t_0,y_0)\| \leq \exp(\int_{t_0}^t \alpha(s)ds)\|y_0\| + \int_{t_0}^t \exp(\int_s^t \alpha(\theta)d\theta)g_1(s)\|y(t,t_0,y_0)\|ds$$
$$+ \int_{t_0}^t \exp(\int_s^t \alpha(\theta)d\theta)g_2(s)\|y(t,t_0,y_0)\|^{\alpha_1}\exp((1-\alpha_1)\int_{t_0}^s \alpha(\theta)d\theta)ds$$
$$+ \sum_{j=1}^L \int_{t_0}^t \exp(\int_s^t \alpha(\theta)d\theta)b_j(s)\int_{t_0}^s c_j(\theta)\|y(t,t_0,y_0)\|d\theta ds$$
$$+ \sum_{k=1}^M \int_{t_0}^t \exp(\int_s^t \alpha(\theta)d\theta)b_k(s)\int_{t_0}^s f_k(\theta)\|y(t,t_0,y_0)\|^{\alpha_1}\exp((1-\alpha_1)\int_{t_0}^\theta \alpha(\theta)d\theta)d\theta ds.$$
(9.26)

Let
$$u(t) = \exp(-\int_{t_0}^t \alpha(s)ds)\|y(t,t_0,x_0)\|, \qquad (9.27)$$

Substitution of $u(t)$ for $\|y(t)\|$, we obtain from (9.27)

$$\|u(t)\| \leq \|y_0\| + \int_{t_0}^t g_1(s)u(s,t_0,y_0)ds + \int_{t_0}^t g_2(s)u(s,t_0,y_0)^{\alpha_1}ds$$
$$+ \sum_{j=1}^L \int_{t_0}^t b_j(s)(\int_{t_0}^s c_j(\theta)\exp(\int_\theta^s -\alpha(\theta)d\theta)u(t,t_0,y_0)d\theta ds$$
$$+ \sum_{k=1}^M \int_{t_0}^t e_k(s)\int_{t_0}^{st} f_k(\theta)\exp(\int_\theta^s -\alpha(\theta)d\theta)u(t,t_0,y_0)^{\alpha_1}d\theta ds. \qquad (9.28)$$

The fact holds that there exists a sufficient smaller $\|y_0\|$ such as

$$1 + (1-\alpha)\|y_0\|^{\alpha-1}\int_{t_0}^t Q_1(s)\exp((\alpha-1)\int_{t_0}^s P_1(\theta)d\theta)ds \geq \varepsilon, \qquad (9.29)$$

for each $\varepsilon > 0$ which is in view of the assumption of Theorem 9.8(3).

Using Theorems 8.1, 9.8(2), (3), and (9.29), we have

$$u(t) \leq \|y_0\|\exp(\int_{t_0}^s P_1(s)ds)$$
$$\cdot \{1 + (1-\alpha_1)\|y_0\|^{\alpha-1}\int_{t_0}^t Q_1(s)\exp((\alpha_1-1)\int_{t_0}^s P_1(\theta)d\theta)ds\}^{\frac{1}{1-\alpha_1}}, t \geq t_0. \qquad (9.30)$$

Further, we obtain

$$\|y(t,t_0,y_0)\| \leq \|y_0\|\exp(\int_{t_0}^t \alpha(s)ds)\exp(\int_{t_0}^t P_1(s)ds)$$
$$\left\{1 + (1-\alpha_1)\|y_0\|^{\alpha-1}\int_{t_0}^t Q_1(s)\exp((\alpha_1-1)\int_{t_0}^s P_1(\theta)d\theta)ds\right\}^{\frac{1}{1-\alpha_1}}. \qquad (9.31)$$

Consider the following facts that $\alpha(s) > 0$, $\int_{t_0}^{\infty} P_1(s)ds < \infty$, for each $t_0 \in \mathbb{I}$,

$$1 + (1 - \alpha_1)\|y_0\|^{\alpha-1} \int_{t_0}^{t} Q_1(s) \exp((\alpha_1 - 1) \int_{t_0}^{s} P_1(\theta)d\theta)ds < 1$$
$$1 + (1 - \alpha_1)\|y_0\|^{\alpha-1} \int_{t_0}^{t} Q_1(s) \exp((\alpha_1 - 1) \int_{t_0}^{s} P_1(\theta)d\theta)ds \geq \varepsilon. \quad (9.32)$$

Let

$$M(t, t_0) := \exp(\int_{t_0}^{t} (\alpha(s) + P_1(s))ds)$$
$$\cdot \left\{ 1 + (1 - \alpha_1)\|y_0\|^{\alpha-1} \int_{t_0}^{t} Q_1(s) \exp((\alpha_1 - 1) \int_{t_0}^{s} P_1(\theta)d\theta)ds \right\}^{\frac{1}{1-\alpha_1}}. \quad (9.33)$$

We conclude that there exists a positive constant M such that

$$M(t, t_0) \leq M. \quad (9.34)$$

Other, we still consider the following inequalities

$$M(t, t_0) \geq 0,$$
$$M(t, t_0) \leq \exp(\int_{t_0}^{t} \alpha(s)ds) \exp(\int_{t_0}^{t} P_1(s)ds).$$

So

$$\lim_{t \to \infty} M(t, t_0) \leq \exp(\int_{t_0}^{t} \alpha(s)ds) \exp(\int_{t_0}^{t} P_1(s)ds) \leq 0.$$

Clearly, we obtain that

$$\lim_{t \to \infty} M(t, t_0) = 0$$

uniformly for $t \geq t_0$ holds.

The proof is completed. □

Definition 9.5 The solution $x(t, t_0, x_0)$ of (9.20) is called exponentially stable behind T if for each $t_0 \in \mathbb{I}, \|x_0\| > 0$, there exists positive constants T, ϕ, λ, such that

$$\|x(t, t_0, x_0)\| \leq \phi \|x_0\| \exp(-\lambda(t - t_0)), \ t > T \gg t_0 > 0.$$

Theorem 9.9 *In Theorem 9.8, the other assumptions hold except the assumption (1), (1)* $\lambda(f_x(t, x)) \leq \alpha(t) < 0$ and*

$$\tau = \lim_{t \to \infty} \sup_{\forall t_0 \in I} \frac{1}{t - t_0} \int_{t_0}^{t} \alpha(s)ds < 0, \quad (9.35)$$

for each $t_0 \in \mathbb{I}$, then the trivial solution of (9.20) is exponentially stable behind T.

Proof Based on Theorem 9.8, we can easily obtain

9.3 Lipschitz Stability on Nonlinear Perturbed Systems

$$\|y(t, t_0, y_0)\| \leq \|y_0\| \exp(\int_{t_0}^{t} (\alpha(s)ds) \exp(\int_{t_0}^{t} (P_1(s)ds)$$
$$\cdot \left\{ 1 + (1 - \alpha_1) \|y_0\|^{\alpha-1} \int_{t_0}^{t} Q_1(s) \exp((\alpha_1 - 1) \int_{t_0}^{s} P_1(\theta)d\theta)ds \right\}^{\frac{1}{1-\alpha_1}}.$$
(9.36)

In view of the assumption of Theorem 9.9(1)*

$$\lambda(f_x(t, x)) \leq \alpha(t) < 0$$

and

$$\tau = \lim_{t \to \infty} \sup_{\forall t_0 \in I} \frac{1}{t-t_0} \int_{t_0}^{t} \alpha(s)ds < 0,$$
(9.37)

for each $t_0 \in \mathbb{I}$.

We know from [2], that there exists some T such that

$$\int_{t_0}^{t} \alpha(s)ds \leq \frac{\tau}{2}(t - t_0), t \geq T \gg t_0 > 0.$$
(9.38)

At the same time, we conclude that there exists a positive constant ϕ such that

$$\exp(\int_{t_0}^{t} \alpha(s)ds) \leq \phi, \forall t \geq t_0,$$
(9.39)

in view of Theorem 9.8(3).

So, we obtain, if $t > T \gg t_0 > 0$ and $\phi_1 = \max(\phi, \phi \varepsilon^{\frac{1}{1-\alpha}})$,

$$\|y(t, t_0, y_0)\| \leq \phi \|y_0\| \exp\left(-\frac{\lambda}{2}(t - t_0) \right), t > T \gg t_0 > 0,$$

which is along with (9.32), (9.38) and (9.39).

Corollary 9.1 *For Theorem 9.8, if*

$$\sup_{t \in \mathbb{I}} \alpha(t) \leq \alpha < 0,$$

where α is some negative constant, then the trivial solution of (9.20) is exponentially stable.

Moreover, the above conclusions are similarly generalized to the linear perturbed system (see [2–4])

$$\dot{y} = A(t)y + f(t, Ty) + g(t, y, Ly)$$
(9.40)

where f, g, T, L are defined the same as (9.20)'s. We omit these discussions.

9.4 Example Analysis

In this section, we still illustrate the above results by the simple example.

Example 9.4 Consider the different equation

$$\begin{aligned}\dot{x} &= -\frac{e^t}{1+2e^t}x, \\ \dot{y} &= -\frac{e^t}{1+2e^t}y + g(t, Ty) + (t, y, Ly),\end{aligned} \quad (9.41)$$

where

$$\|g(t, Ty)\| \leq \exp\left(-\tfrac{1}{2}t\right) \int_{t_0}^{t} c_1(s) \|y(s)\| ds,$$
$$\|h(t, y, Ly)\| \leq g_1(t) \|y(t)\| + g_2(t) \|y(t)\|^2 + \exp\left(-\tfrac{1}{2}t\right) \int_{t_0}^{t} f(s) \|y(s)\|^2 ds,$$

$$\int_{t_0}^{t} c_1(s) ds < \infty,$$
$$\int_{t_0}^{t} g_1(s) ds < \infty,$$
$$\int_{t_0}^{t} g_2(s) ds < \infty,$$

and

$$\int_{t_0}^{t} f(s) ds < \infty.$$

Especially, $\alpha(t) \leq -\frac{1}{2}, b(t) = e(t)$. Combination of the assumption and Corollary 9.10, we obtain that the trivial solution of (9.41) is exponential stable.

9.5 Conclusion

In the chapter, the notions of uniform Lipschitz stability are generalized and the relationship between these notions is analyzed. Further, several sufficient conditions are presented about uniform Lipschitz asymptotical stability of nonlinear systems. These sufficient conditions can similarly be generalized to linear perturbed differential systems. Finally, an example of uniform Lipschitz asymptotic stability of nonlinear differential systems is shown.

Acknowledgements This work is Supported by National Key Research and Development Program of China (2017YFF0207400).

References

1. Alekseev VM. An estimate for the perturbation of the solution of ordinary differential equation (in Russian). Vestinik Moskov Univ Serl Math Mekn. 1961;2:28–36.
2. Brauer F. Perturbations of nonlinear systems of differential equations. J Math Anal Appl. 1966;14:198–206.

References

3. Dannan FM, Elaydi S. Lipschitz stability of nonlinear differential equation II. J Math Anal Appl. 1986;113:562–79.
4. Dannan FM, Elaydi S. Lipschitz stability of nonlinear differential equation. J Math Anal Appl. 1989;143:517–29.
5. Elaydi S, Rao M, Rama M. Lipschitz stability for nonlinear volterra integrodifferential systems. Appl Math Comput. 1988;27(3):191–9.
6. Gallo A, Piccirillo AM. About new analogies of GronwallCBellmanCBihari type inequalities for discontinuous functions and estimated solutions for impulsive differential systems. Nonlinear Anal Theory Methods Appl. 2007;67(5):1550–9.
7. Giovanni A, Sergio V. Lipschitz stability for the inverse conductivity problem. Adv Appl Math. 2005;35(2):207–41.
8. Guo S, Irene M, Si L, Han L. Several integral inequalities and their applications in nonlinear differential systems. Appl Math Comput. 2013;219:4266–77.
9. Hale JK. Ordinary differential equations. New York: Wiley-Interscience; 1969.
10. Jiang F, Meng F. Explicit bounds on some new nonlinear integral inequalities with delay. J Comput Appl Math. 2007;205(1):479–86.
11. Li Y. Boundness, stability and error estimate of the solution of nonlinear different equation. Acta Mathematia Sinica. 1962;12(1):28–36 (in Chinese).
12. Mitropolskiy YA, Iovane G, Borysenko SD. About a generalization of Bellman-Bihari type inequalities for discontinuous functions and their applications. Nonlinear Anal Theory Methods Appl. 2007;66(10):2140–65.
13. Soliman AA. Lipschitz stability with perturbing liapunov functionals. Appl Math Lett. 2004;17(8):939–44.
14. Soliman AA. On Lipschitz stability for comparison systems of differential equations via limiting equation. Appl Math Comput. 2005;163(3):1061–7.
15. Wang W. A generalized retarded Gronwall-like inequality in two variables and applications to BVP. Appl Math Comput. 2007;191(1):144–54.
16. Ye H, Gao J, Ding Y. A generalized Gronwall inequality and its application to a fractional differential equation. J Math Anal Appl. 2007;328(2):1075–81.
17. Yoshizawa T. Stability theory and the existence of periodic solutions and almost periodic solutions. New York: Springer; 1975.

Chapter 10
(c_1, c_1) Stability of a Class of Neutral Differential Equations with Time-Varying Delay

10.1 Introduction

In analyzing many dynamical engineering systems governed by certain differential and integral equations, one often needs some kinds of inequalities, such as Greene inequalities, Gronwall–Bellman inequalities, the Hardy type inequalities and the Volterra type integral inequalities, see [1–17]. These inequalities and their various linear and nonlinear generalizations are crucial in the discussion of the existence, uniqueness, continuation, boundedness, stability, and other qualitative properties of solutions of differential and integral equations. see [1–17]. This field of integral inequalities has developed significantly over the last decade, and has yielded many new results and powerful applications in numerical integration, probability theory and stochastic systems, statistics, information theory, integral operator theory, approximation theory, modern control theory, numerical analysis, and engineering problems.

In this chapter, we shall study (c_1, c_1) stable and (c_1, c_1) asymptotically stable of time-varying delay neutral differential equations based on proposed inequalities method. It is well known that nonlinear differential equation theory has been an active area of research for many years [1–11, 16]. There are many methods or tools to deal with different nonlinear systems, including integral inequalities methods, Lyapunov functional methods. As one of the most useful methods, it is available for studying the asymptotic behavior of the solutions of a nonlinear differential systems, by comparison with a suitable linear systems and by the use of the variation of constants formula in references [1, 2, 5]. We cannot forget the fact that these methods are restricted to nonlinear systems which are small perturbations of the linear systems, or to some nonlinear systems about which information is available, because it involves the variation of constants formula. In applications of time-varying delay neutral differential equations, we present several stability criteria about (c_1, c_1) stable and (c_1, c_1) asymptotically stable, as sufficient conditions, by using the Theorem 8.1 in the Chap. 8 and the parameter variation method in [5]. These results extend and improve the corresponding conclusions appearing in reference [1–4]. More importantly, the above sufficient conditions may similarly be generalized to the above time-varying delay neutral linear differential equations.

10.2 Stability Criteria

In this section, we discuss some sufficient conditions of stability on a type of neutral delay differential system by using the Theorem 8.1 and the parameter variation method in [5], which extend and improve the conclusions in [1–4].

Let $\mathbb{I}^+ = [t_0, +\infty), \mathbb{R}^+ = [0, +\infty)$, we consider the following scalar neutral delay equation:

$$\dot{x}(t) = f(t, x(t)), x(t - \Delta(t)), \dot{x}(t - \Delta(t))), \tag{10.1}$$

where $f : \mathbb{I}^+ \times \mathbb{R} \times \mathbb{R} \times \mathbb{R}, \Delta : \mathbb{I}^+ \to \mathbb{R}^+$.

The initial conditions
$$x(t) = \varphi(t), \forall t \in \mathbb{E}_{t_0}. \tag{10.2}$$

Suppose that $\varphi(t)$ is continuously differentiable in $\mathbb{E}_{t_0}$, the (10.1)'s right side satisfies $f(t, 0, 0, 0) = 0$, and $x(t, \varphi)$ indicates the solution of (10.1) with initial conditions (10.2).

Now we introduce some definitions that we need in the sequel.

Definition 10.1 For every $\varepsilon > 0, t \geq t_0$, if there exists a $\delta > 0$ such that $\rho_\omega(\varphi(t), 0) < \delta$, then $\rho_\Omega(x(t, \varphi), 0) < \varepsilon$ holds, we call that the solution of Eq. (10.1) is (ω, Ω) stable.

Definition 10.2 If the solution of Eq. (10.1) is (ω, Ω) stable, and for any $t \geq t_0$, there exists an $\delta_0 = \delta_0(t_0)$ such that $\rho_\omega(\varphi(t), 0) < \delta$, then $\lim_{t \to +\infty} \rho_\Omega(x(t, \varphi), 0) = 0$ holds, we call that the solution of Eq. (10.1) is (ω, Ω) asymptotically stable.

Here, we need to note that $\rho_\omega(x, y)$ indicates the ω-norm between x and y in some norm space.

Note 10.1 If we choose $\rho_\omega(\varphi(t), 0), \rho_\Omega(x(t, \varphi), 0)$ as follows:

$$\rho_\omega(\varphi(t), 0) = |\varphi(t)| + |\dot{\varphi}(t)|,$$
$$\rho_\Omega(x(t, \varphi), 0) = |x(t, \varphi)| + |\dot{x}(t, \varphi)|,$$

Then, these above stable concepts appearing in Definitions 10.1 and 10.2 mean (c_1, c_1) stable and (c_1, c_1) asymptotically stable, correspondingly.

We omit to outline these definitions here.
Consider the following neutral delay system:

10.2 Stability Criteria

$$\dot{x}(t) = A(t)x(t) + f(t, x(t - \Delta(t)), \dot{x}(t - \Delta(t)), \int_{t_0}^{t} h(s, x(s - \Delta(s)), \dot{x}(s - \Delta(s))ds))), \quad (10.3)$$

where $A(t) \in (a_{ij}(t))^{n \times n}$, $x(t) \in \mathbb{R}^n$, $f \in C(\mathbb{I} \times \mathbb{R}^n \times \mathbb{R}^n, \mathbb{R}^n)$, $h \in (\mathbb{I} \times \mathbb{R}^n \times \mathbb{R}^n, \mathbb{R}^n)$, the delay $\Delta(t)$ is a positive bounded continuous function, and $0 < \Delta_0 \leq \Delta \leq \Delta_1$.

Meanwhile, we assume that the right side of $A(t)x(t) + f(t, y, z, \omega)$ on system (10.3) is defined and continuous when $t \geq t_0$, $||x|| \leq H$, $||y|| \leq H$, $||z|| \leq H$, $||\omega|| \leq H$, H is some constant. In addition, for all $t \geq t_0$, $A(t)$ is bounded and $f(t, 0, 0, 0) = 0$.

Given the initial conditions

$$x(t) = \varphi(t), \quad \dot{x}(t) = \dot{\varphi}(t), \quad t_0 - \Delta \leq t \leq t_0, \quad (10.4)$$

where $\varphi(t)$ and $\dot{\varphi}(t)$ are continuous on $t_0 - \Delta \leq t \leq t_0$.

According to the parameter variation method in the [5] and integrating (10.3), we can get the equivalent solution of (10.3) as follows:

$$x(t) = Y(t, t_0)\varphi(t_0) + \int_{t_0}^{t} Y(t, s) f(s, x(s - \Delta(s)), \dot{x}(s - \Delta(s)), \int_{t_0}^{t} h(\theta, x(\theta - \Delta(\theta)), \dot{x}(\theta - \Delta(\theta)))d\theta)ds, \quad (10.5)$$

where $Y(t, s)$ $(t, s \geq t_0)$ satisfied

$$\frac{\partial(t, s)}{\partial t} = A(t)Y(t, s), \quad Y(s, s) = I. \quad (10.6)$$

We begin by recalling some existed results, which are useful in our subsequent analysis.

Lemma 10.1 [10] *Assume that $h(t)$ and $k(t)$ are nonnegative continuous functions for $t \geq t_0$, then for $\alpha \geq 1$ and all $t \geq t_0$, we have*

$$h(t)^\alpha + k(t)^\alpha \leq (h(t) + k(t))^\alpha. \quad (10.7)$$

The proof procedure is easily obtained from Jensen inequality in [10], so we omit the details.

The following theorems are one of main results in this chapter.

Theorem 10.2 *For system (10.3), assume the following conditions hold for $t \geq t_0$.*
(I) $||Y(t, s)|| \leq \exp(\int_s^t r(\theta)d\theta)$ $(t \geq s \geq t_0)$ and $r(t)$ is a nonpositive bounded function (that is to say that there exists a constant r, which satisfies $r \leq r(t) < 0$);
(II)

$$\|f(t, x(t - \Delta(t)), \dot{x}(t - \Delta(t)), \int_{t_0}^t h(s, x(s - \Delta(s)), \dot{x}(s - \Delta(s))ds)))\|$$
$$\leq g_1(t)(\|x(t - \Delta(t))\| + \|\dot{x}(t - \Delta(t))\|)$$
$$+ \sum_{i=2}^N g_i(s)(\|x(s - \Delta(s))\|^{\alpha_i} + \|\dot{x}(s - \Delta(s))\|^{\alpha_i})$$
$$+ b_1(t) \int_{t_0}^t e^{r(t-s)} c_1(s)(\|x(s - \Delta(s))\| + \|\dot{x}(s - \Delta(s))\|) ds \qquad (10.8)$$
$$+ \sum_{j=2}^L b_j(t) \int_{t_0}^t e^{r(t-s)} c_j(s)((\|x(s - \Delta(s))\|^{\beta_j} + \|\dot{x}(s - \Delta(s))\|^{\beta_j})) ds,$$

where $g_i(t)$ $(i = 1, \cdots, N), b_j(t)$ $(j = 1, \cdots, L)$ both are monotonically nonincreasing positive functions, $c_j(t)$ $(j = 1, \cdots, L)$ are nonnegative continuous functions, $\alpha_1 = 1 < \alpha_2 \leq \alpha_3 \leq \cdots \leq \alpha_N$, $\beta_1 = 1 < \beta_2 \leq \beta_3 \leq \cdots \leq \beta_L$;

(III)
$$\int_{t_0}^{+\infty} g_i(s) e^{(\alpha_i - 1) \int_s^t r(\theta) d\theta} ds < +\infty, \quad (i = 1, \cdots, N),$$
$$\int_{t_0}^{+\infty} b_j(s) c_j(s) ds < +\infty,$$
$$\int_{t_0}^{+\infty} b_j(s) \int_{t_0}^t c_j(s) ds ds < +\infty, \ (c_j = 1, \cdots, L).$$

Then the zero solution of the system (10.3) is (c_1, c_1) stable; Especially if

$$\int_{t_0}^{+\infty} (r(t) + M_4(g_1(t) + b_1(t) \int_{t_0}^t c_1(s) ds)) dt = -\infty$$

holds, where

$$M_1 = \sup_{t_0 \leq t < +\infty} e^{-\int_{t-\Delta(t)}^t r(s) ds} < +\infty,$$
$$M_2 = \frac{M_1}{\Delta_0} \int_{t_0 - \Delta_0}^{t_0} g_1(s) ds + \sum_{i=2}^N \frac{M_1^{\alpha_i}}{\Delta_0} \int_{t_0 - \Delta_0}^{t_0} g_i(s) e^{(\alpha_i - 1) \int_{t_0}^s r(\theta) d\theta} ds, \qquad (10.9)$$
$$M_3 = \max \frac{M_1}{\Delta_0}, \frac{M_1^{\alpha_i}}{\Delta_0}, (i = 2, \cdots, N),$$
$$M_4 = \max((m + M_2), (mM_1 + M_3), M_1, (mM_1^{\alpha_i} + M_3)(i = 2, \cdots, N),$$
$$M_1^{\beta_i} (j = 2, \cdot, L), mM_1, mM_1^{\beta_i} (j = 2, \cdots, L)).$$

then, the zero solution of (10.3) is (c_1, c_1) asymptotically stable.

Proof From the integral system (10.5), we have

$$x(t) = Y(t, t_0) \varphi(t_0) + \int_{t_0}^t Y(t, s) f(s, x(s - \Delta(s)), \dot{x}(s - \Delta(s)),$$
$$\int_{t_0}^t h(\theta, x(\theta - \Delta(\theta)), \dot{x}(\theta - \Delta(\theta))) d\theta) ds.$$

From (10.3) and (10.5), we further have

$$\dot{x} = A(t) Y(t, t_0) \varphi(t_0)$$
$$+ \int_{t_0}^t A(t) Y(t, s) f(s, x(s - \Delta(s)), \dot{x}(s - \Delta(s)), \int_{t_0}^s h(\theta, x(\theta - \Delta(\theta)), \dot{x}(\theta - \Delta(\theta))) d\theta) ds$$
$$+ f(t, x(t - \Delta(t)), \dot{x}(t - \Delta(t)), \int_{t_0}^t h(s, x(s - \Delta(s)), \dot{x}(s - \Delta(s))) ds).$$

By Theorem 10.2 (I) and (II), we obtain

10.2 Stability Criteria

$$\|x(t)\| \leq \|\varphi(t_0)\| \exp(\int_{t_0}^{t} r(s)ds)$$
$$+ \int_{t_0}^{t} \exp(\int_{s}^{t} r(\theta)d\theta) g_1(s)(\|x(s - \Delta(s))\| + \|\dot{x}(s - \Delta(s))\|)ds$$
$$+ \sum_{i=2}^{N} \int_{t_0}^{t} g_i(s) \exp(\int_{s}^{t} r(\theta)d\theta)(\|x(s - \Delta(s))\|^{\alpha_i} + \|\dot{x}(s - \Delta(s))\|^{\alpha_i})ds$$
$$+ \int_{t_0}^{t} b_1(s) e^{\int_{s}^{t} r(\theta)d\theta} \int_{t_0}^{s} \exp(r(s - \theta)) c_1(\theta)(\|x(\theta - \Delta(\theta))\| + \|\dot{x}(\theta - \Delta(\theta))\|)d\theta ds$$
$$+ \sum_{j=2}^{L} \int_{t_0}^{t} b_j(s) e^{\int_{s}^{t} r(\theta)d\theta} \int_{t_0}^{s} e^{r(s-\theta)} c_j(\theta)(\|x(\theta - \Delta(\theta))\|^{\beta_j} + \|\dot{x}((\theta - \Delta(\theta))\|^{\beta_j})d\theta ds,$$

(10.10)

$$\|\dot{x}\| \leq A e^{\int_{t_0}^{t} r(s)ds} \|\varphi(t_0)\|$$
$$+ A \int_{s}^{t} e^{\int_{t_0}^{t} r(\theta)d\theta} g_1(s)(\|x(s - \Delta(s))\| + \|\dot{x}(s - \Delta(s))\|)ds$$
$$+ \sum_{i=2}^{N} \int_{t_0}^{t} A g_i(s) e^{\int_{s}^{t} r(\theta)d\theta}(\|x(s - \Delta(s))\|^{\alpha_i} + \|\dot{x}(s - \Delta(s))\|^{\alpha_i})ds$$
$$+ \int_{t_0}^{t} A e^{\int_{s}^{t} r(\theta)d\theta} b_1(s) \int_{t_0}^{t} e^{r(s-\theta)} c_1(\theta)(\|x(\theta - \Delta(\theta))\| + \|\dot{x}(\theta - \Delta(\theta))\|d\theta ds$$
$$\sum_{j=2}^{L} \int_{t_0}^{t} A e^{\int_{t}^{s} r(\theta)d\theta} b_j(s) \int_{t_0}^{s} e^{r(s-\theta)} c_j(\theta)(\|x(\theta - \Delta(\theta))\|^{\beta_j} + \|\dot{x}(\theta - \Delta(\theta))\|^{\beta_j})d\theta ds$$
$$+ g_1(t)(\|x(t - \Delta(t))\| + \|\dot{x}(t - \Delta(t))\|)$$
$$+ \sum_{i=2}^{N} g_j(t)(\|x(t - \Delta(t))\|^{\alpha_i} + \|\dot{x}(t - \Delta(t))\|^{\alpha_i})$$
$$+ b_1(t) \int_{t_0}^{t} e^{r(t-s)} c_1(s)(\|x(s - \Delta(s))\| + \|\dot{x}(s - \Delta(s))\|)ds$$
$$+ \sum_{j=2}^{L} b_j(t) \int_{t_0}^{t} e^{r(t-s)} c_j(s)(\|x(s - \Delta(s))\|^{\beta_j} + \|\dot{x}(s - \Delta(s))\|^{\beta_j})ds,$$

(10.11)

where $\sup_{t_0 \leq t \leq +\infty} \|A(t)\| \leq A$, A is a positive constant.

Then, combining (10.10) and (10.11), we have

$$(\|x(t)\| + \|\dot{x}(t)\|) \leq m\|\varphi(t_0)\| e^{\int_{t_0}^{t} r(s)ds}$$
$$+ m \int_{t_0}^{t} e^{\int_{s}^{t} r(\theta)d\theta} g_1(s)(\|x(s - \Delta(s))\| + \|\dot{x}(s - \Delta(s))\|)ds$$
$$+ m \sum_{i=2}^{N} \int_{t_0}^{t} e^{\int_{s}^{t} r(\theta)d\theta} g_i(s)(\|x(s - \Delta(s))\|^{\alpha_i} + \|\dot{x}(s - \Delta(s))\|^{\alpha_i})ds$$
$$+ m \int_{t_0}^{t} e^{\int_{s}^{t} r(\theta)d\theta} b_1(s) \int_{t_0}^{s} e^{r(s-\theta)} c_1(\theta)(\|x(\theta - \Delta(\theta))\| + \|\dot{x}(\theta - \Delta(\theta))\|)d\theta ds$$
$$+ m \int_{t_0}^{t} \sum_{j=2}^{L} e^{\int_{s}^{t} r(\theta)d\theta} b_j(s) \int_{t_0}^{s} e^{r(s-\theta)} c_j(\theta)(\|x(\theta - \Delta(\theta))\|^{\beta_j} + \|\dot{x}(\theta - \Delta(\theta))\|^{\beta_j})d\theta ds$$
$$+ g_1(t)(\|x(t - \Delta(t))\| + \|\dot{x}(t - \Delta(t))\|) + \sum_{i=2}^{N} g_i(t)(\|x(t - \Delta(t))\|^{\alpha_i} + \|\dot{x}(t - \Delta(t))\|^{\alpha_i})$$
$$+ b_1(t) \int_{t_0}^{t} e^{r(t-s)} c_1(s)(\|x(s - \Delta(s))\| + \|\dot{x}(s - \Delta(s))\|)ds$$
$$+ \sum_{j=2}^{L} b_j(t) \int_{t_0}^{t} e^{r(t-s)} c_j(s)(\|x(s - \Delta(s))\|^{\beta} + \|\dot{x}(s - \Delta(s))\|^{\beta_j})ds,$$

(10.12)

where $m = 1 + A$.

Equation (10.12)'s both sides divide by $e^{\int_{t_0}^{t} r(s)ds}$, we have

$$(\|x(t)\| + \|\dot{x}(t)\|)\exp(-\int_{t_0}^{t} r(s)ds) \leq m\|\varphi(t_0)\|$$
$$+m\int_{t_0}^{t} e^{-\int_{t_0}^{s} r(\theta)d\theta} g_1(s)(\|x(s - \Delta(s))\| + \|\dot{x}(s - \Delta(s))\|)ds \qquad (10.13)$$
$$+m\sum_{i=2}^{N}\int_{t_0}^{t} e^{-\int_{t_0}^{s} r(\theta)d\theta} g_i(s)(\|x(s - \Delta(s))\|^{\alpha_i} + \|\dot{x}(s - \Delta(s))\|^{\alpha_i})ds$$

$$+m\int_{t_0}^{t} e^{-\int_{t_0}^{s} r(\theta)d\theta} b_1(s)\int_{t_0}^{s} e^{r(s-\theta)} c_1(\theta)(\|x(\theta - \Delta(\theta))\| + \|\dot{x}(\theta - \Delta(\theta))\|)d\theta ds$$
$$+m\int_{t_0}^{t}\sum_{j=2}^{L} e^{-\int_{t_0}^{s} r(\theta)d\theta} b_j(s)\int_{t_0}^{s} e^{r(s-\theta)} c_j(\theta)(\|x(\theta - \Delta(\theta))\|^{\beta_j} + \|\dot{x}(\theta - \Delta(\theta))\|^{\beta_j})d\theta ds$$
$$+g_1(t)(\|x(t - \Delta(t))\| + \|\dot{x}(t - \Delta(t))\|)e^{-\int_{t_0}^{t} r(s)ds}$$
$$+\sum_{i=2}^{N} g_i(t)(\|x(t - \Delta(t))\|^{\alpha_i} + \|\dot{x}(t - \Delta(t))\|^{\alpha_i})e^{-\int_{t_0}^{t} r(s)ds}$$
$$+b_1(t)\int_{t_0}^{t} e^{r(t-s)} c_1(s)(\|x(s - \Delta(s))\| + \|\dot{x}(s - \Delta(s))\|)ds\, e^{-\int_{t_0}^{t} r(s)ds}$$
$$+\sum_{j=2}^{L} b_j(t)\int_{t_0}^{t} e^{r(t-s)} c_j(s)(\|x(s - \Delta(s))\|^{\beta} + \|\dot{x}(s - \Delta(s))\|^{\beta_j})ds\, e^{-\int_{t_0}^{t} r(s)ds}.$$

Using Lemma 10.1, Eq. (10.13) can be written as

$$(\|x(t)\| + \|\dot{x}(t)\|)\exp(-\int_{t_0}^{t} r(s)ds) \leq m\|\varphi(t_0)\|$$
$$+m\int_{t_0}^{t} e^{-\int_{t_0}^{s} r(\theta)d\theta} g_1(s)(\|x(s - \Delta(s))\| + \|\dot{x}(s - \Delta(s))\|)ds$$
$$+m\sum_{i=2}^{N}\int_{t_0}^{t} e^{-\int_{t_0}^{s} r(\theta)d\theta} g_i(s)(\|x(s - \Delta(s))\|^{\alpha_i} + \|\dot{x}(s - \Delta(s))\|^{\alpha_i})ds$$
$$+m\int_{t_0}^{t} e^{-\int_{t_0}^{s} r(\theta)d\theta} b_1(s)\int_{t_0}^{s} e^{r(s-\theta)} c_1(\theta)(\|x(\theta - \Delta(\theta))\| + \|\dot{x}(\theta - \Delta(\theta))\|)d\theta ds$$
$$+m\int_{t_0}^{t}\sum_{j=2}^{L} e^{-\int_{t_0}^{s} r(\theta)d\theta} b_j(s)\int_{t_0}^{s} e^{r(s-\theta)} c_j(\theta)(\|x(\theta - \Delta(\theta))\|^{\beta_j} + \|\dot{x}(\theta - \Delta(\theta))\|^{\beta_j})d\theta ds$$
$$+g_1(t)(\|x(t - \Delta(t))\| + \|\dot{x}(t - \Delta(t))\|)e^{-\int_{t_0}^{t} r(s)ds}$$
$$+\sum_{i=2}^{N} g_i(t)(\|x(t - \Delta(t))\|^{\alpha_i} + \|\dot{x}(t - \Delta(t))\|^{\alpha_i})e^{-\int_{t_0}^{t} r(s)ds}$$
$$+b_1(t)\int_{t_0}^{t} e^{r(t-s)} c_1(s)(\|x(s - \Delta(s))\| + \|\dot{x}(s - \Delta(s))\|)ds\, e^{-\int_{t_0}^{t} r(s)ds}$$
$$+\sum_{j=2}^{L} b_j(t)\int_{t_0}^{t} e^{r(t-s)} c_j(s)(\|x(s - \Delta(s))\|^{\beta} + \|\dot{x}(s - \Delta(s))\|^{\beta_j})ds\, e^{-\int_{t_0}^{t} r(s)ds}$$

$$\leq m\|\varphi(t_0)\| + m\int_{t_0}^{t} e^{-\int_{s-\Delta(s)}^{s} r(\theta)d\theta} g_1(s)(\|x(s - \Delta(s))\| + \|\dot{x}(s - \Delta(s))\|)e^{-\int_{t_0}^{s-\Delta(s)} r(\theta)d\theta} ds$$
$$+m\sum_{i=2}^{N}\int_{t_0}^{t} e^{(\alpha_i - 1)\int_{t_0}^{s} r(\theta)d\theta} g_i(s) e^{-\alpha_i \int_{s-\Delta(s)}^{s} r(\theta)d\theta}$$
$$*[(\|x(s - \Delta(s))\| + \|\dot{x}(s - \Delta(s))\|)e^{-\int_{t_0}^{s-\Delta(s)} r(\theta)d\theta}]^{\alpha_i} ds$$
$$+m\int_{t_0}^{t} b_1(s)\int_{t_0}^{s} e^{-\int_{\theta}^{s}(r-r(\tau)d\tau} c_1(\theta) e^{-\int_{\theta-\Delta(\theta)}^{\theta} r(\theta)d\theta}$$
$$*(\|x(\theta - \Delta(\theta))\| + \|\dot{x}(\theta - \Delta(\theta))\|)e^{\int_{t_0}^{\theta-\Delta(\theta)} r(s)ds} d\theta ds$$

10.2 Stability Criteria

$$+ m \sum_{j=2}^{L} \int_{t_0}^{t} b_j(s) \int_{t_0}^{s} e^{\int_{-\theta}^{s}(r-r(\tau))d\tau} c_j(\theta) e^{(\beta_j-1)\int_{t_0}^{\theta} r(s)ds} e^{-\beta_j \int_{\theta-\Delta(\theta)}^{\theta} r(s)ds}$$
$$* [(\|x(\theta-(\theta))\| + \|\dot{x}(\theta-(\theta))\| e^{-\beta_j \int_{t_0}^{\theta-(\theta)} r(s)ds}]^{\beta_j})) d\theta ds$$
$$+ g_1(t)(\|x(t-(t))\| + \|\dot{x}(t-(t))\|)e^{-\int_{t_0}^{t-\Delta(t)} r(s)ds} e^{-\int_{t-\Delta(t)}^{t} r(s)ds}$$
$$+ \sum_{i=2}^{N} g_i(t)[(\|x(t-(t))\| + \|\dot{x}(t-(t))\|)e^{-\int_{t_0}^{t} r(s)ds}]^{\alpha_i}) e^{-\alpha_i \int_{t-\Delta(t)}^{t} r(s)ds} e^{(\alpha_i-1)\int_{t_0}^{t} r(s)ds}$$

$$+ b_1(t) \int_{t_0}^{t} e^{\int_{s}^{t}(r-r(\theta))d\theta} c_1(s)$$
$$* [(\|x(s-(s))\| + \|\dot{x}(s-(s))\|) e^{\int_{t_0}^{s-\Delta(s)} r(\theta)d\theta}] e^{-\int_{s-\Delta(s)}^{s} r(\theta)d\theta}] ds$$
$$+ \sum_{j=2}^{L} b_j(t) \int_{t_0}^{t} e^{\int_{s}^{t}(r-r(\theta))d\theta} c_j(s) \tag{10.14}$$
$$* [(\|x(s-(s))\| + \|\dot{x}(s-(s))\|) e^{-\int_{t_0}^{s-\Delta(s)} r(\theta)d\theta}]^{\beta_j}) e^{-\beta_j \int_{s-\Delta(s)}^{s} r(\theta)d\theta} e^{(\beta_j-1) \int_{t_0}^{s} r(\theta)d\theta} ds.$$

According to $\Delta(t) \leq \Delta$ and Theorem 10.2 (I), we can get $-\int_{t-\Delta(t)}^{t} r(s)ds$ is bounded. Based on the facts that

$$M_1 = \sup_{t_0 \leq t < +\infty} e^{-\int_{t-\Delta(t)}^{t} r(s)ds} < +\infty, \tag{10.15}$$

combining (10.14) and (10.15), we have

$$\begin{aligned}
u(t) \leq & m\phi + mM_1 \int_{t_0}^{t} g_1(s)u(s-\Delta(s))ds \\
& + \sum_{i=2}^{N} mM_1^{\alpha_i} \int_{t_0}^{t} g_i(s) e^{(\alpha_i-1)\int_{t_0}^{s} r(\theta)d\theta} u(s-\Delta(s))^{\alpha_i} ds \\
& + mM_1 \int_{t_0}^{t} b_1(s) \int_{t_0}^{s} c_1(\theta) u(\theta-\Delta(\theta)) d\theta ds \\
& + \sum_{j=2}^{L} \int_{t_0}^{t} mM_1^{\beta_j} b_j(s) \int_{t_0}^{s} c_j(\theta) e^{(\beta_j-1)\int_{t_0}^{\theta} r(s)ds} u(s-\Delta(s))^{\beta_j} d\theta ds \\
& + M_1 g_1(t) u(t-\Delta(t)) + \sum_{i=2}^{N} M_1^{\alpha_i} g_i(t) e^{(\alpha_i-1)\int_{t_0}^{t} r(s)ds} \\
& + M_1 b_1(t) \int_{t_0}^{t} c_1(s) u(s-\Delta(s))ds \\
& + \sum_{j=2}^{L} M_1^{\beta_j} b_j(s) \int_{t_0}^{t} c_j(\theta) e^{(\beta_i-1)\int_{t_0}^{s} r(s)ds} u(s-\Delta(s))^{\beta_j} ds.
\end{aligned} \tag{10.16}$$

where

$$u(t) = (\|x(t)\| + \|\dot{x}(t)\|) e^{-\int_{t_0}^{t} r(s)ds},$$
$$\phi = \max(\phi_1, \phi_2),$$
$$\phi_1 = \sup_{t_0-\Delta \leq t < +\infty} (\|\phi(t)\| + \|\dot{\phi}(t)\|),$$
$$\phi_2 = \max\{\sup_{t_0-\Delta \leq t < +\infty} (\|\phi(t)\| + \|\dot{\phi}(t)\|^{\alpha_i}), (i = 2, \cdots, N),$$
$$\sup_{t_0-\Delta \leq t < +\infty} (\|\phi(t)\| + \|\dot{\phi}(t)\|)^{\beta_j}, (j = 2, \cdots, L).$$

Since $\xi - \Delta(\xi) \leq t_0$, we have

$$u(\xi - \Delta(\xi)) = (\|\phi(\xi - \Delta(\xi))\| + \|\dot{\phi}(\xi - \Delta(\xi))\|)e^{-\int_{t_0}^{\xi - \Delta(\xi)} r(\theta)d\theta}, \quad (10.17)$$

Combining the above results, we obtain $u(\xi - \Delta(\xi)) \leq \phi_1$.

Let
$$\|u(t)\| = \max\{\phi_1, \max(u(\xi)), t_0 \leq \xi \leq t\},$$

obviously, $\|u(t)\|$ is a monotone nondecreasing function.

Further, we obtain from (10.16),

$$u(t) \leq m\phi + mM_1 \int_{t_0}^{t} g_1(s)\|u(s-\Delta(s))\|ds$$

$$+ \sum_{i=2}^{N} mM_1^{\alpha_i} \int_{t_0}^{t} g_i(s) e^{(\alpha_i - 1)\int_{t_0}^{s} r(\theta)d\theta} u(s - \Delta(s))^{\alpha_i} ds$$

$$+ mM_1 \int_{t_0}^{t} b_1(s) \int_{t_0}^{s} c_1(\theta) u(\theta - \Delta(\theta)) d\theta ds$$

$$+ \sum_{j=2}^{L} \int_{t_0}^{t} mM_1^{\beta_j} b_j(s) \int_{t_0}^{s} c_j(\theta) e^{(\beta_j - 1)\int_{t_0}^{\theta} r(s)ds} u(s - \Delta(s))^{\beta_j} d\theta ds \quad (10.18)$$

$$+ M_1 b_1(t)\|u(t - \Delta(t))\| + \sum_{i=2}^{N} M_1^{\alpha_i} g_i(t)\|u(t - \Delta(t))\|^{\alpha_i} e^{(\alpha_i - 1)\int_{t_0}^{t} r(s)ds}$$

$$+ M_1 b_1(t) \int_{t_0}^{t} c_1(s)\|u(s - \Delta(s))\|ds$$

$$+ \sum_{j=2}^{L} M_1^{\beta_j} b_j(t) \int_{t_0}^{t} c_j(s) e^{(\beta_j - 1)\int_{t_0}^{s} r(\theta)d\theta} \|u(s - \Delta(s))\|^{\beta_j} ds.$$

On the other hand, we have, from Theorem 10.2 (I) and (II),

$$M_1 g_1(t)\|u(t - \Delta(t))\| + \sum_{i=2}^{N} M_1^{\alpha_1} g_i(s)\|u(t - \Delta(t))\|^{\alpha_i} e^{(\alpha_i - 1)\int_{t_0}^{t} r(s)ds}$$

$$\leq M_1 g_1(t)\|u(t - \Delta_0)\| + \sum_{i=2}^{N} M_1^{\alpha_1} g_i(t)\|u(t - \Delta(t))\|^{\alpha_i} e^{(\alpha_i - 1)\int_{t_0}^{t} r(s)ds}$$

$$\leq \frac{M_1}{\Delta_0} \int_{t_0 - \Delta_0}^{t_0} g_1(s)\|u(s)\|ds + \sum_{i=2}^{N} \frac{M_1^{\alpha_i}}{\Delta_0} \int_{t_0 - \Delta_0}^{t_0} g_i(s) e^{(\alpha_i - 1)\int_{t_0}^{s} r(\theta)d\theta} \|u(s)\|^{\alpha_i} ds$$

$$+ \frac{M_1}{\Delta_0} \int_{t_0}^{t} g_1(s)\|u(s)\|ds + \sum_{i=2}^{N} \frac{M_1^{\alpha_i}}{\Delta_0} \int_{t_0}^{t} g_i(s) e^{(\alpha_i - 1)\int_{t_0}^{s} r(\theta)d\theta} \|u(s)\|^{\alpha_i} ds$$

$$\leq M_2 \phi + \int_{t_0}^{t} M_3 g_1(s)\|u(s)\|ds + \sum_{i=2}^{N} \int_{t_0}^{t} M_3 g_i(s) e^{(\alpha_i - 1)\int_{t_0}^{s} r(\theta)d\theta} \|u(s)\|^{\alpha_i} ds, \quad (10.19)$$

where

10.2 Stability Criteria

$$M_2 = \frac{M_1}{\Delta_0}\int_{t_0-\Delta_0}^{t_0} g_1(s)ds + \sum_{i=2}^{N}\frac{M_1^{\alpha_i}}{\Delta_0}\int_{t_0-\Delta_0}^{t_0} g_i(s)e^{(\alpha_i-1)\int_{t_0}^{s}r(\theta)d\theta}ds,$$

$$M_3 = \max\left\{\frac{M_1}{\Delta_0}, \frac{M_1^{\alpha_i}}{\Delta_0}, (i=2,\cdots,N)\right\},$$

$$M_1 b_1(t)\int_{t_0}^{t} c_1(s)\|u(s-\Delta(s))\|ds + \sum_{j=2}^{L} M_1^{\beta_j} b_j(t)\int_{t_0}^{t} c_j(s)e^{(\beta_j-1)\int_{t_0}^{s}r(\theta)d\theta}\|u(s-\Delta(s))\|^{\beta_j}ds$$

$$\leq M_1\int_{t_0}^{t} b_1(s)c_1(s)\|u(s)\|ds + \sum_{j=2}^{L} M_1^{\beta_j}\int_{t_0}^{t} b_j(s)c_j(s)e^{(\beta_j-1)\int_{t_0}^{s}r(\theta)d\theta}\|u(s)\|^{\beta_j}ds.$$
(10.20)

Combining (10.18), (10.19) and (10.20) can be written as

$$u(t) \leq (m+M_2)\phi + \int_{t_0}^{t}[(mM_1+M_3)g_1(s) + M_1 b_1(s)c_1(s)]\|u(s)\|ds$$
$$+ \int_{t_0}^{t}\sum_{i=2}^{N}(mM_1^{\alpha_i}+M_3)g_i(s)e^{(\alpha_i-1)\int_{t_0}^{s}\alpha(\theta)d\theta}\|u(s)\|^{\alpha_i}ds$$
$$+ \int_{t_0}^{t}\sum_{j=2}^{L} M_1^{\beta_j} b_j(s)c_j(s)e^{(\beta_i-1)\int_{t_0}^{s}r(\theta)d\theta}\|u(s)\|^{\beta_j}ds \qquad (10.21)$$
$$+ mM_1\int_{t_0}^{t} b_j(s)\int_{t_0}^{s} c_1(\theta)\|u(\theta)\|d\theta ds$$
$$+ \sum_{j=2}^{L} mM_1^{\beta_j}\int_{t_0}^{t} b_j(s)\int_{t_0}^{s} c_j(\theta)e^{(\beta_i-1)\int_{t_0}^{\theta}r(s)ds}\|u(\theta)\|^{\beta_j}d\theta ds.$$

Noting the following facts such as a monotone nondecreasing function $u(t)$ in (10.21) which is not less than $(m+M_2)\phi$ and the definition of $u(t)$, Eq. (10.21) can be rewritten as

$$\|u(t)\| \leq M_4\phi + M_4\int_{t_0}^{t}(g_1(s)+b_1(s)c_1(s))\|u(s)\|ds$$
$$+M_4\int_{t_0}^{t}\sum_{i=2}^{N} g_i(s)e^{(\alpha_i-1)\int_{t_0}^{s}r(\theta)d\theta}\|u(s)\|^{\alpha_i}ds$$
$$+M_4\int_{t_0}^{t}\sum_{j=2}^{L} b_j(s)c_j(s)e^{(\beta_j-1)\int_{t_0}^{s}r(\theta)d\theta}\|u(s)\|^{\beta_j}ds \qquad (10.22)$$
$$+M_4\int_{t_0}^{t} b_1(s)\int_{t_0}^{s} c_1(\theta)\|u(\theta)\|d\theta ds$$
$$+M_4\sum_{j=2}^{L}\int_{t_0}^{t} b_j(s)\int_{t_0}^{s} c_j(\theta)e^{(\beta_i-1)\int_{t_0}^{\theta}r(s)ds}\|u(\theta)\|^{\beta_j}d\theta ds,$$

where

$$M_4 = \max\{(m+M_2), (mM_1+M_3), M_1, (mM_1^{\alpha_i}+M_3)(i=2,\cdots,N),$$
$$M_1^{\beta_j}(j=2,\cdots,L), mM_1, mM_1^{\beta_j}(j=2,\cdots,L)\}$$

$$G_n(s) = \begin{cases} M_4 g_i(s)e^{(\alpha_i-1)\int_{t_0}^{s}r(\theta)d\theta}, \\ \quad \text{if } k_n=\alpha_i, k_n\neq\beta_j, (i=2,\cdots,N, j=2,\cdots,L); \\ M_4 b_j(s)c_j(s)e^{(\beta_i-1)\int_{t_0}^{s}r(\theta)d\theta}, \\ \quad \text{if } k_n\neq\alpha_i, k_n=\beta_j, (i=2,\cdots,N, j=2,\cdots,L); \\ M_4(g_i(s)e^{(\alpha_i-1)\int_{t_0}^{s}r(\theta)d\theta} + b_j(s)c_j(s)e^{(\beta_i-1)\int_{t_0}^{s}r(\theta)d\theta}), \\ \quad \text{if } k_n=\alpha_i=\beta_j, (i=2,\cdots,N; j=2,\cdots,L), \end{cases}$$

and k_n satisfies $k_2 \leq k_3 \leq \cdots \leq k_M$ and

$$\max(L, N) \leq M \leq L + N - 1, k_2 = \min(\alpha_2, \beta_2), k_M = \max(\alpha_L, \beta_N),$$
$$B_j(s) = M_4 b_j(s), C_j(s) = c_j(s) e^{(\beta_i - 1)\int_{t_0}^s r(\theta)d\theta} (j = 2, \cdots, L).$$

Then, (10.22) can be written as

$$\|u(t)\| \leq u_0 + \int_{t_0}^t G_1(s)\|u(s)\|ds + \sum_{i=2}^N \int_{t_0}^t G_n(s)\|u(s)\|^{k_n}ds$$
$$+ \int_{t_0}^t B_1(s)\int_{t_0}^s C_1(\theta)\|u(\theta)\|d\theta ds + \sum_{j=2}^L \int_{t_0}^t B_j(s)\int_{t_0}^s C_j(\theta)\|u(\theta)\|^{\beta_j}d\theta ds,$$
(10.23)

where $u_0 = M_4 \phi$.

Clearly, the following facts hold that

$$\bar{\alpha} = \max(k_M, \beta_L) = \max(\max(\alpha_N, \beta_L), \beta_L) = \max(\alpha_N, \beta_L),$$
$$\underline{\beta} = \min(k_2, \beta_2) = \min(\min(\alpha_2, \beta_2), \beta_2) = \min(\alpha_2, \beta_2).$$

In view of assumptions of Theorem 10.2,

$$A(t) = \sum_{n=1}^M G_n(t) + \sum_{j=1}^L B_j(t)\int_{t_0}^t C_j(s)ds + \sum_{j=1}^L \int_{t_0}^t B_j(t)\int_{t_0}^t C_j(s)ds,$$
$$= G_1(t) + \sum_{n=1}^M G_n(t) + B_1(t)\int_{t_0}^t C_j(s)ds + 2\sum_{j=2}^L B_j(t)\int_{t_0}^t C_j(s)ds,$$
$$B(t) = \sum_{n=2}^M G_n(t) + \sum_{j=2}^L B_j(t)\int_{t_0}^t C_j(s)ds.$$

and Theorem 10.2 (III)

$$\int_{t_0}^{+\infty} g_i(s) e^{(\alpha_i - 1)\int_{t_0}^s r(\theta)d\theta}ds < +\infty, \quad (i = 1, \cdots, N),$$
$$\int_{t_0}^{+\infty} b_j(s)c_j(s))ds < +\infty, \int_{t_0}^{+\infty} b_j(s)\int_{t_0}^{+\infty} c_j(s)ds < +\infty, (j = 1\cdots, L),$$

further, we have

$$\int_{t_0}^{+\infty} A(t)dt = \int_{t_0}^{+\infty} G_1(s)ds + \sum_{n=2}^M \int_{t_0}^{+\infty} G_n(s)d + \int_{t_0}^{+\infty} B_1(s)\int_{t_0}^s C_1(\theta)d\theta ds$$
$$+ 2\sum_{j=2}^L \int_{t_0}^{+\infty} B_j(s)\int_{t_0}^s C_j(\theta)d\theta ds$$
$$< +\infty.$$
(10.24)

We know that ϕ in u_0 is decided by the initial values, so there exists a sufficient smaller ϕ such that the following inequality

10.2 Stability Criteria

$$1 + (1-\bar{\alpha})u_0^{(\bar{\alpha}-1)} \int_{t_0}^t B(s)\exp((\bar{\alpha}-1)\int_{t_0}^s A(\theta)d\theta)ds > 0$$

holds.

It is easily to find that $||u(t)||$ is bounded from the proof of the Theorem 8.1, so we assume $||u(t)|| \leq M_0$, where M_0 is a positive constant.

Based on Theorem 8.1, let

$$C(t) = M_4 g_1(t) + b_1(t)c_1(t) + b_1(t)\int_{t_0}^t c_1(s)ds,$$

$$D(t) = M_4 \sum_{n=2}^M G_n(t) M_0^{k_n - \underline{\beta}} + \sum_{j=2}^L M_4 b_j(t) \int_{t_0}^t c_j(s) e^{(\beta_j - 1)\int_{t_0}^s r(\theta)d\theta} ds M_0^{\beta_j - \underline{\beta}}.$$

For the same reasons, we have

$$\int_{t_0}^{+\infty} C(s)ds < +\infty, \int_{t_0}^{+\infty} D(s)ds < +\infty,$$
$$1 + (1-\underline{\beta})u_0^{(\underline{\beta}-1)} \int_{t_0}^t D(s)\exp((\underline{\beta}-1)\int_{t_0}^s C(\theta)d\theta)ds > 0. \quad (10.25)$$

Further, we have

$$||u(t)|| \leq u_0 \exp(\int_{t_0}^t C(s)ds)$$
$$\{1 + (1-\underline{\beta})u_0^{(\underline{\beta}-1)} \int_{t_0}^t D(s)\exp[(\underline{\beta}-1)\int_{t_0}^s C(\theta)d\theta]ds\}^{\frac{1}{1-\underline{\beta}}}. \quad (10.26)$$

Now, in view of the fact $u(\xi - \Delta(\xi)) \leq \phi_1$, (10.26) can be written as

$$||x(t)|| \leq u_0 \exp(\int_{t_0}^t (r(s) + C(s))ds)$$
$$\{1 + (1-\underline{\beta})u_0^{(\underline{\beta}-1)} \int_{t_0}^t D(s)\exp[(\underline{\beta}-1)\int_{t_0}^s C(\theta)d\theta]ds\}^{\frac{1}{1-\underline{\beta}}}. \quad (10.27)$$

We obtain the desired result along with the definitions of (c_1, c_1) stable and (c_1, c_1) asymptotically stable. The proof is completed. □

Theorem 10.3 *For Theorem 10.2, if $\alpha_i = i (i = 2, \cdots, N)$, the Theorem 1 and Theorem 2 in [2] is obtained.*

Theorem 10.4 *For Theorem 10.2, if $b_j(t) \equiv 0 (j = 1, \cdots, L)$ or $c_j(t) \equiv 0 (j = 1, \cdots, L)$, many important results such that Theorem 4.5.1, Theorem 4.5.2 in [3, 4] are obtained.*

The above conditions may similarly be generalized to the above time-varying delay neutral linear differential equations based on Theorem 8.1. We omit these details.

10.3 Numerical Examples

Example 10.1 Consider the following complex system:

$$\dot{x}(t) = A(t)x(t) + f(t, x(t - \Delta(t)), \dot{x}(t - \Delta(t)), \int_{t_0}^{t} h(s, x(s - \Delta(s)), \dot{x}(s - \Delta(s))ds))), \quad (10.28)$$

where $A(t) = \begin{bmatrix} -5 & -6 \\ 1 & 0 \end{bmatrix}$ and

$$f(t, x(t - \Delta(t)), \dot{x}(t - \Delta(t)), \int_{t_0}^{t} h(s, x(s - \Delta(s)), \dot{x}(s - \Delta(s))ds)$$
$$= \begin{bmatrix} e^{-1.5t} & 2e^{-1.5t} \\ 0 & e^{-1.5t} \end{bmatrix} \begin{bmatrix} x_1(t - \Delta(t)) \\ x_2(t - \Delta(t)) \end{bmatrix} + \begin{bmatrix} e^{-1.4t} & 2e^{-1.4t} \\ 0 & e^{-1.4t} \end{bmatrix} \begin{bmatrix} \dot{x}_1(t - \Delta(t)) \\ \dot{x}_2(t - \Delta(t)) \end{bmatrix}$$
$$+ \begin{bmatrix} e^{-0.8t} & 2e^{-0.8t} \\ 0 & e^{-0.8t} \end{bmatrix} \begin{bmatrix} x_1^3(t - \Delta(t)) + x_1(t - \Delta(t))x_2^2(t - \Delta(t)) \\ x_2(t - \Delta(t))x_1^2(t - \Delta(t)) + x_2^3(t - \Delta(t)) \end{bmatrix}$$
$$+ \begin{bmatrix} e^{-0.7t} & 2e^{-0.7t} \\ 0 & e^{-0.7t} \end{bmatrix} \begin{bmatrix} \dot{x}_1^3(t - \Delta(t)) + \dot{x}_1(t - \Delta(t))\dot{x}_2^2(t - \Delta(t)) \\ \dot{x}_2(t - \Delta(t))\dot{x}_1^2(t - \Delta(t)) + \dot{x}_2^3(t - \Delta(t)) \end{bmatrix}$$
$$+ \begin{bmatrix} e^{-0.5t} & 2e^{-0.5t} \\ 0 & e^{-0.5t} \end{bmatrix} \int_{t_0}^{t} \begin{bmatrix} e^{r(t-s)}e^{-1.2s} & 2e^{r(t-s)}e^{-1.2s} \\ 0 & e^{r(t-s)}e^{-1.2s} \end{bmatrix} \begin{bmatrix} x_1(s - \Delta(s)) \\ x_2(s - \Delta(s)) \end{bmatrix} ds$$
$$+ \begin{bmatrix} e^{-0.4t} & 2e^{-0.4t} \\ 0 & e^{-0.4t} \end{bmatrix} \int_{t_0}^{t} \begin{bmatrix} e^{r(t-s)}e^{-s} & 2e^{r(t-s)}e^{-s} \\ 0 & e^{r(t-s)}e^{-s} \end{bmatrix} \begin{bmatrix} \dot{x}_1(s - \Delta(s)) \\ \dot{x}_2(s - \Delta(s)) \end{bmatrix} ds$$
$$+ \begin{bmatrix} e^{-0.9t} & 2e^{-0.9t} \\ 0 & e^{-0.9t} \end{bmatrix} \int_{t_0}^{t} \begin{bmatrix} e^{r(t-s)}e^{-s} & 2e^{r(t-s)}e^{-s} \\ 0 & e^{r(t-s)}e^{-s} \end{bmatrix} \begin{bmatrix} x_1^3(s - \Delta(s)) + x_1(s - \Delta(s))x_2^2(s - \Delta(s)) \\ x_2(s - \Delta(s))x_1^2(s - \Delta(s)) + x_2^3(s - \Delta(s)) \end{bmatrix} ds$$
$$+ \begin{bmatrix} e^{-t} & 2e^{-t} \\ 0 & e^{-t} \end{bmatrix} \int_{t_0}^{t} \begin{bmatrix} e^{r(t-s)}e^{-0.8s} & 2e^{r(t-s)}e^{-0.8s} \\ 0 & e^{r(t-s)}e^{-0.8s} \end{bmatrix} \begin{bmatrix} \dot{x}_1^3(s - \Delta(s)) + \dot{x}_1(s - \Delta(s))\dot{x}_2^2(s - \Delta(s)) \\ \dot{x}_2(s - \Delta(s))\dot{x}_1^2(t - \Delta(s)) + \dot{x}_2^3(s - \Delta(s)) \end{bmatrix} ds.$$

We obtain

$$\|f(t, x(t - \Delta(t)), \dot{x}(t - \Delta(t)), \int_{t_0}^{t} h(s, x(s - \Delta(s)), \dot{x}(s - \Delta(s))ds)\|$$
$$\leq e^{-1.4t}(\|x(t - \Delta(t))\| + \|\dot{x}(t - \Delta(t))\|)$$
$$+ e^{-0.7t}(\|x(t - \Delta(t))\|^3 + \|\dot{x}(t - \Delta(t))\|^3)$$
$$+ e^{-0.4t} \int_{t_0}^{t} e^{r(t-s)} e^{-s}(\|x(s - \Delta(s))\| + \|\dot{x}(s - \Delta(s))\|) ds$$
$$+ e^{-0.9t} \int_{t_0}^{t} e^{r(t-s)} e^{-0.8s}(\|x(s - \Delta(s))\|^3 + \|\dot{x}(s - \Delta(s))\|^3) ds.$$

Comparing with Lemma 10.1, we can know that

$$N = L = 2, g_1(t) = e^{-1.4t}, g_2(t) = e^{-0.7t},$$
$$b_1(t) = e^{-0.4t}, b_2(t) = e^{-0.9t},$$
$$c_1(t) = e^{-t}, c_2(t) = e^{-0.8t},$$
$$\alpha_2 = \beta_2 = 3, r(t) = r = -2, m = 1 + A = -1.$$

It is obvious that

10.3 Numerical Examples

Fig. 10.1 Trajectories of x_1 starting on $x_1 \times x_2 = 0, -0.7 \leq x_1 \leq 0.7$

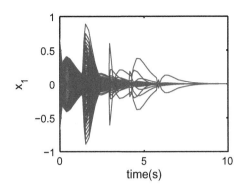

$$\int_{t_0}^{+\infty} g_1(s) e^{(\alpha_2 - 1) \int_s^t r(\theta) d\theta} ds < +\infty,$$
$$\int_{t_0}^{+\infty} g_2(s) e^{(\alpha_2 - 1) \int_s^t r(\theta) d\theta} ds < +\infty,$$
$$\int_{t_0}^{+\infty} b_j(s) c_j(s) ds < +\infty,$$
$$\int_{t_0}^{+\infty} b_j(s) \int_{t_0}^t c_j(s) ds\, ds < +\infty.$$

We suppose $\Delta(t) = \sin^2 t + 1$, $t_0 = 0$, then $\Delta = 2, \Delta_0 = 1$ and

$$M_1 = \sup_{t_0 \leq t < +\infty} e^{-\int_{t-\Delta(t)}^t r(s) ds} = \sup_{t_0 \leq t < +\infty} e^{2\Delta(t)} = e^4 < +\infty,$$
$$M_2 = \frac{M_1}{\Delta_0} \int_{t_0 - \Delta_0}^{t_0} g_1(s) ds + \frac{M_1^{\alpha_2}}{\Delta_0} \int_{t_0 - \Delta_0}^{t_0} g_2(s) e^{(\alpha_2 - 1) \int_{t_0}^s r(\theta) d\theta} ds$$
$$= e^4 \int_{-1}^0 e^{-1.4s} ds + e^{12} \int_{-1}^0 e^{-0.7s} e^{2 \int_0^s -2d\theta} ds$$
$$= \tfrac{5}{7} e^4 (1 - e^{1.4}) - \tfrac{10}{47} e^{12} (1 - e^{4.7}),$$
$$M_3 = \max \left\{ \frac{M_1}{\Delta_0}, \frac{M_1^{\alpha_2}}{\Delta_0} \right\} = \max \{e^4, e^{12}\} = e^{12},$$
$$M_4 = \max \left\{ (m + M_2), (mM_1 + M_3), M_1, (mM_1^{\alpha_2} + M_3), M_1^{\beta_2}, mM_1, mM_1^{\beta_2} \right\}$$
$$= \max \{M_2 - 1, M_3 - M_1, M_1, M_3 - M_1^3, M_1^3, -M_1, -M_1^3\}$$
$$= M_2 - 1$$
$$= \tfrac{5}{7} e^4 (1 - e^{1.4}) - \tfrac{10}{47} e^{12} (1 - e^{4.7}) - 1.$$

Further, we have

$$\int_{t_0}^{+\infty} (r(t) + M_4(g_1(t) + b_1(t) \int_{t_0}^t c_1(s) ds)) dt$$
$$= \int_{t_0}^{+\infty} (-2 + M_4(e^{-1.4t} + e^{-0.4t} \int_{t_0}^t e^{-s} ds)) dt$$
$$= \int_{t_0}^{+\infty} (-2 + (\tfrac{5}{7} e^4 (1 - e^{1.4}) - \tfrac{10}{47} e^{12} (1 - e^{4.7}) - 1) e^{-0.4t}) dt$$
$$= -\infty.$$

According to the Lemma 10.1, the zero solution of the system is (c_1, c_1) asymptotically stable.

The results are shown in following figures:

Fig. 10.2 Trajectories of x_2 starting on $x_1 \times x_2 = 0, -0.4 \leq x_2 \leq 0.4$

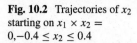

Fig. 10.3 Phase trajectories starting on $x_1 \times x_2 = 0, -0.7 \leq x_1 \leq 0.7, -0.4 \leq x_2 \leq 0.4$

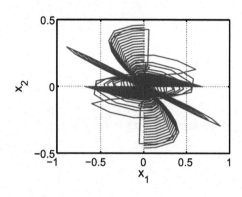

Fig. 10.4 Trajectories of x_1 starting on the $x_2 = \pm 0.5 x_1, -0.4 \leq x_1 \leq 0.4$

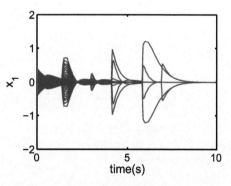

(i) Figures 10.1, 10.2, and 10.3 show the trajectories which start on the axis, where $-0.7 < x_1 < 0.7, x_2 = 0$ and $-0.4 < x_2 < 0.4, x_1 = 0$.

From Fig. 10.1, 10.2, 10.3, 10.4, 10.5, 10.6, 10.7, 10.8, and 10.9, we obtain that the zero of (10.28) is (c_1, c_1) asymptotically stable, but from Fig. 10.10, the trajectories, which start on the $x_1^2 + x_2^2 = 1$ and $x_1 = -1, -0.9, -0.8, \cdots, -0.1, 0, 0.1, \cdots, 0.8, 0.9, 1$ are diffuse. So, we conclude that the zero of (10.28) is locally (c_1, c_1) asymptotically stable, not (c_1, c_1) asymptotically stable in large.

10.3 Numerical Examples

Fig. 10.5 Trajectories of x_2 starting on the $x_2 = \pm 0.5x_1, -0.4 \leq x_1 \leq 0.4$

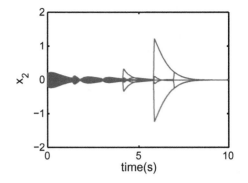

Fig. 10.6 Phase trajectories starting on the $x_2 = \pm 0.5x_1, -0.4 \leq x_1 \leq 0.4$

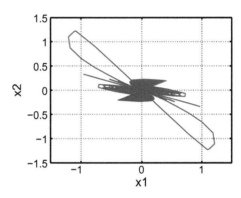

Fig. 10.7 Trajectories of x_1 starting on the $x_2 = x_1^2, -0.4 \leq x_1 \leq 0.4$

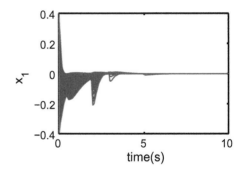

(ii) Figures 10.4, 10.5 and 10.6 show the trajectories which start on the lines of $x_2 = \pm 0.5x_1$, where $-0.4 \leq x_1 \leq 0.4$.

(iii) Figures 10.7, 10.8 and 10.9 show the trajectories which start on the curve of $x_2 = x_1{}^2$, where $-0.4 \leq x_1 \leq 0.4$.

(iv) Figure 10.10 shows the trajectories which start on the $x_1^2 + x_2^2 = 1$, where $x_1 = -1, -0.9, -0.8, \cdots, -0.1, 0, 0.1, \cdots, 0.8, 0.9, 1$.

Fig. 10.8 Trajectories of x_2 starting on the $x_2 = x_1^2, -0.4 \leq x_1 \leq 0.4$

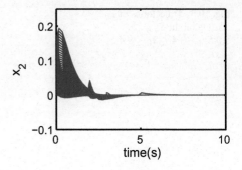

Fig. 10.9 Phase trajectories starting on the $x_2 = x_1^2, -0.4 \leq x_1 \leq 0.4$

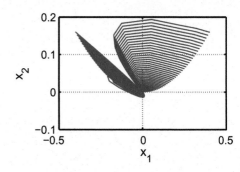

Fig. 10.10 Phase trajectories starting on the $x_1^2 + x_2^2 = 1, -1 \leq x_1 \leq 1$

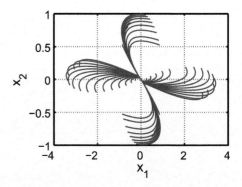

10.4 Conclusion

In this chapter, the notions of (ω, Ω) stable and (ω, Ω) asymptotically stable, especially (c_1, c_1) stable and (c_1, c_1) asymptotically stable are presented. Several sufficient conditions about (c_1, c_1) stable and (c_1, c_1) asymptotically stable of time-varying delay neutral differential equations are established by the proposed integral inequalities. The above sufficient conditions may similarly be generalized to the above time-varying delay neutral linear differential equations. Finally, a complex numerical example is presented to illustrate the main result effectively.

Acknowledgements This work is Supported by National Key Research and Development Program of China (2017YFF0207400).

References

1. Guo S, Irene M, Si L, Han L. Several integral inequalities and their applications in nonlinear differential systems. Appl Math Comput. 2013;219:4266–77.
2. Hong J. A result on stability of time-varying delay differential equation. Acta Math Sin. 1983;26(3):257–61.
3. Si L. Boundness, stability of the solution of time-varying delay neutral differential equation. Acta Math Sin. 1974;17(3):197–204.
4. Si L. Stability of delay neutral differential equations. Huhhot: Inner Mongolia Educational Press; 1994. p. 106–41.
5. Alekseev V M. An estimate for the perturbation of the solution of ordinary differential equation. Vestnik Moskovskogo Universiteta. Seriya I. Matematika, Mekhanika, 1961;2:28-36. in Russian.
6. Li Y. Boundness, stability and error estimate of the solution of nonlinear different equation. Acta Math Sin. 1962;12(1):28–36 In Chinese.
7. Brauer F. Perturbations of nonlinear systems of differential equations. J Math Anal Appl. 1966;14:198–206.
8. Elaydi S, Rao M, Rama M. Lipschitz stability for nonlinear Volterra integro differential systems. Appl Math Comput. 1988;27(3):191–9.
9. Giovanni A, Sergio V. Lipschitz stability for the inverse conductivity problem. Adv Appl Math. 2005;35(2):207–41.
10. Hale JK. Ordinary differential equations. Interscience, New York: Wiley; 1969.
11. Jiang F, Meng F. Explicit bounds on some new nonlinear integral inequalities with delay. J Comput Appl Math. 2007;205(1):479–86.
12. Soliman AA. Lipschitz stability with perturbing Liapunov functionals. Appl Math Lett. 2004;17(8):939–44.
13. Soliman AA. On Lipschitz stability for comparison systems of differential equations via limiting equation. Appl Math Comput. 2005;163(3):1061–7.
14. Wang W. A generalized retarded Gronwall-like inequality in two variables and applications to BVP. Appl Math Comput. 2007;191(1):144–54.
15. Ye H, Gao J, Ding Y. A generalized Gronwall inequality and its application to a fractional differential equation. J Math Anal Appl. 2007;328(2):1075–81.
16. Mitropolskiy YA, Iovane G, Borysenko SD. About a generalization of bellman-bihari type inequalities for discontinuous functions and their applications. Nonlinear Anal Theory Methods and Appl. 2007;66(10):2140–65.
17. Guo S, Han L. Generalization of integral inequalities and (c_1, c_1) stability of a type of neutral differential equations with time-varying delays. J Syst Eng Electron. 2017;28(2):347–60.

Part IV
Applications in Nonlinear Mechanical and Electrical Systems

Part IV

Chapter 11
The Chaos Synchronization, Encryption for a Type of Three-Stage Communication System

11.1 Introduction

In the recent years, the application of chaotic systems has attracted lots of attention in various fields of science and technology, such as biological systems [1], chemicals [2–5], economics [6, 7], and secure communication [8–11]. In 1990, Pecora and Carroll [12, 13] first proposed the idea for chaos synchronization which was verified by the electric current later. In 1963, Lorenz found the first chaotic attractor [15]. In 1999, Chen found another chaotic attractor [16]. In 2001, *Lü* found a new chaotic system [17], and later, *Lü* et al produced a unified chaotic system [14, 18] which correspondingly contained the Lorenz and Chen systems as two extremes. The unified chaotic system bridges the gap between the Lorenz system and Chen system. Since the unified chaotic system has only one parameter and is always chaotic for the whole interval of $\alpha \in [0, 1]$, it is very convenient for us to apply it for the secure communication. There are also a lot of researches using the unified parameter system in recent decades and has achieved great success [19–21].

One of the most effective methods of secure communication is based on chaotic synchronization [22]. In this way, three message encoding methods such as chaotic masking [23, 24], chaotic shift keying [25], and chaotic modulation [26, 27] have been developed. In particular, for the chaotic masking method, the message signal is added to the output of the chaotic system in the transmitter terminal. So the sending information signal will be a noise-like signal, which contains the original message signal. This signal will be subtracted from the output of the chaotic system in the receiver terminal and the message signal will be recovered. As the message signal is only floating on the output of the chaotic system, it will be relatively easy to be decoded. In the chaotic shift keying method, the transmitted signal is obtained by switching among several chaotic generators in the transmitter terminal according to the binary sequence. The shift keying method has its security flaws since the binary message is encrypted by shifting the parameter sets among fixed sets of values. Hence, the encrypted information can be revealed by detecting the statistical properties of the intercepted information signal. In chaotic modulation, the information signal modifies the states or the parameters of the chaotic system and the generated chaotic signal inherently contains the information of the transmitted signal [28–32]. So it will be relatively safer than the chaotic masking and the chaotic shift keying method. But in real situation, the single-stage secure communication system will be relatively easier for the illegal receiver to decode. Here, we present the three-stage unified chaotic system to carry out the transmission. As the chaotic system is very sensitive to initial conditions, it is very hard to predict even the single-stage chaotic system, so the proposed three-stage chaotic system in this chapter will have very strong security and practicability.

This chapter is organized as follows: In Sect. 11.2, we give out the principle diagram and the working process of the system. In Sect. 11.3, we prove that the system will realize synchronization quickly. In addition, the numerical simulation is carried out in Sect. 11.4. Finally the conclusions are collected in Sect. 11.5.

11.2 Problem Statement

The unified chaotic system is described as below:

$$\begin{cases} \dot{x} = (25a + 10)(y - x), \\ \dot{y} = (28 - 35a)x - xz + (29a - 1)y, \\ \dot{z} = xy - \frac{1}{3}(8 + a)z, \end{cases} \tag{11.1}$$

where $\alpha \in [0, 1]$ obviously, when $\alpha = 0$, the system (11.1) belongs to the original Lorenz system; for $\alpha = 1$, system (11.1) belongs to the original Chen system; When $\alpha = 0.8$, the system (11.1) is the critical $Lü$ system. In fact, the system (11.1) bridges the gap between the Lorenz system and Chen system. Since the system (11.1) is always chaotic for the whole interval of $\alpha \in [0, 1]$, it is very convenient for us to apply it for the secure communication. The specific attractors of the unified parameter chaotic system are shown in Fig. 11.1.

As the unified parameter chaotic system has only one parameter, we will use it to improve the transmission and safety performance of the single-stage chaotic system, and here we present a new three-stage chaotic communication system based on PC method, as it is shown in Fig. 11.2, where all the systems in both transmitter and receiver terminal are the unified parameter chaotic system.

There are three-stage chaotic systems in both the transmitter and the receiver terminal. As to the transmitter terminal, the first-stage main system is (x_1, x_2, x_3); The second-stage subsystem is (y_1, y_2, y_3) and is driven by x_1 of the main system; The third-stage subsystem is (z_1, z_2, z_3) and is driven by y_2 of the second-stage subsystem. The working process of the transmitter terminal is defined as below, first the message signal $m(t)$ is added to $z_1(t)$ of the third-stage subsystem (z_1, z_2, z_3) and we will get $s_1(t) = m(t) + z_1(t) + x(t)$, then $s_1(t)$ is added to y_1 of the second-stage subsystem, then we will get $s_2(t)$, the output of the second-stage subsystem, at last

11.2 Problem Statement

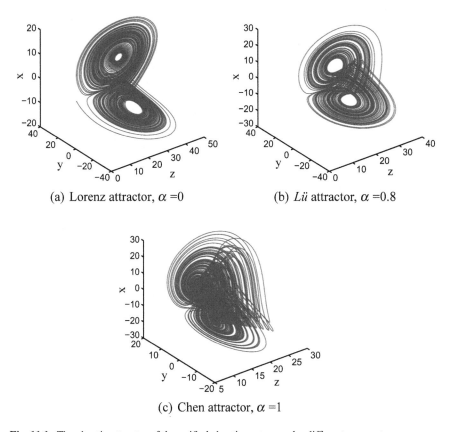

Fig. 11.1 The chaotic attractor of the unified chaotic system under different parameters

after $s_2(t)$ passing the Linear Module L, we will get $s_3(t)$ and it is added to x_1 of the main system (x_1, x_2, x_3). The transmission information signal will be $x_1(t)$.

For the receiver terminal, the first-stage main system is (x'_1, x'_2, x'_3) and is driven by the signal information $x_1(t)$, which is also the information signal transmitted in the channel; The second-stage subsystem is (y'_1, y'_2, y'_3) and is driven by x'_2 of the main system; The third-stage subsystem is (z'_1, z'_2, z'_3) and is driven by $\hat{s}_2(t)$ of the Linear Module L^{-1}. The working process of the receiver terminal is defined as below, first the information signal is added to the main system (x'_1, x'_2, x'_3) and the first recovery module, after the first recovery module we will have $\hat{s}_3(t)$, the recovery signal of $s_3(t)$. After the Linear Module L^{-1} we will have $\hat{s}_2(t)$, the recovery signal of $s_2(t)$. After $\hat{s}_2(t)$ passing through the second recovery module, we will have $\hat{s}_1(t)$, the recovery signal of $s_1(t)$. As the conditional Lyapunov Index of (y_2, y_3), (z_1, z_3), (x'_2, x'_3), (y'_1, y'_3) and (z'_1, z'_3) are not positive, the synchronization between each subsystem is feasible. At last we will get the final recovery signal $\hat{m}(t) = \hat{s}_1(t) - z'_1(t) - x'(t)$.

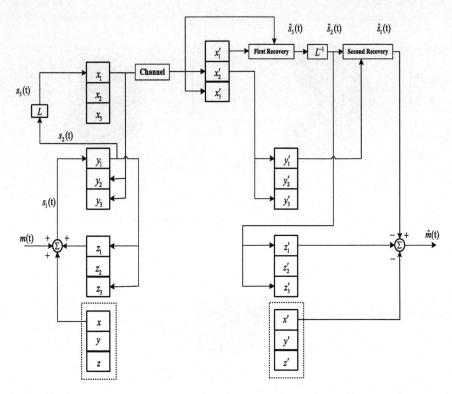

Fig. 11.2 The principle diagram of the three-stage chaotic communication system based on PC method

In order to prevent the encryption signal from being decoded easily, we add the interference system (x, y, z) in the transmitter end and (x', y', z') in the receiver end. (x, y, z) and (x', y', z') have the same system parameter and initial condition. According to Fig. 11.2, we will have the state equations of each part.

The state equation of the main system of the transmitter terminal:

$$\begin{cases} \dot{x}_1 = (25a + 10)(x_2 - x_1) + s_3(t), \\ \dot{x}_2 = (28 - 35a)x_1 - x_1 x_3 + (29a - 1)x_2, \\ \dot{x}_3 = x_1 x_2 - \frac{1}{3}(8 + a)x_3. \end{cases} \quad (11.2)$$

The state equation of the second-stage subsystem of the transmitter terminal:

$$\begin{cases} \dot{y}_1 = (25a + 10)(y_2 - y_1), \\ \dot{y}_2 = (28 - 35a)x_1 - x_1 y_3 + (29a - 1)y_2, \\ \dot{y}_3 = x_1 y_2 - \frac{1}{3}(8 + a)y_3. \end{cases} \quad (11.3)$$

The state equation of the third-stage subsystem of the transmitter terminal:

11.2 Problem Statement

$$\begin{cases} \dot{z}_1 = (25a + 10)(y_1 - z_1), \\ \dot{z}_2 = (28 - 35a)z_1 - z_1 z_3 + (29a - 1)z_2, \\ \dot{z}_3 = z_1 y_1 - \frac{1}{3}(8 + a)z_3. \end{cases} \quad (11.4)$$

For Fig. 11.2, $m(t)$ is the message signal, $s_1(t) = m(t) + z_1(t) + x(t)$. L is the linear module with internal function $f(x) = cx + d$. c and d are the variables that we will discuss in Sect. 11.4. From the principle diagram, we will also have the state equation in the receiver end.

The state equation of the main system of the receiver terminal:

$$\begin{cases} \dot{x}'_1 = (25a + 10)(x'_2 - x'_1), \\ \dot{x}'_2 = (28 - 35a)x_1 - x_1 x_3 + (29a - 1)x'_2, \\ \dot{x}'_3 = x_1 x'_2 - \frac{1}{3}(8 + a)x'_3. \end{cases} \quad (11.5)$$

The state equation of the second-stage subsystem of the receiver terminal:

$$\begin{cases} \dot{y}'_1 = (25a + 10)(x'_2 - y'_1), \\ \dot{y}'_2 = (28 - 35a)y'_1 - y'_1 y_3 + (29a - 1)y'_2, \\ \dot{y}'_3 = y'_1 x'_2 - \frac{1}{3}(8 + a)y'_3. \end{cases} \quad (11.6)$$

The state equation of the third-stage subsystem of the receiver terminal:

$$\begin{cases} \dot{z}'_1 = (25a + 10)(\hat{s}_2(t) - z'_1), \\ \dot{z}'_2 = (28 - 35a)z'_1 - z'_1 z'_3 + (29a - 1)z'_2, \\ \dot{z}'_3 = z'_1 \hat{s}_2(t) - \frac{1}{3}(8 + a)z'_3. \end{cases} \quad (11.7)$$

Figures 11.3 and 11.4 show the internal algorithm of the first and the second recovery module.

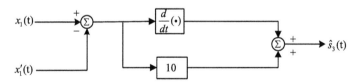

Fig. 11.3 The first recovery module

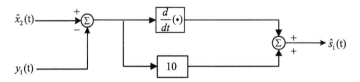

Fig. 11.4 The first recovery module

We have give the overall description of the principle diagram of the three-stage chaotic communication system, and in the next section, we will present the proof that the message signal will be successfully recovered.

11.3 Proof of the Feasibility of the Recovery Signal

From the above analysis, we can get the basic theory of the system that the message signal is mixed to the chaotic drive signal to be the pretend transmitted signal, then the information signal is transmitted to the receiver and is recovered by decoding in the synchronized receiver. In this section, we will give the proof that the message signal will be recovered accurately and quickly in the receiver terminal.

First, we will prove $\hat{s}_3(t) = s_3(t)$, from the first recovery module we can get

$$\hat{s}_3(t) = \dot{x}_1 - \dot{x}_1' + 10(x_1 - \dot{x}_1). \tag{11.8}$$

Form Eqs. (11.2) and (11.5), we have

$$s_3(t) = \dot{x}_1 - \dot{x}_1' + 10(x_1 - \dot{x}_1) - 10(x_2 - x_2'). \tag{11.9}$$

As (11.2) and (11.5) is in synchronization, we have $x_2 = x_2'$, then we can easily have

$$s_3(t) = \dot{x}_1 - \dot{x}_1' + 10(x_1 - \dot{x}_1) = \hat{s}_3(t). \tag{11.10}$$

Second, we wish to prove $\hat{s}_2(t) = s_2(t)$, as we already prove $\hat{s}_3(t) = s_3(t)$ in Eq. (11.10) and L^{-1} is the inverse function of L, we will clearly get

$$\hat{s}_2(t) = s_2(t). \tag{11.11}$$

Thirdly, we will prove $\hat{s}_1(t) = s_1(t)$, from the second recovery module, we have

$$\hat{s}_1(t) = \dot{\hat{s}}_2 - \dot{y}_1' + 10(\hat{s}_2 - y_1'). \tag{11.12}$$

Based on the Eqs. (11.3) and (11.6), we have

$$s_1(t) = \dot{y}_1 - \dot{y}_1' + 10(y_1 - y_1') - 10(y_2 - x_2'). \tag{11.13}$$

As the systems (11.3) and (11.5) are in synchronization, we have $y_2 = x_2'$, as we have proven $\hat{s}_2(t) = s_2(t)$ in Eq. (11.11), and from the principle diagram we can see that $s_2(t) = y_1(t)$, so it is obvious that $\hat{s}_2(t) = y_1(t)$.

Then Eq. (11.12) can be rewritten as

$$\hat{s}_1(t) = \dot{y}_1 - \dot{y}_1' + 10(y_1 - y_1'). \tag{11.14}$$

11.3 Proof of the Feasibility of the Recovery Signal

And Eq. (11.13) can be rewritten as

$$s_1(t) = \dot{y}_1 - \dot{y}'_1 + 10(y_1 - y'_1). \tag{11.15}$$

It is obvious that $\hat{s}_1(t) = s_1(t)$.

Finally, we will prove that the message signal $m(t)$ can be successfully recovered, that is $\hat{m}(t) = m(t)$. From Fig. 11.2 we further have

$$\hat{m}(t) = \hat{s}_1(t) - z'_1(t) - x'(t). \tag{11.16}$$

As we have proven that $\hat{s}_1(t) = s_1(t)$ in Eq. (11.15), and that Eq. (11.4) and (11.7) is in synchronization, we have $z_1(t) = z'_1(t)$, then Eq. (11.16) can be rewritten as

$$\hat{m}(t) = \hat{s}_1(t) - z'_1(t) - x'(t) = s_1(t) - z_1(t) - x(t) = m(t). \tag{11.17}$$

In a word, we have proven that the message signal $m(t)$ can be successfully recovered, we will present the numerical simulation in the next section.

11.4 Numerical Simulation

Before transmission, we first send out four binary signal, after passing through the parameter control module, we will get the four dynamic parameters for this transmission, in which a is the parameter for the unified chaotic system, c and d are the parameters for the linear system, b is the parameter for the interference system. The algorithm of the parameter control module is presented in the below table (Table 11.1):

Table 11.1 The algorithm in the parameter control module

Binary Sequence	Parameter a	Parameter b	Parameter c	Parameter d
0000	0.0	0.5	1.0	0.5
0001	0.1	0.6	1.1	0.6
0010	0.2	0.7	1.2	0.7
0011	0.3	0.8	1.3	0.8
0100	0.4	0.9	1.4	0.9
0101	0.5	1.0	1.5	1.0
0110	0.6	0.1	1.6	0.1
0111	0.7	0.2	1.7	0.2
1000	0.8	0.3	1.8	0.3
1001	0.9	0.4	1.9	0.4
1010	1.0	0.5	2.0	0.5

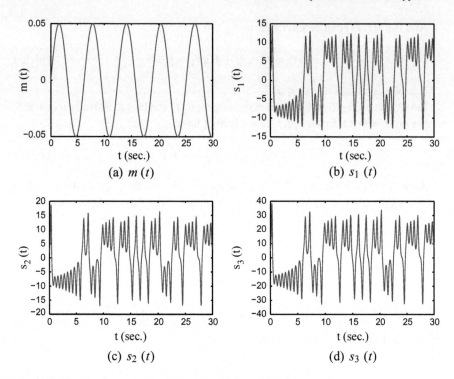

Fig. 11.5 The signal waveforms of the transmitter terminal when $\alpha = 0$

When the binary sequence is 0000, the simulation results of the transmitter terminal are shown in Fig. 11.5, where Fig. 11.5a is the message signal $m(t) = 0.05sin(t)$, Fig. 11.5b denotes $s_1(t)$, Fig. 11.5c denotes $s_2(t)$ and Fig. 11.5d denotes $s_3(t)$.

In the receiver terminal, the simulation results are shown in Fig. 11.6, where Fig. 11.6a denotes $\hat{s}_3(t)$ from the first recovery module, Fig. 11.6b denotes $\hat{s}_2(t)$ after the linear module L^{-1}, Fig. 11.6c denotes $\hat{s}_1(t)$ from the second recovery module, Fig. 11.6d denotes the recovered message signal $\hat{m}(t)$.

Figure 11.7 depicts the error curve of the systems, where $e_1(t) = s_1(t) - \hat{s}_1(t)$, $e_2(t) = s_2(t) - \hat{s}_2(t)$, $e_3(t) = s_3(t) - \hat{s}_3(t)$, $e_4(t) = m(t) - \hat{m}(t)$.

When the binary sequence is 0011, the simulation results of the transmitter terminals are shown in Fig. 11.8, where Fig. 11.8a denotes the message signal $m(t) = 0.05sin(t)$, Fig. 11.8b denotes $s_1(t)$, Fig. 11.8c denotes $s_2(t)$ and Fig. 11.8d denotes $s_3(t)$.

11.4 Numerical Simulation

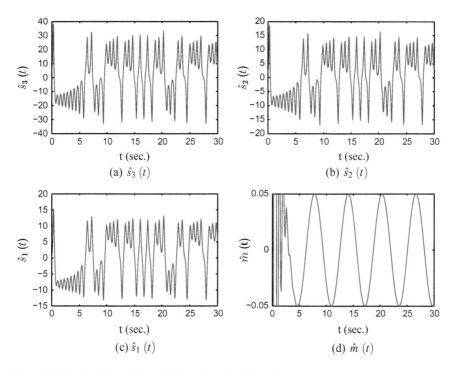

Fig. 11.6 The signal waveforms of the receiver terminal when $\alpha = 0$

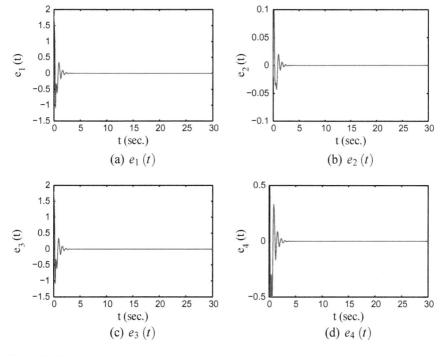

Fig. 11.7 The error curve of the system

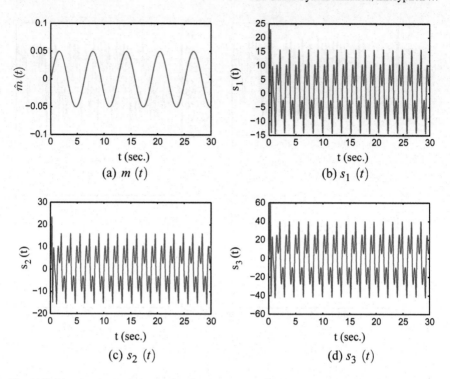

Fig. 11.8 The signal waveforms of the transmitter terminal when $\alpha = 0.3$

In the receiver terminal, the simulation results are shown in Fig. 11.9, where Fig. 11.9(a) denotes $\hat{s}_3(t)$ from the first recovery module, Fig. 11.9b denotes $\hat{s}_2(t)$ after the linear module L^{-1}, Fig. 11.9c denotes $\hat{s}_1(t)$ from the second recovery module, and Fig. 11.9d denotes the recovered message signal $\hat{m}(t)$.

11.5 Conclusion

If the illegal receiver wants to decode the message signal from the transmission signal $x_1(t)$, it has to first predict the main system (x'_1, x'_2, x'_3) and then to predict the second-stage subsystem (y'_1, y'_2, y'_3), which will cause lots of error. As the chaotic system is very sensitive to initial conditions, and it is very hard to use the error signal to predict the third-stage subsystem (z'_1, z'_2, z'_3), too. What is more, at every time, the parameters of the unified chaotic system, the linear system, and the interference system are dynamically time-varying and the correspondingly algorithm in the parameter control module can be changed regularly. As it is different from the Chaotic Masking method, the message signal is modulated into the chaotic system in the transmitter terminal instead of simply being added to the output of the chaotic system, and it is very

11.5 Conclusion

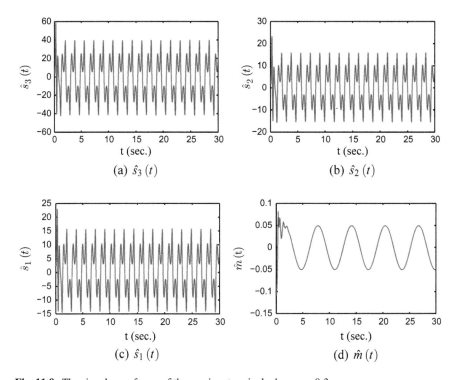

Fig. 11.9 The signal waveforms of the receiver terminal when $\alpha = 0.3$

hard for the illegal receiver to decode the legal message signal if the algorithms on the parameter control modules, the type of the interference systems, and the initial conditions do not be learned. As a result, the proposed three-stage chaotic secure communication system has strong security and flexibility.

Acknowledgements This work is Supported by National Key Research and Development Program of China (2017YFF0207400).

References

1. Boccaletti S, Kurths J, Osipov G, Valladares DL, Zhou CS. The synchronization of chaotic systems. Phys Rep. 2002;302:777–88.
2. Henryk M. Characteristic time series and operation region of the system of two tank reactors (CSTR) with variable division of recirculation stream. Chaos Solitons Fractals. 2006;27:279–85.
3. Henryk M. Chaotic dynamics of a cascade of plug flow tubular reactors (PFTRs) with division of recirculating stream. Chaos Solitons Fractals. 2005;23:1211–9.
4. Tomasz S, Toshio O. A numerical analysis of chaos in the double pendulum. Chaos Solitons Fractals. 2006;29:417–22.

5. Sunita G, Brahampal S. Complex dynamic behavior in a food web consisting of two preys and a predator. Chaos Solitons Fractals. 2005;24:789–801.
6. Zhang J, Li X, Chu Y, Yu J, Chang Y. Hopf bifurcations, lyapunov exponents and control of chaos for a class of centrifugal flywheel governor system. Chaos Solitons Fractals. 2009;39(5):2150–68.
7. Chu YD, Zhang JG, Li XF, Chang YX, Luo GW. Chaos and chaos synchronization for a non-autonomous rotational machine systems. Nonlinear Anal (RWA). 2008;9:1378–93.
8. Qi G. Michaël AW, Barend JW, Chen G. A chaotic secure communication scheme based on observer. Chaos Solitons Fractals. 2007;10:1–6.
9. Park JH. Adaptive control for modified projective synchronization of a four-dimensional chaotic system with uncertain parameters. J Comput Appl Math. 2008;213:288–93.
10. Yu Y. Adaptive synchronization of a unified chaotic system. Chaos Solitons Fractals. 2008;36:329–33.
11. Wang H, Han ZZ, Zhang W, Xie QY. Synchronization of unified chaotic systems with uncertain parameters based on the CLF. Nonlinear Anal RWA. 2009;10:715–22.
12. Pecora LM, Carroll TL. Synchronization in chaotic systems. Phys Rev lett. 1990;64(8):821.
13. Pecora LM, Carroll TL. Driving systems with chaotic signals. Phys Rev A. 1991;44(4):2374.
14. Lü J, Chen G, Cheng D et al. Bridge the gap between the Lorenz system and the Chen system. Int J Bifurc Chaos 2002;12(12): 2917–2926.
15. Lorenz EN. Deterministic nonperiodic flow. J Atmos Sci. 1963;20(2):130–41.
16. Chen G, Ueta T. Yet another chaotic attractor. Int J Bifurc Chaos. 1999;9(07):1465–6.
17. Lü J, Chen G. A new chaotic attractor coined. Int J Bifurc Chaos. 2002;12(03): 659–661.
18. Tao CH, Lu JA, Lu JH. The feedback synchronization of a unified chaotic system. Acta Phys Sin. 2002;51(7):1497–501.
19. Smaoui N, Karouma A, Zribi M. Secure communications based on the synchronization of the hyperchaotic Chen and the unified chaotic systems. Commun Nonlinear Sci Numer Simul. 2011;16(8):3279–93.
20. Li XF, Leung ACS, Han XP, et al. Complete (anti-) synchronization of chaotic systems with fully uncertain parameters by adaptive control. Nonlinear Dyn. 2011;63(1–2):263–75.
21. Wu X, Li Y, Zhang J. Synchronisation of unified chaotic systems with uncertain parameters in finite time. Int J Model Identif Control. 2012;17(4):295–301.
22. Sheikhan M, Shahnazi R, Garoucy S. Synchronization of general chaotic systems using neural controllers with application to secure communication. Neural Comput Appl. 2013;22(2):361–73.
23. Chee CY, Xu D. Secure digital communication using controlled projective synchronisation of chaos. Chaos Solitons Fractals. 2005;23(3):1063–70.
24. Zhu F. Observer-based synchronization of uncertain chaotic system and its application to secure communications. Chaos Solitons Fractals. 2009;40(5):2384–91.
25. Kolumban G, Kennedy MP, Chua LO. The role of synchronization in digital communications using chaos. I. Fundamentals of digital communications. IEEE Trans Circuits Syst I: Fundam Theory Appl. 1997;44(10):927–936.
26. Bai EW, Lonngren KE, Ucar A. Secure communication via multiple parameter modulation in a delayed chaotic system. Chaos Solitons Fractals. 2005;23(3):1071–6.
27. Wang XY, Wang MJ. A chaotic secure communication scheme based on observer. Commun Nonlinear Sci Numer Simul. 2009;14(4):1502–8.
28. Wu XJ, Wang H, Lu HT. Hyperchaotic secure communication via generalized function projective synchronization. Nonlinear Anal Real World Appl. 2011;12(2):1288–99.
29. Yang LX, Zhang J. Synchronization of three identical systems and its application for secure communication with noise perturbation. Int Conf Inf Eng Comput Sci ICIECS 2009. IEEE. 2009;2009:1–4.
30. Grzybowski JMV, Rafikov M, Balthazar JM. Synchronization of the unified chaotic system and application in secure communication. Commun Nonlinear Sci Numer Simul. 2009;14(6):2793–806.

31. Chen S. Lü J. Synchronization of an uncertain unified chaotic system via adaptive control. Chaos Solitons Fractals. 2002;14(4):643–7.
32. Lu J, Wu X. Lü J. Synchronization of a unified chaotic system and the application in secure communication. Phys Lett A. 2002;305(6):365–70.

Chapter 12
The Numerical Solutions and Their Applications in 2K-H Planetary Gear Transmission Systems

12.1 Introduction

Planetary gear transmission is widely used in both civil and national defense industry because of its small volume, light weight, large speed ratio, and high efficiency. Clearance, which is inevitable during the gear mesh due to lubrication, installation and wear, will cause strong vibration and larger dynamic load under high speed light load mode, so that the service life and reliability of gear transmission will be affected. In the practical gear transmission system, disastrous accident is often caused by abnormal vibration which can only be explained by nonlinear vibration theory. In recent years, many researchers, both in and abroad, have done lots of research on dynamic characteristics of planetary gear transmission [1–19]. But many of them use linear vibration theory during modeling and analysis without considering the influence of clearance to the dynamic characteristic of the system [4, 7]. Now many researchers consider the nonlinear factors, i.e., not only the dynamic loads, but nonlinear dynamic characteristics, including chaos characteristics [6–9].

The nonlinear dynamic model and its normalized equation of 2 K-H is presented in the chapter, numerical method of solving the equation is discussed, and particularly the discretion largest Lyapunov Exponent is presented, too. Various chaotic characteristics of the system are collected, such as the largest Lyapunov Exponent, Poincaré section, FFT spectrum and the phase locus.

This chapter is organized as follows. In Sect. 12.2, classical Range Kutta(RK) methods are introduced and analyzed. And some important algorithms are outlined, In Sect. 12.3, the nonlinear dynamic model of 2K-H planetary gear transmission system and its normalized equation are presented. and a new method to get its largest Laypunov exponent is described. The numerical simulation results are shown in Sect. 12.4 and the chaos characteristics are presented. Finally the conclusions are collected in Sect. 12.5.

© Springer Nature Singapore Pte Ltd. 2018
S. Guo and L. Han, *Stability and Control of Nonlinear Time-varying Systems*,
https://doi.org/10.1007/978-981-10-8908-4_12

12.2 The Numerical Integration Algorithms

In contrast to the analytical solutions presented previously, the numerical solutions that we will compute are ordered pairs of numbers. i.e., The numerical integration is then step-by-step algorithm going from the solution point (t_i, y_i) to the point (t_{i+1}, y_{i+1}). This stepping procedure can be represented mathematically by a Taylor series

$$y_{i+1} = y_i + \frac{dy_i}{dt}h + \frac{d^2 y_i}{dt^2} + \ldots, \tag{12.1}$$

where $h = t_{i+1} - t_i$.

We can truncate this series after the linear term in h

$$y_{i+1} = y_i + \frac{dy_i}{dt}h, \tag{12.2}$$

and use this approximation to step along the solution from y_0 to y_1 (with $i = 0$), then from y_1 to y_2 (with $i = 1$), etc. This is famous Euler's method.

Application of the Taylor series method with taking derivatives becomes cumbersome. The Runge Kutta method can be used to fit the numerical ODE solution exactly to an arbitrary number of the terms in the underlying Taylor series without differentiating the ODE. The other important characteristics of a numerical integration algorithm are away of estimating the truncation error ε, so that the integration step h, can be adjusted to achieve a solution with a prescribed accuracy.

The modified Euler method or the extended Euler method is as follows:

$$y_{i+1} = y_i + \left(\frac{dy_{i+1}}{dt} + \frac{dy_i}{dt}\right) \frac{1}{2!} h. \tag{12.3}$$

Equation (12.3) can be written as a two-step algorithm

$$\begin{cases} y_{i+1}^p = y_i + \frac{dy_i}{dt}h, & (12.4a) \\ \varepsilon_i = (\frac{dy_{i+1}^p}{dt} - \frac{dy_i}{dt})\frac{h}{2!}, & (12.4b) \\ y_{i+1}^c = y_i + \frac{dy_i}{dt}h + \varepsilon_i, & (12.4c) \end{cases}$$

the above algorithm of (12.4) is called a predictor–corrector method.

Now, we introduce the RK notation and define k_1 and k_2 (termed Range Kutta constants)

$$\begin{cases} k_1 = f(y_i, t_i)h, \\ k_2 = f(y_i + k_1, t_i + H)h, \end{cases} \tag{12.5}$$

So, the Euler method can be written as $y_{i+1} = y_i + k_1$, the modified Euler method can be written as $y_{i+1} = y_i + \frac{k_1 + k_2}{2}$.

12.2 The Numerical Integration Algorithms

Equation (12.5) can be conveniently written in RK notation

$$\begin{cases} y_{i+1}^p = y_i + k_1, & (12.6a) \\ \varepsilon_i = \dfrac{k_2 - k_1}{2!}, & (12.6b) \\ y_{i+1}^c = y_i + k_1 + \dfrac{k_2 - k_1}{2} h = y_i + \dfrac{k_1 + k_2}{2}. & (12.6c) \end{cases}$$

The second-order RK method is actually a family of second-order methods; a particular member of this family is selected by choosing an arbitrary constant in the general second-order RK formulas. The origin of these formula is illustrated by the following development.

$$\begin{cases} y_{i+1} = y_i + c_1 k_1 + c_2 k_2, & (12.7a) \\ k_1 = f(y_i, t_i)h, & (12.7b) \\ k_2 = f(y_i + a_2 k_1(y_i, t_i), t_i + a_2 h)h = f(y_i + a_2 k_1(y_i, t_i), t_i + a_2 h)h, & (12.7c) \end{cases}$$

and c_1, c_2, a_2 are constants to be determined.

Further, we have

$$\begin{cases} k_2 = [f(y_i, t_i) + f_y(y_i, t_i)a_2 f(y_i, t_i)h + f_t(y_i, t_i)a_2 h]h + o(h^3), \\ y_{i+1} = y_i + (c_1 + c_2) f(y_i, t_i)h + c_2[f_y(y_i, t_i)a_2 f(y_i, t_i) + f_t(y_i, t_i)a_2]h^2 + o(h^3) \\ = y_i + f(y_i, t_i)h + [f_y(y_i, t_i)f(y_i, t_i) + f_t(y_i, t_i)]\dfrac{h^2}{2!} + o(h^3). \end{cases}$$
(12.8)

Thus, we conclude

$$\begin{cases} c_1 + c_2 = 1, \\ c_2 a_2 = \tfrac{1}{2}. \end{cases} \quad (12.9)$$

For example, we can easily obtain the following pairs of (c_2, c_1, a_2):

$$\begin{cases} (c_2, c_1, a_2) = (\tfrac{1}{2}, \tfrac{1}{2}, 1), \\ (c_2, c_1, a_2) = (1, 0, \tfrac{1}{2}), \\ (c_2, c_1, a_2) = (\tfrac{3}{4}, \tfrac{1}{4}, \tfrac{2}{3}). \end{cases} \quad (12.10)$$

The third-order stepping formula is

$$y_{i+1} = y_i + c_1 k_1 + c_2 k_2 + c_3 k_3, \quad (12.11)$$

whose RK constants are

$$\begin{cases} k_1 = f(y_i, t_i)h. \\ k_2 = f(y_i + a_2k_1, t_i + a_2h)h. \\ k_3 = f(y_i + b_3k_1 + (a_3 - b_3)k_2, t_i + a_3h)h. \end{cases} \quad (12.12)$$

Based on matching the stepping formula, and the Taylor series with the term $(\frac{d^3y}{dt^3})\frac{h^3}{3!}$, the following four algebraic equations define the six constants $c_1, c_2, c_3, a_2, a_3, b_3$:

$$\begin{cases} c_1 + c_2 + c_3 = 1, & (12.13a) \\ c_2a_2 + c_3a_3 = \frac{1}{2}, & (12.13b) \\ c_2a_2^2 + c_3a3^2 = \frac{1}{3}, & (12.13c) \\ c_3(a_3 - b_3)a_2 = \frac{1}{6}, & (12.13d) \end{cases}$$

Table 12.1 Algorithm 12.1. (1, 2) equation

calculation	(1, 2) equation
Stage 1.	$k_1 = f(y_i, t_i)h;$
Stage 2.	$k_2 = f(y_i + k_i, t_i + h)h;$
$O(h^1)$	$y_{1+1}^{(1)} = y_1 + k_1;$
$O(h^2)$	$y_{i+1}^{(2)} = y_{i+1}^{(1)} + e_{i+1}^{(1)};$
Estimation Error	$e_{i+1}^{(1)} = \frac{1}{2}(k_2 - k_1);$
step t	$t_{i+1} = t_i + h;$

Table 12.2 Algorithm 12.2. on (2, 3) equation

calculation	(2, 3) equation with fixed step
Stage 1.	$k_1 = f(y_i, t_i)h$
Stage 2.	$k_2 = f(y_i + \frac{2}{3}k_1, t_i + \frac{2}{3}h)h$
Stage 3.	$k_3 = f(y_i + \frac{2}{3}k_2, t_i + \frac{2}{3}h)h$
$O(h^2)$	$y_{i+1}^{(2)} = y_i + \frac{1}{4}k_1 + \frac{3}{4}k_2$
$O(h^3)$	$y_{i+1}^{(3)} = y_i + \frac{1}{4}k_1 + \frac{3}{8}k_2 + \frac{3}{8}k_3$
Estimation Error	$e_{1+1}^{(2)} = y_{i+1}^{(3)} - y_{i+1}^{(2)}$
step t	$t_{i+1} = t_i + h$

12.2 The Numerical Integration Algorithms

Table 12.3 Algorithm 12.3. on (2, 3) equation

calculation	(2, 3) equation with variable step
Stage 1.	$k_1 = f(y_i, t_i)h$
Stage 2.	$k_2 = f(y_i + \frac{1}{2}k_1, t_i + \frac{1}{2}h)h$
Stage 3.	$k_3 = f(y_i - k_1 + 2k_2, t_i + h)h$
$O(h^2)$	$y_{i+1}^{(2)} = y_i + \frac{1}{2}(k_1 + k_3)$
$O(h^3)$	$y_{i+1}^{(3)} = y_{i+1}^{(2)} + e_{i+1}^{(2)}$
Estimation Error	$e_{i+1}^{(2)} = \frac{1}{3}(-k_1 + 2k_2 - k_3)$
step t	$t_{i+1} = t_i + h$

Table 12.4 Algorithm 12.4. on (3, 4) equation

calculation	(3, 4) equation with variable step
Stage 1.	$k_1 = f(y_i, t_i)h$
Stage 2.	$k_2 = f(y_i + k_1, t_i + \frac{1}{3}h)\frac{1}{3}h$
Stage 3.	$k_3 = f(y_i + \frac{1}{2}(k_1 + k_2), t_i + \frac{1}{3}h)\frac{1}{3}h$
Stage 4.	$k_4 = f(y_i + \frac{3}{8}k_1 + \frac{9}{8}k_3, t_i + \frac{1}{2}h)\frac{1}{3}h$
Stage 5.	$k_5 = f(y_i + \frac{3}{2}k_1 - \frac{9}{2}k_3 + 6k_4, t_i + h)\frac{1}{3}h$
$O(h^3)$	$y_{i+1}^{(3)} = y_i + \frac{1}{2}(k_1 + 4k_4 + k_5)$
$O(h^4)$	$y_{i+1}^{(4)} = y_{i+1}^{(3)} + e_{i+1}^{(3)}$
Estimation Error	$e_{i+1}^{(3)} = \frac{1}{5}(-k_1 + \frac{9}{2}k_3 - 4k_4 + \frac{1}{2}k_5)$
step t	$t_{i+1} = t_i + h$

For the higher order stepping formula cases, we omit these discussion. Here, we outline some important and classical pairs algorithms including the higher order stepping formula cases as below (Tables 12.1, 12.2, 12.3, 12.4, 12.5, 12.6, 12.7, 12.8, 12.9, 12.10, 12.11, 12.12, 12.13, 12.14, 12.15, 12.16 and 12.17) (See [1–19]).

The Algorithm 12.16 and Algorithm 12.17 are recommended for those problems with particularly stringent accuracy requirements.

12.3 Nonlinear Dynamics of 2K-H Planetary Gear Transmission Systems

First, we introduce the following nonlinear 2K-H planetary gear transmission system.

The sketch diagram of 2K-H planetary gear transmission is shown in Fig. 12.1. Here, the gears are assumed to be straight spur gear and each planet gear has the same

Table 12.5 Algorithm 12.5. on (3, 4) equation

calculation	(3, 4) equation with variable step
Stage 1.	$k_1 = f(y_i, t_i)h$
Stage 2.	$k_2 = f(y_i - 0.4k_1, t_i - 0.4h)h$
Stage 3.	$k_3 = f(y_i + 0.668489588k_1 - 0.2434895833k_2, t_i + 0.425h)h$
Stage 4.	$k_4 = f(y_i - 2.323685857k_1 + 1.125483559k_2 + 2.198202298k_3, t_i + h)h$
$O(h^3)$	$y_{i+1}^{(3)} = y_i + 0.03968253968k_2 + 0.7729468599k_3$
	$+ 0.18737060041k_4$
$O(h^4)$	$y_{i+1}^{(4)} = y_{i+1}^{(3)} + e_{i+1}^{(3)}$
Estimation Error	$e_{i+1}^{(3)} = -y_{i+1}^{(3)} + y_i + 0.03431372549k_1 + 0.027056277706k_2 + 0.7440130202k_3$
	$+ 0.1946169772k_4$
step t	$t_{i+1} = t_i + h$

Table 12.6 Algorithm 12.6. on (3, 4) equation

calculation	(3, 4) equation with variable step
Stage 1.	$k_1 = f(y_i, t_i)h$
Stage 2.	$k_2 = f(y_i + \frac{1}{3}k_1, t_i + \frac{1}{3}h)h$
Stage 3.	$k_3 = f(y_i + \frac{1}{6}(k_1 + k_2), t_i + \frac{1}{3}h)h$
Stage 4.	$k_4 = f(y_i + \frac{1}{8}(k_1 + 3k_3), t_i + \frac{1}{2}h)h$
Stage 5.	$k_5 = f(y_i + \frac{1}{2}k_1 - \frac{3}{2}k_3 + 2k_4, t_i + h)h$
$O(h^3)$	$y_{i+1}^{(3)} = y_i + \frac{1}{2}k_1 - \frac{3}{2}k_3 + 2k_4)$
$O(h^4)$	$y_{i+1}^{(4)} = y_{i+1}^{(3)} + e_{i+1}^{(3)}$
Estimation Error	$e_{i+1}^{(3)} = -\frac{1}{3}k_1 + \frac{3}{2}k_3 - \frac{4}{3}k_4 + \frac{1}{6}k_5$
step t	$t_{i+1} = t_i + h$

physical and geometrical parameters, and lumped mass model is used and neglect the influence of friction during gear mesh.

The dynamic model of 2K-H planetary gear transmission system (i.e., 3 planet gears) is shown in Fig. 12.1. There are $N + 2$ coordinates in all, θ_s denotes the angular displacement of the sth sun gear, θ_{pi} denotes the angular displacement of the ith planet gear, $(i = 1, 2, \ldots, N)$, θ_c denotes the angular displacement of the cth planet carrier. In this chapter, s, pi, c, r are used to represent the correspondingly sun gear, planet carrier, and gear ring. and spi is used to represent the meshing parameter

12.3 Nonlinear Dynamics of 2K-H Planetary Gear Transmission Systems

Table 12.7 Algorithm 12.7. on (4, 5) equation

calculation	(4, 5) equation with variable step
Stage 1.	$k_1 = f(y_i, t_i)h$
Stage 2.	$k_2 = f(y_i + 0.0005k_1, t_i + 0.0005h)h$
Stage 3.	$k_3 = f(y_i - 80.89939470k_1 + 81.18439470k_2, t_i + 0.285h)h$
Stage 4.	$k_4 = f(y_i + 2113.327899k_1 - 2117.778035k_2 + 5.442136522k_3,$ $t_i + 0.992h)h$
Stage 5.	$k_5 = f(y_i + 2249.757677k_1 - 2254.489040k_2 + 5.739991965k_3$ $-0.008629230728k_4, t_i + h)h$
$O(h^4)$	$y_{i+1}^{(4)} = y_i - 131.2823524k_1 + 131.4998223k_2 + 0.4837620276k_3$ $+ 0.2987680554k_4$
$O(h^4)$	$y_{i+1}^{(5)} = y_{i+1}^{(4)} + e_{i+1}^{(4)}$
Estimation Error	$e_{i+1}^{(4)} = -y_{i+1}^{(4)} + y_i + 65.80784286k_1 - 65.94767173k_2 0.7959885276k_3$ $+ 4.715404915k_4 - 4.371564570k_5$
step t	$t_{i+1} = t_i + h$

Table 12.8 Algorithm 12.8. on (4, 5) equation

calculation	(4, 5) equation with variable step
Stage 1.	$k_1 = f(y_i, t_i)h$
Stage 2.	$k_2 = f(y_i + \frac{1}{2}k_1, t_i + \frac{1}{2}h)h$
Stage 3.	$k_3 = f(y_i + \frac{1}{4}(k_1 + k_2), t_i + \frac{1}{2}h)h$
Stage 4.	$k_4 = f(y_i - k_2 + 2k_3, t_i + h)h$
Stage 5.	$k_5 = f(y_i + \frac{1}{27}(7k_1 + 10k_2 + k_4), t_i + \frac{2}{3}h)h$
Stage 6.	$k_6 = f(y_i + \frac{1}{625}(28k_1 - 125k_2 + 546k_3 + 54k_4 - 378k_5), t_i + \frac{1}{5})h$
$O(h^4)$	$y_{i+1}^{(4)} = y_i + \frac{1}{6}(k_1 + 4k_3 + 4k_4)$
$O(h^4)$	$y_{i+1}^{(5)} = y_{i+1}^{(4)} + e_{i+1}^{(4)}$
Estimation Error	$e_{i+1}^{(4)} = \frac{1}{336}(-42k_1 - 224k_3 - 21k_4 + 162k_5 + 126k_6)$
step t	$t_{i+1} = t_i + h$

between sun gear and the ith planet gear, rpi is used to represent the meshing parameter between sun gear and gear ring.

From Fig. 12.1, the dynamics differential equation of the system will be obtained through Lagrange Equation as follows.

Table 12.9 Algorithm 12.9. on (4, 5) equation

calculation	(4, 5) equation with variable step
Stage 1.	$k_1 = f(y_i, t_i)h$
Stage 2.	$k_2 = f(y_i + \frac{1}{4}k_1, t_i + \frac{1}{4}h)h$
Stage 3.	$k_3 = f(y_i + \frac{1}{32}(3k_1 + 9k_2), t_i + \frac{3}{8}h)h$
Stage 4.	$k_4 = f(y_i + \frac{1}{2197}(1932k_1 - 7200k_2 + 7296k_3), t_i + \frac{12}{13}h)h$
Stage 5.	$k_5 = f(y_i + \frac{439}{216}k_1 - 8k_2 + \frac{3680}{513}k_3 - \frac{845}{4104}k_4), t_i + h)h$
Stage 6.	$k_6 = f(y_i + \frac{8}{27}k_1 + 2k_2 - \frac{3544}{2565}k_3 + \frac{1859}{4104}k_4 - \frac{11}{40}k_5), t_i + \frac{1}{2}h)h$
$O(h^4)$	$y_{i+1}^{(4)} = y_i + \frac{25}{216}k_1 + \frac{1408}{2565}k_3 + \frac{2197}{4104}k_4 - \frac{1}{5}k_5$
$O(h^5)$	$y_{i+1}^{(5)} = y_i + \frac{16}{135}k_1 + \frac{6656}{12825}k_3 + \frac{28561}{56430}k_4 - \frac{9}{50}k_5 + \frac{2}{55}k_6$
Estimation Error	$e_{i+1}^{(4)} = y_{i+1}^{(5)} - y_{i+1}^{(4)}$
step t	$t_{i+1} = t_i + h$

Table 12.10 Algorithm 12.10. on (4, 5) equation

calculation	(4, 5) equation with variable step
Stage 1.	$k_1 = f(y_i, t_i)h$
Stage 2.	$k_2 = f(y_i + \frac{1}{5}k_1, t_i + \frac{1}{5}h)h$
Stage 3.	$k_3 = f(y_i + \frac{3}{40}k_1 + \frac{9}{40}k_2), t_i + \frac{3}{10}h)h$
Stage 4.	$k_4 = f(y_i + \frac{3}{10}k_1 - \frac{9}{10}k_2 + \frac{6}{5}k_3), t_i + \frac{3}{5}h)h$
Stage 5.	$k_5 = f(y_i - \frac{11}{54}k_1 + \frac{5}{2}k_2 - \frac{70}{27}k_3 + \frac{35}{27}k_4), t_i + h)h$
Stage 6.	$k_6 = f(y_i + \frac{1631}{55296}k_1 + \frac{175}{512}k_2 + \frac{575}{13824}k_3 + \frac{44275}{110592}k_4 + \frac{253}{4096}k_5, t_i + \frac{7}{8}h)h$
$O(h^4)$	$y_{i+1}^{(4)} = y_i + \frac{2825}{27648}k_1 + \frac{18575}{48384}k_3 + \frac{13525}{55296}k_4 + \frac{277}{14336}k_5 + \frac{1}{4}k_6$
$O(h^5)$	$y_{i+1}^{(5)} = y_i + \frac{37}{378}k_1 + \frac{250}{612}k_3 + \frac{125}{595}k_4 + \frac{512}{1771}k_6$
Estimation Error	$e_{i+1}^{(4)} = y_{i+1}^{(5)} - y_{i+1}^{(4)}$
step t	$t_{i+1} = t_i + h$

$$\begin{cases} J_s \ddot{\theta}_s + \sum_{i=1}^{N}(P_{spi} + D_{spi})r_{bs} = T_D(t), \\ J_{pi}\ddot{\theta}_{pi} - (D_{spi} + D_{rpi} + P_{spi} + P_{rpi})r_{bpi} = 0, \\ (J_c + \sum_{i=1}^{N} M_{pi}r_c)\ddot{\theta}_c - \sum_{i=1}^{N}(D_{spi} + D_{rpi} + P_{spi} + P_{rpi})r_c \cos\alpha = -T_L(t), (i = 1, 2 \ldots N), \end{cases}$$
(12.14)

where J is rotational inertia of each gear, M represents the mass of each gear, r_b is the base radius, r_c is the radius of the planet carrier.

12.3 Nonlinear Dynamics of 2K-H Planetary Gear Transmission Systems

Table 12.11 Algorithm 12.11. on (2, 4) equation

calculation	(2, 4) equation with variable step
Stage 1.	$k_1 = f(y_i, t_i)h$
Stage 2.	$k_2 = f(y_i + \frac{1}{2}k_1, t_i + \frac{1}{2}h)h$
Stage 3.	$k_3 = f(y_i + \frac{1}{2}k_2, t_i + \frac{1}{2}h)h$
Stage 4.	$k_4 = f(y_i + k_3, t_i + h)h$
$O(h^2)$	$y_{i+1}^{(2)} = y_i + k_2$
$O(h^4)$	$y_{i+1}^{(4)} = y_{i+1}^{(2)} + e_{i+1}^{(2)}$
Estimation Error	$e_{i+1}^{(2)} = \frac{1}{6}(k_1 - 4k_2 + 2k_3 + k_4)$
step t	$t_{i+1} = t_i + h$

Table 12.12 Algorithm 12.12. on (2, 4) equation

calculation	(2, 4) equation with variable step
Stage 1.	$k_1 = f(y_i, t_i)h$
Stage 2.	$k_2 = f(y_i + \frac{1}{3}k_1, t_i + \frac{1}{3}h)h$
Stage 3.	$k_3 = f(y_i - \frac{1}{3}k_1 + k_2, t_i + \frac{2}{3}h)h$
Stage 4.	$k_4 = f(y_i + k_1 + -k_2 + k_3, t_i + h)h$
$O(h^2)$	$y_{i+1}^{(2)} = y_i - \frac{1}{2}k_1 + \frac{3}{2}k_2$
$O(h^4)$	$y_{i+1}^{(4)} = y_{i+1}^{(2)} + e_{i+1}^{(2)}$
Estimation Error	$e_{i+1}^{(2)} = \frac{1}{8}(5k_1 - 9k_2 + 3k_3 + k_4)$
step t	$t_{i+1} = t_i + h$

Table 12.13 Algorithm 12.13. on (2, 4) equation

calculation	(2, 4) equation with variable step
Stage 1.	$k_1 = f(y_i, t_i)h$
Stage 2.	$k_2 = f(y_i + \frac{1}{2}k_1, t_i + \frac{1}{2}h)h$
Stage 3.	$k_3 = f(y_i - (\frac{1}{2} - \frac{1}{\sqrt{2}})k_1 + (1 - \frac{1}{\sqrt{2}})k_2, t_i + \frac{1}{2}h)h$
Stage 4.	$k_4 = f(y_i - \frac{1}{\sqrt{2}}k_2 + (1 + \frac{1}{\sqrt{2}})k_3, t_i + h)h$
$O(h^2)$	$y_{i+1}^{(2)} = y_i + k_2$
$O(h^4)$	$y_{i+1}^{(4)} = y_{i+1}^{(2)} + e_{i+1}^{(2)}$
Estimation Error	$e_{i+1}^{(2)} = \frac{1}{6}(k_1 - 2(2 + \frac{1}{\sqrt{2}})k_2 + 2(1 + \frac{1}{\sqrt{2}})k_3 + k_4)$
step t	$t_{i+1} = t_i + h$

Table 12.14 Algorithm 12.14. on (2, 4) equation

calculation	(2, 4) equation with variable step
Stage 1.	$k_1 = f(y_i, t_i)h$
Stage 2.	$k_2 = f(y_i + \frac{1}{2}k_1, t_i + \frac{1}{2}h)h$
Stage 3.	$k_3 = f(y_i - \frac{1}{2}k_1 + k_2, t_i + \frac{1}{2}h)h$
Stage 4.	$k_4 = f(y_i + \frac{1}{2}k_2 + \frac{1}{2}k_3, t_i + h)h$
$O(h^2)$	$y_{i+1}^{(2)} = y_i + k_2$
$O(h^4)$	$y_{i+1}^{(4)} = y_{i+1}^{(2)} + e_{i+1}^{(2)}$
Estimation Error	$e_{i+1}^{(2)} = \frac{1}{6}(k_1 - 3k_2 + k_3 + k_4)$
step t	$t_{i+1} = t_i + h$

Table 12.15 Algorithm 12.15. on (2, 4) equation

calculation	(2, 4) equation with variable step
Stage 1.	$k_1 = f(y_i, t_i)h$
Stage 2.	$k_2 = f(y_i + \frac{2}{5}k_1, t_i + \frac{2}{5}h)h$
Stage 3.	$k_3 = f(y_i + 0.29697760924775360 k_1 + 0.15875964497103583 k_2,$ $t_i + 0.45573725421878943h)h$
Stage 4.	$k_4 = f(y_i + 0.21810038822592047 k_1 - 3.0509651486929308 k_2$ $+3.8328647604670103 k_3, t_i + h)h$
$O(h^2)$	$y_{i+1}^{(2)} = y_i - \frac{1}{4}k_1 + 1.25 k_2$
$O(h^4)$	$y_{i+1}^{(4)} = y_{i+1}^{(2)} + e_{i+1}^{(2)}$
Estimation Error	$e_{i+1}^{(2)} = 0.42476028226269037 k_1 - 1.8014806628787329 k_2$ $+1.2055355993965235 k_3 + 0.17118478121951903 k_4)$
step t	$t_{i+1} = t_i + h$

$$r_c = \frac{(r_{sb} + r_{pb})}{\cos \alpha}, \qquad (12.15)$$

where α is the engaging angle of gear pair, $T_D(t)$ is input torque excitation function, $T_L(t)$ is load torque excitation function, P represents the elastic force of gear pair, and D represents the viscous force of gear pair.

Let the backlash of the meshing gear pair be $2b$, then some parameters in the Eq. (12.14) will be obtained:

12.3 Nonlinear Dynamics of 2K-H Planetary Gear Transmission Systems

Table 12.16 Algorithm 12.16. on (2, 5) equation

calculation	(4, 5) equation with variable step
Stage 1.	$k_1 = f(y_i, t_i)h$
Stage 2.	$k_2 = f(y_i + \frac{1}{2}k_1, t_i + \frac{1}{2}h)h$
Stage 3.	$k_3 = f(y_i + \frac{1}{4}(k_1 + k_2), t_i + \frac{1}{2}h)h$
Stage 4.	$k_4 = f(y_i - k_2 + 2k_3, t_i + h)h$
Stage 5.	$k_5 = f(y_i + \frac{1}{27}(7k_1 + 10k_2 + k_4), t_i + \frac{2}{3}h)h$
Stage 6.	$k_6 = f(y_i + \frac{1}{625}(28k_1 - 125k_2 + 546k_3 + 54k_4 - 378k_5), t_i + \frac{1}{5}h)h$
$O(h^2)$	$y_{i+1}^{(2)} = y_i + \frac{1}{2}k_2$
$O(h^5)$	$y_{i+1}^{(5)} = y_{i+1}^{(2)} + e_{i+1}^{(2)}$
Estimation Error	$e_{i+1}^{(2)} = \frac{1}{336}(14k_1 - 336k_2 + 35k_4 + 162k_5 + 125k_6)$
step t	$t_{i+1} = t_i + h$

Table 12.17 Algorithm 12.17. on (2, 5) equation

calculation	(4, 5) equation with variable step
Stage 1.	$k_1 = f(y_i, t_i)h$
Stage 2.	$k_2 = f(y_i + \frac{1}{3}k_1, t_i + \frac{1}{3}h)h$
Stage 3.	$k_3 = f(y_i + \frac{1}{25}(4k_1 + 6k_2), t_i + \frac{2}{3}h)h$
Stage 4.	$k_4 = f(y_i + \frac{1}{4}(k_1 - 12k_2 + 15k_3), t_i + h)h$
Stage 5.	$k_5 = f(y_i + \frac{1}{81}(6k_1 + 90k_2 - 50k_3 + 8k_4), t_i + \frac{2}{3}h)h$
Stage 6.	$k_6 = f(y_i + \frac{1}{75}(6k_1 + 36k_2 + 10k_3 + 8k_4), t_i + \frac{4}{5}h)h$
$O(h^2)$	$y_{i+1}^{(2)} = y_i - \frac{1}{2}k_1 + \frac{3}{2}k_2$
$O(h^5)$	$y_{i+1}^{(5)} = y_{i+1}^{(2)} + e_{i+1}^{(2)}$
Estimation Error	$e_{i+1}^{(2)} = \frac{1}{192}(119k_1 - 288k_2 + 125k_3 - 81k_5 + 125k_6)$
step t	$t_{i+1} = t_i + h$

$$\begin{cases} P_{spi} = K_{spi}(t) f(\theta_s r_{bs} - \theta_{pi} r_{bpi} - \theta_c r_c \cos\alpha - e_{spi}(t), b_{spi}), \\ P_{rpi} = K_{rpi}(t) f(\theta_{pi} r_{bpi} - \theta_c r_c \cos\alpha - e_{rpi}(t), b_{rpi}), \end{cases} \quad (12.16)$$

$$\begin{cases} D_{spi} = C_{spi}(t) f(\dot\theta_s r_{bs} - \dot\theta_{pi} r_{bpi} - \dot\theta_c r_c \cos\alpha - \dot e_{spi}(t)), \\ D_{rpi} = C_{rpi}(t) f(\dot\theta_{pi} r_{bpi} - \dot\theta_c r_c \cos\alpha - \dot e_{rpi}(t)) \end{cases}, \quad (12.17)$$

Fig. 12.1 Dynamic model of 2K-H planetary gear transmission

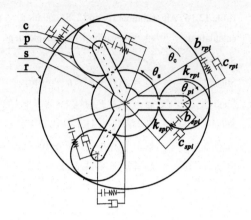

where $f(x, b)$ is clearance nonlinear function as

$$f(x, b) = \begin{cases} x - b & x > b, \\ 0 & -b \leq x \leq b, \\ x + b & x < -b, \end{cases} \quad (12.18)$$

$e(t)$ is composition error of meshing gear pair as

$$e(t) = \sum_{i=1}^{M} e_i \sin(i\omega t + \phi_i), \quad (12.19)$$

$K(t)$ is the meshing stiffness of gear pair along the active line as

$$K(t) = K_0 + \sum_{i=1}^{\infty} [p_i \cos(i\omega t) + q_i \sin(i\omega t)], \quad (12.20)$$

$C(t)$ is the damping factor of meshing gear pair as

$$C = 2\zeta \sqrt{\bar{K}/(1/M_1^{eq} + 1/M_2^{eq})}. \quad (12.21)$$

Second, we shall consider its normalized procedures.

For Eq. (12.14), the $N + 2$ coordinates are not mutually independent, actually the system has $N + 1$ degrees of freedom and each variables in the equation has clear physical significance. But when considering the performance of the system, too many parameters are not convenient.

As the normalized equation only considers the dynamical characteristics in form and does not depend on the specific physical meanings, not only can it present the linear displacement, but also it can express the angular displacement of the motion.

12.3 Nonlinear Dynamics of 2K-H Planetary Gear Transmission Systems

As a result, during studying the system, the constraints of specific parameters can be separated and it will be easy to be generalized.

In order to get the normalized equation, defining the displacement nominal scale r_{bc} and time nominal scale ω_n as

$$\tau = \omega_n t, \tag{12.22}$$

$$\omega_n = \sqrt{\bar{K}_{sp1}(1/M_s^{eq} + 1/M_{p1}^{eq})}, \tag{12.23}$$

and defining a planet carrier base circle r_{bc} as

$$r_{bc} = r_c \cos\alpha, \tag{12.24}$$

and at the same time, defining a new set of generalized coordinates as

$$x = \{X_{sp1}, X_{sp2}, \ldots, X_{spN}, X_{sc}\}^T, \tag{12.25}$$

further, we have

$$\begin{cases} X_{spi} = \dfrac{\theta_s r_{bs} - \theta_{pi} r_{bpi} - \theta_c r_{bc} - e_{spi}(\tau)}{b_c}, \\ X_{sc} = \dfrac{\theta_s r_{bs} - 2\theta_c r_{bc}}{b_c}. \end{cases} \tag{12.26}$$

Then differential equations (12.14) can be transformed to the following $N + 1$ normalized equations as

$$\begin{cases}
\ddot{X}_{spi} + \dfrac{1}{\omega_n^2}((\dfrac{1}{M_s^{eq}} + \dfrac{1}{M_c^{eq}})\sum_{i=1}^{N} K_{spi}(\tau)f(X_{spi}, \bar{b}_{spi}) + \dfrac{1}{M_{pi}^{eq}}K_{spi}(\tau)f(X_{spi}, \bar{b}_{spi})) + \\
\dfrac{1}{\omega_n^2}(\dfrac{1}{M_c^{eq}}\sum_{i=1}^{N} K_{rpi}(\tau) - \dfrac{1}{M_{pi}^{eq}}K_{rpi}(\tau))f(X_{sc} - X_{spi} - \bar{e}_{spi}(\tau) - \bar{e}_{rpi}(\tau), \bar{b}_{rpi}) + \\
\dfrac{1}{\omega_n}((\dfrac{1}{M_s^{eq}} + \dfrac{1}{M_c^{eq}})\sum_{i=1}^{N} C_{spi}(\tau)\dot{X}_{spi} + \dfrac{1}{M_{pi}^{eq}}C_{spi}(\tau)\dot{X}_{spi}) + \\
\dfrac{1}{\omega_n}(\dfrac{1}{M_c^{eq}}\sum_{i=1}^{N} C_{rpi}(\tau) - \dfrac{1}{M_{pi}^{eq}}C_{rpi}(\tau))(\dot{X}_{sc} - \dot{X}_{spi} - \dot{\bar{e}}_{spi}(\tau) - \dot{\bar{e}}_{rpi}(\tau)) + \\
= \dfrac{1}{\omega_n^2 b_c}(\dfrac{T_D(\tau)}{M_s^{eq} r_{bs}} + \dfrac{T_L(\tau)}{M_c^{eq} r_{bc}}) - \ddot{\bar{e}}_{spi}(\tau), \\
\ddot{X}_{sc} + \dfrac{1}{\omega_n^2}(\dfrac{1}{M_s^{eq}} + \dfrac{2}{M_c^{eq}})\sum_{i=1}^{N} K_{spi}(\tau)f(X_{spi}, \bar{b}_{spi}) + \dfrac{2}{M_c^{eq}\omega_n^2}\sum_{i=1}^{N} K_{rpi}(\tau)f(X_{sc} - \\
X_{spi} - \bar{e}_{spi}(\tau) - \bar{e}_{rpi}(\tau), \bar{b}_{rpi}) + \dfrac{1}{\omega_n}(\dfrac{1}{M_s^{eq}} + \dfrac{2}{M_c^{eq}})\sum_{i=1}^{N} C_{spi}(\tau)\dot{X}_{spi} + \\
\dfrac{2}{M_c^{eq}\omega_n}\sum_{i=1}^{N} C_{rpi}(\tau)(\dot{X}_{sc} - \dot{X}_{spi} - \dot{\bar{e}}_{spi}(\tau) - \dot{\bar{e}}_{rpi}(\tau)) = \dfrac{1}{\omega_n^2 b_c}(\dfrac{T_D(\tau)}{M_s^{eq} r_{bs}} + \dfrac{2T_L(\tau)}{M_c^{eq} r_{bc}}).
\end{cases} \tag{12.27}$$

where

$$M_c^{eq} = \frac{J_c + \sum_{i=1}^{N} M_{pi} r_c}{r_{bc}^2}, \quad (12.28)$$

$$\bar{b}_{spi} = b_{spi}/b_c,$$
$$\bar{b}_{rpi} = b_{rpi}/b_c, \quad (12.29)$$

$$\Omega = \omega/\omega_n, \quad (12.30)$$

$\bar{e}_{spi}(\tau)$ and $\bar{e}_{rpi}(\tau)$ are the fundamental frequency of Fourier series of meshing error:

$$\begin{cases} \psi_{spi} = \dfrac{2\pi \Delta l_{1i}}{p_b}, \\ \psi_{spi} = \dfrac{2\pi \Delta l'_{1i}}{p_b}, \end{cases} \quad (12.31)$$

ϕ is the starting phase angle; ψ_{spi} and ψ_{rpi} represent the phase angles of external meshing and internal meshing, where

$$\begin{cases} \psi_{spi} = \dfrac{2\pi \Delta l_{1i}}{p_b}, \\ \psi_{spi} = \dfrac{2\pi \Delta l'_{1i}}{p_b}, \end{cases} \quad (12.32)$$

γ_{sr} is the phase angle between external meshing and internal meshing, where

$$\gamma_{sr} = \frac{2\pi \Delta l_1^1}{pb}. \quad (12.33)$$

The system's analysis based on the Algorithm 12.13 on (2, 4) equation will be discussed. Here we adapt the following transform:

$$Y = (y, \dot{y})^T, \dot{Y} = (\dot{y}, \ddot{y})^T, \quad (12.34)$$

Then the Eq. (12.27) should be reduced to the following form:

$$\begin{cases} \dot{y} = f(y, t), y \in \mathbb{R}^n, t > t_0, \\ y(t_0) = y_0. \end{cases} \quad (12.35)$$

It is well known that for any given time t_d, choosing a large enough positive integer N, the discrete value of $[t_0, t_d]$ will be obtained:

12.3 Nonlinear Dynamics of 2K-H Planetary Gear Transmission Systems

$$t_k = t_0 + k\Delta t, \, \Delta t = (t_d - t_0)/N, \, k = 0, 1, \ldots N. \tag{12.36}$$

Getting the integral of Eq. (12.27) within each time interval, according to the mean value theorem for integration, there exists $s_k \in [t_k, t_{k+1}]$ such that

$$\mathbf{y}(t_{k+1}) = \mathbf{y}(t_k) + \int_{t_k}^{t_{k+1}} f(\psi(s), s)ds = \mathbf{y}(t_k) + f(\psi(s_k), s_k)\Delta t, \, k = 0, 1, \ldots N. \tag{12.37}$$

It is very clear that, if the approximate solution of $s_k (k = 1, 2, \ldots, N)$ is obtained, then the corresponding approximate solution of $x_k (k = 1, 2, \ldots, N)$ within time interval of $[t_0, t_d]$ will be obtained. Now, we consider the following first-order differential equations through the RK integration method:

$$\begin{cases} y'_0 = f_0(t, y_0, y_1, \cdots, y_{n-1}), \, y_0(t_0) = y_{00}, \\ y'_1 = f_1(t, y_0, y_1, \cdots, y_{n-1}), \, y_1(t_0) = y_{10}, \\ \cdots\cdots\cdots, \\ y'_{n-1} = f_{n-1}(t, y_0, y_1, \cdots, y_{n-1}), \, y_{n-1}(t_0) = y_{n-1,0}, \end{cases} \tag{12.38}$$

Using Algorithm 12.13.on (2, 4) equation, we have

$$\begin{cases} k_{0i} = hf_i(t_j, y_{0j}^{(0)}, y_{1j}^{(0)}, \cdots, y_{n-1,j}^{(0)}), \\ k_{1i} = hf_i(t_j + \frac{h}{2}, y_{0j}^{(1)}, y_{1j}^{(1)}, \cdots, y_{n-1,j}^{(1)}), \\ k_{2i} = hf_i(t_j + \frac{h}{2}, y_{0j}^{(2)}, y_{1j}^{(2)}, \cdots, y_{n-1,j}^{(2)}), \\ k_{3i} = hf_i(t_j + h, y_{0j}^{(3)}, y_{1j}^{(3)}, \cdots, y_{n-1,j}^{(3)}), \\ (i = 0, 1, \ldots, n-1), \end{cases} \tag{12.39}$$

where

$$\begin{cases} y_{ij}^{(1)} = y_{ij}^{(0)} + \frac{1}{2}(k_{0i} - 2q_i^{(0)}), \\ y_{ij}^{(2)} = y_{ij}^{(1)} + \left(1 - \sqrt{\frac{1}{2}}\right)(k_{1i} - q_i^{(1)}), \\ y_{ij}^{(3)} = y_{ij}^{(2)} + \left(1 + \sqrt{\frac{1}{2}}\right)(k_{2i} - q_i^{(2)}), \\ y_{ij}^{(4)} = y_{ij}^{(3)} + \frac{1}{6}(k_{3i} - 2q_i^{(3)}), \\ (i = 0, 1, \ldots, n-1), \end{cases} \tag{12.40}$$

$$\begin{cases} q_i^{(1)} = q_i^{(0)} + 3\left[\frac{1}{2}(k_{0i} - 2q_i^{(0)})\right] - \frac{1}{2}k_{0i}, \\ q_i^{(2)} = q_i^{(1)} + 3\left[\left(1 - \sqrt{\frac{1}{2}}\right)(k_{1i} - 2q_i^{(1)})\right] - \left(1 - \sqrt{\frac{1}{2}}\right)k_{1i}, \\ q_i^{(3)} = q_i^{(2)} + 3\left[\left(1 + \sqrt{\frac{1}{2}}\right)(k_{2i} - 2q_i^{(2)})\right] - \left(1 + \sqrt{\frac{1}{2}}\right)k_{2i}, \\ q_i^{(4)} = q_i^{(3)} + 3\left[\frac{1}{6}(k_{3i} - 2q_i^{(3)})\right] - \frac{1}{2}k_{3i}, \\ (i = 0, 1, \ldots, n-1), \end{cases} \qquad (12.41)$$

and $y_{ij}^{(0)}$ is the function value of $y_i(t_j)$ in t_j, passing one integral step later, $y_{ij}^{(4)}$ is the function value $y_i(t_j + 1)$ in $t_j + 1$, and the initial value of $q_i^{(0)}$ is 0, and passing each integral step later, $q_i^{(4)}$ is the next step value of $q_i^{(0)}$.

In a word, its main procedures are as follows.

The first step is to calculate $y_{ij}^{(h)}$, $(i = 0, 1, \ldots, n-1)$ from $y_{i,j-1}$ with passing step length h;

The second step is to calculate $y_{ij}^{(h/2)}$, $(i = 0, 1, \ldots, n-1)$ from $y_{i,j-1}$ with passing step length $h/2$. When the following condition

$$\max_{0 \le i \le n-1} \|y_{ij}^{(h/2^m)} - y_{ij}^{(h)}\| < \varepsilon, \qquad (12.42)$$

holds, then calculating stops and let $y_i(t_j) = y_{ij}^{(h/2)} (i = 0, 1, \ldots, n-1)$. Otherwise continuing the next smaller step length. Repeating the above process until the following condition

$$\max_{0 \le i \le n-1} \left| y_{ij}^{(h/2^m)} - y_{ij}^{(h/2^{m-1})} \right| < \varepsilon \qquad (12.43)$$

holds, finally we obtain the following equation:

$$y_i(t_j) = y_{ij}^{h/2^m}, (i = 0, 1, \ldots, n-1), \qquad (12.44)$$

where ε is the given positive constant in advance.

Third, we consider the largest Lyapunov exponent of the Eq. (12.27).

It is well known that the largest Lyapunov exponent is now the most reliable standard to tell whether a system is in chaos or not. If the largest Lyapunov exponent of the system is greater than zero, then the system is chaotic; and if it is less than zero, then the movement of the system is periodic. Specially, if the largest Lyapunov exponent is equal to zero, the system is in quasiperiodic movement. The system is sectionally continuous because of the existence of clearance, so the Jacobi matrix of the nonlinear system does not exist at the boundary points. Another method has to be found to calculate the largest Lyapunov exponent of the system,

Based on the definition of the Lyapunov Exponent, we deduce the methods on Largest Lyapunov Exponent as follows.

For any sufficiently small disturbance, $\delta Y_1(t_0)$ in t_0 passes a enough small time period, $\delta Y(t)$ will change according to the rule of $\|\delta Y_1(t_0)\|e^{\sigma(t-t_0)}$, where a certain

12.3 Nonlinear Dynamics of 2K-H Planetary Gear Transmission Systems

constant σ, which depends on t_0, will be computed as follows:

$$||\delta Y(t)|| = ||\delta Y(t_0)||e^{\sigma(t-t_0)}. \tag{12.45}$$

Further, the value of σ will be obtained as

$$\sigma = \frac{1}{t-t_0}\ln\left[\frac{||\delta Y(t)||}{||\delta Y(t_0)||}\right]. \tag{12.46}$$

As the Lyapunov Exponent describes the exponential average divergence rate of adjacent phase trajectory, different σ in different moments t_0 should be calculated respectively and the corresponding average value of each σ will be the Largest Lyapunov Exponent.

The basic computing procedure is described as follows.

First, given an initial disturbance vector $\mathbf{d}_0$, we make the length of $\mathbf{d}_0$ equals to a standard length, at least ensure the computer ability with running 1000 times steps for computational accuracy.

Second, the initial vector $\mathbf{d}_0$ is transformed into $\mathbf{d}_{01}$ after a time period of Δt, in order to prevent overflow, we ever ensure the vector direction of $\mathbf{d}_{01}$ unchange and usually adjusting its step length no more than less the initial standard length, then a next standard vector $\mathbf{d}_1$ will be obtained. Obviously, $\mathbf{d}_1$ is more close to the fastest growth direction of the disturbance than $\mathbf{d}_0$.

Third, after passing another time period of Δt, $\mathbf{d}_1$ is transformed into $\mathbf{d}_{11}$. We continue to ensure the direction of $\mathbf{d}_{11}$ is unchanged and usually adjusting its length no more than the initial standard length, then another new standard vector $\mathbf{d}_2$ will be obtained. Obviously, $\mathbf{d}_2$ is more close to the fastest growth direction of the disturbance than $\mathbf{d}_1$.

Fourthly, repeating Steps 2 and 3, passing a enough time period of $K \cdot \Delta t$, the desirable vector series $\mathbf{d}_k$ will be obtained.

Finally, let

$$\begin{aligned}\lambda_1 &= \lim_{\tau\to\infty}\frac{1}{t-t_0}\ln\left[\frac{||\delta X(t)||}{||\delta X(t_0)||}\right]\\ &= \lim_{k\to\infty}\frac{1}{k\Delta t}\ln\left[\frac{||\mathbf{d}_{01}||}{||\mathbf{d}_0||}\frac{||\mathbf{d}_{11}||}{||\mathbf{d}_1||}\frac{||\mathbf{d}_{21}||}{||\mathbf{d}_2||}\cdots\frac{||\mathbf{d}_{k1}||}{||\mathbf{d}_k||}\right]\\ &= \lim_{k\to\infty}\frac{1}{k\Delta t}\sum_{i=1}^{k}\ln\left[\frac{||\mathbf{d}_{i1}||}{||\mathbf{d}_i||}\right]\\ &= \lim_{k\to\infty}\frac{1}{k\Delta t}\sum_{i=1}^{k}\ln\left[\frac{||\mathbf{d}_{i1}||}{||\mathbf{d}_0||}\right],\end{aligned} \tag{12.47}$$

after k is big enough, and λ_1 will converge to a constant value, which will be the desirable largest Lyapunov exponent.

12.4 Numerical Simulation

Let the displacement nominal scale $bc = 1\mu m$. The periodic movement, quasiperiodic movement, and chaotic movement of the system under different parameters will be discussed in the following section.

Periodic Movement

When the damping ratio $\xi = 0.07$, exciting frequency $\Omega = 0.2$, the response of the system is periodic 1 movement without impact, the simulation results are shown in Figs. 12.2 and 12.3, Fig. 12.2a is the phase plane and the black dot comes from the Poincaré section, Fig. 12.2b is the Poincaré section, Fig. 12.2c, d are the FFT spectrum which locates in the points of $n \cdot \Omega$, and Fig. 12.3 is the largest Lyapunov exponent that converges to a negative value -0.03742.

Chaotic 2 Movement

When the damping ratio $\xi = 0.07$, exciting frequency $\Omega = 0.51$, the response of the system is periodic 2 movement, the simulation results are shown in Figs. 12.4 and 12.5. From the pictures, we can see that the Poincaré section contains two discrete points, the FFT spectrum shows that the response spectrum value is located in the discrete points of $n \cdot \Omega/2$ and the largest Lyapunov exponent converges to a negative value -0.00686.

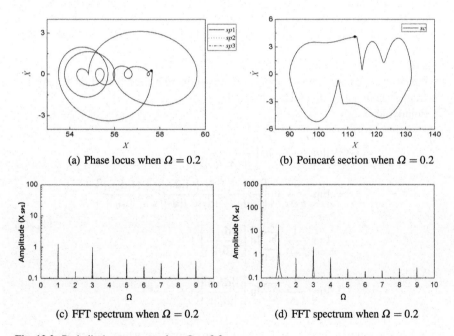

(a) Phase locus when $\Omega = 0.2$

(b) Poincaré section when $\Omega = 0.2$

(c) FFT spectrum when $\Omega = 0.2$

(d) FFT spectrum when $\Omega = 0.2$

Fig. 12.2 Periodic 1 movement when $\Omega = 0.2$

12.4 Numerical Simulation 245

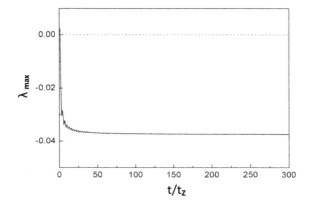

Fig. 12.3 Largest Lyapunov Exponent when $\Omega = 0.2$

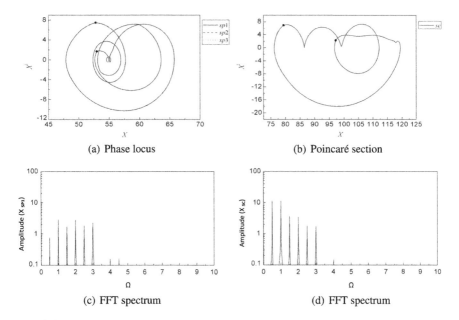

Fig. 12.4 Periodic 2 movement when $\Omega = 0.51$

Chaotic 4 Movement

When the damping ratio $\xi = 0.07$, exciting frequency $\Omega = 0.57$, the response of the system is periodic 4 movement, the simulation results are shown in Figs. 12.6 and 12.7. From the pictures, we can see that the Poincaré section contains four discrete points, the FFT spectrum shows that the response spectrum value is located in the discrete points of $n \cdot \Omega/4$, and the largest Lyapunov exponent converges to a negative value -0.03742.

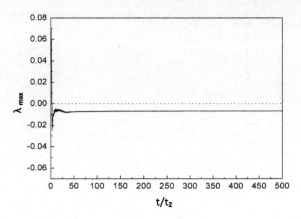

Fig. 12.5 Largest Lyapunov Exponent when $\Omega = 0.51$

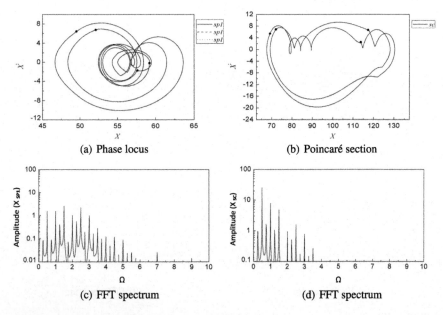

Fig. 12.6 Periodic 4 movement when $\Omega = 0.57$

Quasiperiodicity Movement

When the damping ratio $\xi = 0.00442$, exciting frequency $\Omega = 1.1$, the response of the system is quasiperiodicity movement. The simulation results are shown in Figs. 12.8, 12.9, 12.10, and 12.11. Figure 12.8 is the phase plane, Fig. 12.10 is the Poincaré section, Fig. 12.11 is the FFT spectrum, and Fig. 12.9 is the largest Lyapunov exponent that converges to a positive value 0.000112, while the actual value should be 0. So the calculation error is inevitable.

12.4 Numerical Simulation

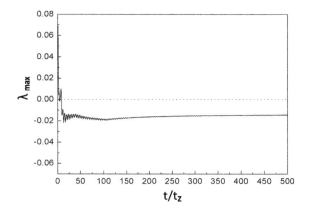

Fig. 12.7 Largest Lyapunov Exponent when $\Omega = 0.57$

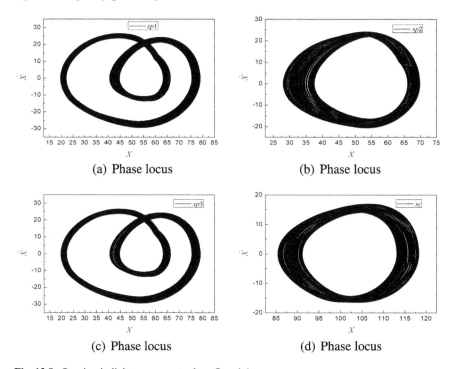

Fig. 12.8 Quasiperiodicity movement when $\Omega = 1.1$

Chaos Movement

When the damping ratio $\xi = 0.007$, exciting frequency $\Omega = 0.64$, the response of the system is chaotic movement. Figure 12.14 shows that the largest Lyapunov exponent of the system that converges to a positive value 0.01745, and Fig. 12.12 shows the phase plane and the FFT spectrum. In Fig. 12.13, the first two pictures are the Poincaré

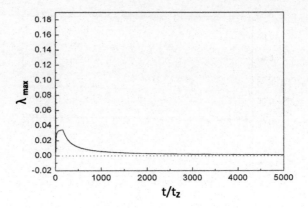

Fig. 12.9 Largest Lyapunov Exponent when $\Omega = 1.1$

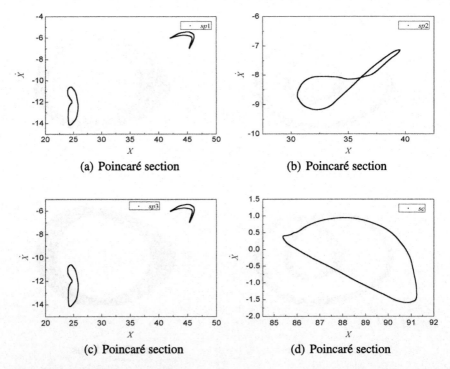

Fig. 12.10 Quasiperiodicity movement when $\Omega = 1.1$

section, and successively enlarge the small rectangle section in the first two pictures, we will have the next four pictures.

12.4 Numerical Simulation

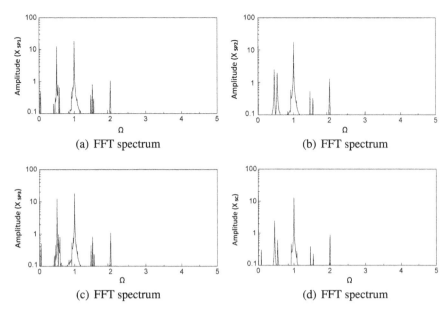

Fig. 12.11 Quasiperiodicity movement when $\Omega = 1.1$

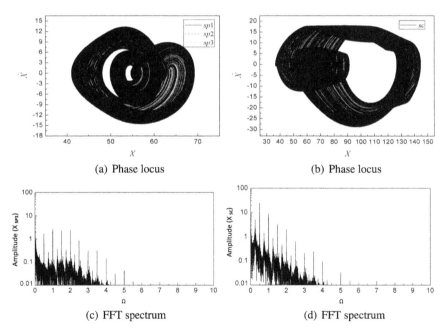

Fig. 12.12 Chaos movement when $\Omega = 0.64$

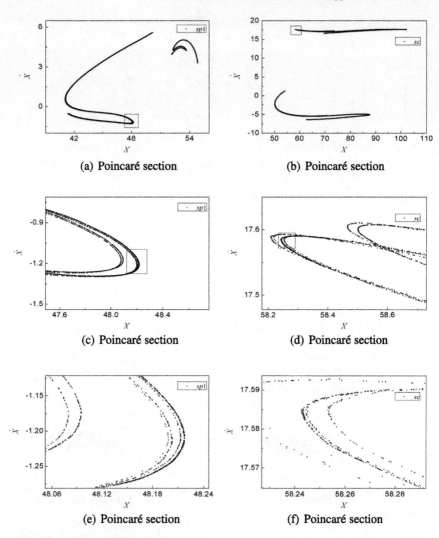

Fig. 12.13 Chaos movement when $\Omega = 0.64$

12.5 Conclusion

The nonlinear dynamic model of 2 K-H planetary gear transmission is presented in this chapter based on lumped mass model, which takes into consideration of clearance, meshing stiffness and composition error of meshing gear pair. Numerical method of solving the equation is discussed and a new discrete method to calculate the largest Lyapunov exponent of the system is deduced. Experimental simulation is carried out and the periodic motion, quasiperiodic motion, and chaotic motions of the system are obtained with different parameters.

12.5 Conclusion

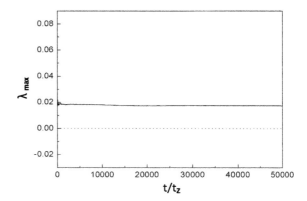

Fig. 12.14 Largest Lyapunov Exponent when $\Omega = 0.64$

Acknowledgements This work is Supported by National Key Research and Development Program of China (2017YFF0207400).

References

1. Parker RG, Lin J, Mesh phasing relationships in planetary and epicyclic gears. ASME, International Design Engineering Technical Conferences and Computers and Information in Engineering Conference. American Society of Mechanical Engineers. 2003;2003:525–34.
2. Kahraman A. Natural modes of planetary gear trains. J Sound Vib. 1994;173(1):125–30.
3. Lin J, Parker RG. Analytical characterization of the unique properties of planetary gear free vibration. J Vib Acoust. 1999;121(3):316–21.
4. Al-Shyyab A, Kahraman A. Non-linear dynamic analysis of a multi-mesh gear train using multi-term harmonic balance method: sub-harmonic motions. J Sound Vib. 2005;279(1):417–51.
5. Sondkar P, Kahraman A. A dynamic model of a double-helical planetary gear set. Mech Mach Theory. 2013;70:157–74.
6. Parker RG, Vijayakar SM, Imajo T. Non-linear dynamic response of a spur gear pair: modelling and experimental comparisons. J Sound Vib. 2000;237(3):435–55.
7. Al-Shyyab A, Alwidyan K, Jawarneh A, et al. Non-linear dynamic behaviour of compound planetary gear trains: model formulation and semi-analytical solution. Proc Inst Mech Eng Part K: J Multi-body Dyn. 2009;223(3):199–210.
8. Lin J, Parker RG. Planetary gear parametric instability caused by mesh stiffness variation. J Sound Vib. 2002;249(1):129–45.
9. Parker RG. A physical explanation for the effectiveness of planet phasing to suppress planetary gear vibration. J Sound Vib. 2000;236(4):561–73.
10. Gao H, Zhang Y. Nonlinear behavior analysis of geared rotor bearing system featuring confluence transmission. Nonlinear Dyn. 2014;1–15.
11. De Jess Rubio J, Prez-Cruz JH. Evolving intelligent system for the modelling of nonlinear systems with dead-zone input. Appl Soft Comput. 2014;14:289–304.
12. Lu JW, Chen H, Zeng FL, et al. Influence of system parameters on dynamic behavior of gear pair with stochastic backlash. Meccanica. 2014;49(2):429–40.
13. Chaari F, Haddar M. Modeling of gear transmissions dynamics in non-stationary conditions., Cyclostationarity: theory and methodsBerlin: Springer International Publishing; 2014. p. 109–24.

14. Asemidis LD, Sackellares JC. The evolution with time of the spatial distribution of the Largest Lyapunov exponent on the human epileptic cortex. Measuring chaos in the human brain; 1991;49–82.
15. Li S, Wu Q, Zhang Z. Bifurcation and chaos analysis of multistage planetary gear train. Nonlinear Dyn. 2014;75(1–2):217–33.
16. Cho S, Jin M, Kuc TY, et al. Control and synchronization of chaos systems using time-delay estimation and supervising switching control. Nonlinear Dyn. 2014;75(3):549–60.
17. Banerjee T, Biswas D, Sarkar BC. Complete and generalized synchronization of chaos and hyperchaos in a coupled first-order time-delayed system. Nonlinear Dyn. 2013;71(1–2):279–90.
18. Ma YD, Lu JG, Chen WD, et al. Robust stability bounds of uncertain fractional-order systems. Fract Calc Appl Anal. 2014;17(1):136–53.
19. Van Zon R, Van Beijeren H, Dellago C. Largest Lyapunov exponent for many particle systems at low densities. Phys Rev Lett. 1998;80(10):2035.

Appendix A
Some Theory about Lyapunov Stability

A.1 Lyapunov Function and K-Class Function

Suppose that $W(x) \in C[\mathbb{R}^n, \mathbb{R}^n]$, $W(0) = 0$, $V(t, x) \in C[\mathbb{I} \times \mathbb{R}^n, \mathbb{R}^n]$, $V(t, 0) = 0$, $\mathbb{I} = [t_0, \infty)$.

Definition A.1 The function $W(x)$ is said to be positive (negative) definite if $W(x) \geq 0 (W(x) \leq 0)$ and $W(x) = 0$ if and only if $x = 0$; The function $W(x)$ is said to be radially unbounded, positive definite if $W(x)$ is positive definite and $W(x) \to \infty$ as $\|x\| \to \infty$; The function $V(t, x)$ is said to be positive (negative) definite if $V(t, x) \geq W(x)$; The function $V(t, x)$ is said to be negative definite if $-V(t, x)$ is positive definite.

The positive (negative) definite functions are usually called Lyapunov functions.

Definition A.2 The function $\varphi \in [\mathbb{R}^+, \mathbb{R}^+]$ (where $\mathbb{R}^+ := [0, +\infty)$), φ is continuous, strictly monotone increasing, and $\varphi(0) = 0$, we call φ a $K-$ class function, denoted by $\varphi \in \mathbb{K}$; If $\varphi \in \mathbb{K}$ and $\lim_{r \to \infty} \varphi(r) = +\infty$, then $\varphi(r)$ is called a radially unbounded $K-$ class function, denoted by $\varphi \in \mathbb{KR}$.

We easily outline the following results.

Lemma A.1 *Given a positive definite function $W(x)$, there exist two functions $\varphi_1, \varphi_2 \in \mathbb{K}$, such that*

$$\varphi_1(x) \leq W(x) \leq \varphi_1;$$

Given a radically unbounded $K-$ class function $W(x)$, there must exist two functions $\varphi_1, \varphi_2 \in \mathbb{KR}$, such that

$$\varphi_1(x) \leq W(x) \leq \varphi_1.$$

A.2 Dini Derivative

Suppose that $f(t) \in C[\mathbb{I}, \mathbb{R}^1], \mathbb{I} = [t_0, \infty)$, for any $t \in \mathbb{I}$, the following four derivatives

$$D^+ f(t) = \varlimsup_{h \to 0^+} \tfrac{1}{h}[f(t+h) - f(t)],$$
$$D_+ f(t) = \varliminf_{h \to 0^+} \tfrac{1}{h}[f(t+h) - f(t)],$$
$$D^- f(t) = \varlimsup_{h \to 0^-} \tfrac{1}{h}[f(t+h) - f(t)],$$
$$D_- f(t) = \varliminf_{h \to 0^-} \tfrac{1}{h}[f(t+h) - f(t)],$$

are called right upper derivative, right lower derivative, left upper derivative, and left lower derivative of $f(t)$ at t, respectively, they are all called Dini Derivative.

Lemma A.2 *If $f(t) \in C[\mathbb{I}, \mathbb{R}^1], \mathbb{I} = [t_0, \infty)$, then $f(t)$ is monotone nondecreasing on $\mathbb{I}$ if and only if $D^+ f(t) \geq 0$ for $I = [t_0, \infty)$, (or $D_+ f(t) \geq 0$, or $D^- f(t) \geq 0$, or $D_- f(t) \geq 0$).*

A.3 Definitions of Lyapunov Stability

We consider physical systems which can be described by the following ordinary differential equation:

$$\tfrac{dy}{dt} = g(t, y), \tag{A.1}$$

where $\Omega \subset \mathbb{R}^n, 0 \in \Omega, g \in C[\mathbb{I} \times \Omega, \mathbb{R}^n]$. Assume that the solution of cauchy problem of (A.1) is unique. Suppose $\tilde{y} = \varphi(t)$ is a particular solution of (A.1), we substitute the transformation $x = y - \varphi(t)$ into the system (A.1), then we obtain

$$\tfrac{dx}{dt} = g(t, x + \varphi(t)) := f(t, x). \tag{A.2}$$

Definition A.3 The zero solution $x = 0$ of (A.2) is said to be stable, if $\forall \varepsilon > 0, \forall t_0 > 0, \exists \delta > 0$, such that $\forall x_0, \|x_0\| < \delta(\varepsilon, t_0)$ implies $\|x(t, t_0, x_0)\| < \varepsilon$;

The zero solution $x = 0$ of (A.2) is said to be unstable, if $\exists \varepsilon_0, \exists t_0, \forall \delta > 0, \exists x_0, \|x_0\| < \delta$, but $\exists t_1 \geq t_0$ such that $\|x(t, t_0, x_0)\| \geq \varepsilon$;

The zero solution $x = 0$ of (A.2) is said to be uniformly stable with respect to t_0, if $\forall \varepsilon > 0, \exists \delta(\varepsilon) > 0$ ($\delta(\varepsilon)$ is independent of t_0), such that $\|x_0\| < \delta(\varepsilon)$ implies $\|x(t, t_0, x_0)\| < \varepsilon$;

The zero solution $x = 0$ of (A.2) is said to be attractive, if $\forall t_0 \in I, \forall \varepsilon > 0, \exists \sigma(t_0) > 0, \exists T(\varepsilon, t_0, x_0) > 0, \|x_0\| < \sigma(t_0)$ implies $\|x(t, t_0, x_0)\| < \varepsilon$ for $t \geq t_0 + T$. i.e., $\lim_{h \to \infty} \|x(t, t_0, x_0)\| = 0$;

The zero solution $x = 0$ of (A.2) is said to be uniformly attractive with respect to x_0, if $\forall t_0 \in I, \forall \varepsilon > 0, \exists \sigma(t_0) > 0, \exists T(\varepsilon, t_0) > 0$ ($T(\varepsilon, t_0)$ is independent of x_0), $\|x_0\| < \sigma(t_0)$ implies $\|x(t, t_0, x_0)\| < \varepsilon$ for $t \geq t_0 + T$. i.e., $\lim_{h \to \infty} \|x(t, t_0, x_0)\| = 0$. $x = 0$ is also said to be equi-attractive;

The zero solution $x = 0$ of (A.2) is said to be uniformly attractive with respect to x_0, t_0, if $x = 0$ is equi-attractive, σ does not depend on t_0, and T does not depend on x_0, t_0, i.e., $\|x(t, t_0, x_0)\| < \varepsilon$, for $\|x_0\| < \sigma$, and $t \geq t_0 + T(\varepsilon)$.

Definition A.4 The zero solution $x = 0$ of (A.2) is respectively said to be asymptotically stable, equi-asymptotically stable, quasi-asymptotically stable if
(a) the zero solution $x = 0$ of (A.2) is stable;
(b) $x = 0$ is attractive, equi-attractive, uniformly attractive, respectively.

Definition A.5 The zero solution $x = 0$ of (A.2) is respectively said to be uniformly asymptotically stable, globally uniformly asymptotically stable, respectively, if
(a) the zero solution $x = 0$ of (A.2) is uniformly stable;
(b) $x = 0$ is uniformly attractive, globally uniformly attractive, and all solutions are uniformly bounded.

A.4 M matrix

Definition A.6 The matrix $A = (a_{ij})$ is said to be nonsingular M matrix, if
(1) $a_{ii} > 0, (i = 1, 2, \ldots, n), a_{ij} \leq 0, (j \neq j, i = 1, 2, \ldots, n)$;
(2)
$$\nabla_i = \begin{vmatrix} a_{11} & \cdots & a_{1i} \\ \vdots & \ddots & \vdots \\ a_{i1} & \cdots & a_{ii} \end{vmatrix} > 0, i = 1, \ldots, n.$$

There are many equivalent conditions for M matrix as follows:

(1) $a_{ii} > 0, (i = 1, 2, \ldots, n), a_{ij} \leq 0, (j \neq j, i = 1, 2, \ldots, n)$, and $A^{-1} \geq 0$, i.e., A^{-1} is a nonnegative matrix;

(2) $a_{ii} > 0, (i = 1, 2, \ldots, n), a_{ij} \leq 0, (j \neq j, i = 1, 2, \ldots, n)$, and there exists a constant $c_i > 0$ such that $\sum_{j=1}^{n} c_j a_{ij} > 0$ (there exist a positive solution $x \in \mathbb{R}^n$ such that $Ax = 0$);

(3) $a_{ii} > 0, (i = 1, 2, \ldots, n), a_{ij} \leq 0, (j \neq j, i = 1, 2, \ldots, n)$, and there exists a constant $c_j > 0$ such that $\sum_{j=1}^{n} c_j a_{ij} > 0$ (there exist a positive solution $x \in \mathbb{R}^n$ such that $A^T x = 0$);

(4) $a_{ii} > 0, (i = 1, 2, \ldots, n), a_{ij} \leq 0, (j \neq j, i = 1, 2, \ldots, n), -A$ is a Hurwitz matrix.

(5) $a_{ii} > 0, (i = 1, 2, \ldots, n), a_{ij} \leq 0, (j \neq j, i = 1, 2, \ldots, n)$, the spectral radius of $G = I - D^{-1}A$ is less than 1, where $D^{-1} = diag(a_{11}^{-1}, \ldots, a_{nn}^{-1})_{n \times n}$.

A.5 Some Results on Lyapunov Stability

We consider an n-dimensional differential system

$$\frac{dx}{dt} = f(t, x), \tag{A.3}$$

where $x \in \mathbb{R}^n, f \in C[\mathbb{I} \times \mathbb{R}^n, \mathbb{R}^n], \mathbb{I} = [t_0, +\infty)$. Suppose that the solution of the initial value problem of (A.3) is unique.

Lemma A.3 *If there exists a function* $V(t, x) \in C[\mathbb{I} \times \mathbb{R}^n, \mathbb{R}^1], \mathbb{I} = [t_0, +\infty)$, *satisfying*

$$\varphi_1(x) \leq V(t, x), \varphi_1 \in \mathbb{K},$$
$$\frac{dV}{dt}|_{(A.3)} \leq 0,$$

then the zero solution of (A.3) is stable.

Lemma A.4 *If there exists a function* $V(t, x) \in C[\mathbb{I} \times \mathbb{R}^n, \mathbb{R}^1], \mathbb{I} = [t_0, +\infty)$, *satisfying*

$$\varphi_1(x) \leq V(t, x) \leq \varphi_1, \varphi_1, \varphi_2 \in \mathbb{K},$$
$$\frac{dV}{dt}|_{(A.3)} \leq 0,$$

then the zero solution of (A.3) is uniform stable.

Lemma A.5 *If there exists a function* $V(t, x) \in C[\mathbb{I} \times \mathbb{R}^n, \mathbb{R}^1], \mathbb{I} = [t_0, +\infty)$, *satisfying*

$$\varphi_1(x) \leq V(t, x) \leq \varphi_1, \varphi_1, \varphi_2 \in \mathbb{K},$$
$$\frac{dV}{dt}|_{(A.3)} = -\varphi(x), \varphi \in \mathbb{K},$$

then the zero solution of (A.3) is globally uniformly asymptotically stable.

Appendix A: Some Theory about Lyapunov Stability

A.6 Two Important Results

A.6.1 LaSalle Invariant Principle

Consider the n-dimensional autonomous system

$$\frac{dx}{dt} = f(x), \; f(0) = 0, \tag{A.4}$$

where $x \in \mathbb{R}^n$, $f^T \in C[\mathbb{R}^n, \mathbb{R}^n]$. Assume that the solution of (A.4) is unique.

Lemma A.6 *Suppose $\mathbb{D}$ is a compact set, $\forall x_0 \in \mathbb{D}$, $x(t, t_0, x_0) \in \mathbb{D}$, i.e., $\mathbb{D}$ is a positive invariant set, and there exists $V(x) \in C[\mathbb{D}, \mathbb{R}^1]$ such that*

$$\frac{dx}{dt}\big|_{(A.4)} \le 0.$$

Let $\mathbb{E} = \{x \mid \frac{dx}{dt}\big|_{(A.4)} = 0, x \in D\}$, $\mathbb{M} \subset \mathbb{E}$ is the maximum invariant set, then $\sum_{t \to +\infty} x(t, t_0, x_0) \in \mathbb{M}$. In particular, if $\mathbb{M} = 0$, then the zero solution of (A.4) is asymptotically stable.

A.6.2 Comparison Principle

Consider the n-dimensional nonautonomous system

$$\frac{dx}{dt} = f(t, x), \; f(t, 0) = 0, \tag{A.5}$$

where $x \in \mathbb{R}^n$, $f^T \in C[\mathbb{I} \times \mathbb{R}^n, \mathbb{R}^n]$. Assume that the solution of (A.5) is unique, and at the same time, consider a scale differential equation

$$\frac{d\varphi}{dt} = g(t, \varphi), \tag{A.6}$$

where $g^T \in C[\mathbb{I} \times \mathbb{R}^+, \mathbb{R}^1]$. $g(t, 0) = 0$ if and only if $\varphi = 0$. If there exists a positive definite function $V(t, x) \in C[\mathbb{I} \times \mathbb{R}^n, \mathbb{R}^+]$, which satisfies the Lipschitz condition for x, and $V(t, 0) = 0$, further

$$D^+ V(t, x)\big|_{(A.5)} = g(t, V),$$

then the following conclusions hold.

(1) The stability of the zero solution (A.6) implies the corresponding one of the zero solution (A.5);
(2) If V has infinitesimal upper bound, then the uniform stability of the zero solution (A.6) implies the corresponding one of the zero solution (A.5);

(3) The asymptotic stability of the zero solution (A.6) implies the corresponding one of the zero solution (A.5);

(4) If V has infinitesimal upper bound, then the uniform asymptotic stability of the zero solution (A.6) implies the corresponding one of the zero solution (A.5).

Glossary

0 − 1 **Algebra-Geometry Structure**
Polytopic Uncertainty
Norm Boundary Uncertainty
Commuting Matrices
Hurwitz Stable
Closed Trajectory
Stable Focus
Unstable Focus
Unstable Node
Stable Node
Saddle
Fuzzy Observer
Fuzzy Controller
Common Hurwitz Matrices
Lyapunov Function
Lie Algebra
Lurie Nonlinear System
Absolute Stable
Solvable Condition
Differential Inclusion
Convex Polyhedron
Polyhedron Lyapunov Functional
Integral Inequalities
Uniform Lipschitz Stable
Uniform Lipschitz Asymptotical Stable
(c_1, c_1) Stable
(c_1, c_1) Asymptotical Stable
The Parameter Variation Method
Chaos Synchronization
Chaos Encryption
Unified Chaos System

Glossary

2K-H Planetary Gear
Largest Lyapunov Exponent
Poincare Section
Gill integration
Fast Fourier Transformation

Printed by Printforce, the Netherlands